INTERNATIONAL
REVIEW OF CYTOLOGY

VOLUME 66

INTERNATIONAL

Review of Cytology

EDITED BY

G. H. BOURNE
St. George's University School of Medicine
St. George's, Grenada
West Indies

J. F. DANIELLI
Worcester Polytechnic Institute
Worcester, Massachusetts

ASSISTANT EDITOR
K. W. JEON
Department of Zoology
University of Tennessee
Knoxville, Tennessee

VOLUME 66

1980

ACADEMIC PRESS   *A Subsidiary of Harcourt Brace Jovanovich, Publishers*
New York   London   Toronto   Sydney   San Francisco

ACADEMIC PRESS, INC.
111 Fifth Avenue, New York, New York 10003

United Kingdom Edition published by
ACADEMIC PRESS, INC. (LONDON) LTD.
24/28 Oval Road, London NW1 7DX

LIBRARY OF CONGRESS CATALOG CARD NUMBER: 52–5203

ISBN 0–12–364466–6

PRINTED IN THE UNITED STATES OF AMERICA

80 81 82 83   9 8 7 6 5 4 3 2 1

Contents

Dynein: The Mechanochemical Coupling Adenosine Triphosphatase of Microtubule-Based Sliding Filament Mechanisms

FRED D. WARNER AND DAVID R. MITCHELL

Structure and Function of Phycobilisomes: Light Harvesting Pigment Complexes in Red and Blue-Green Algae

ELISABETH GANTT

Structural Correlates of Gap Junction Permeation

CAMILLO PERACCHIA

The Kinetics and Metabolism of the Cells of Hibernating Animals during Hibernation

S. G. KOLAEVA, L. I. KRAMAROVA, E. N. ILYASOVA, AND F. E. ILYASOV

CELLSIM: Cell Cycle Simulation Made Easy

CHARLES E. DONAGHEY

The Formation of Axonal Sprouts in Organ Culture and Their Relationship to Sprouting *in Vivo*

I. R. DUCE AND P. KEEN

When Sperm Meets Egg: Biochemical Mechanisms of Gamete Interaction

BENNETT M. SHAPIRO AND E. M. EDDY

Perisinusoidal Stellate Cells (Fat-Storing Cells, Interstitial Cells, Lipocytes), Their Related Structure in and around the Liver Sinusoids, and Vitamin A-Storing Cells in Extrahepatic Organs

KENJIRO WAKE

List of Contributors

Numbers in parentheses indicate the pages on which the authors' contributions begin.

CHARLES E. DONAGHEY (171), *Industrial Engineering Department, University of Houston, Houston, Texas 77004*

I. R. DUCE (211), *Department of Zoology, University of Nottingham, Nottingham, England*

E. M. EDDY (257), *Department of Biological Structure, University of Washington, Seattle, Washington 98195*

ELISABETH GANTT (45), *Radiation Biology Laboratory, Smithsonian Institution, Rockville, Maryland 20852*

F. E. ILYASOV (147), *Laboratory of the Biophysics of Living Structures, Institute of Biological Physics, Academy of Sciences of the USSR, Biological Center of the Academy of Sciences of the USSR, Pushchino, Moscow Region, USSR*

E. N. ILYASOVA (147), *Laboratory of the Biophysics of Living Structures, Institute of Biological Physics, Academy of Sciences of the USSR, Biological Center of the Academy of Sciences of the USSR, Pushchino, Moscow Region, USSR*

P. KEEN (211), *Department of Pharmacology, University of Bristol, Bristol, England*

S. G. KOLAEVA (147), *Laboratory of the Biophysics of Living Structures, Institute of Biological Physics, Academy of Sciences of the USSR, Biological Center of the Academy of Sciences of the USSR, Pushchino, Moscow Region, USSR*

L. I. KRAMAROVA (147), *Laboratory of the Biophysics of Living Structures, Institute of Biological Physics, Academy of Sciences of the USSR, Biological Center of the Academy of Sciences of the USSR, Pushchino, Moscow Region, USSR*

DAVID R. MITCHELL (1), *Department of Biology, Biological Research Laboratories, Syracuse University, Syracuse, New York 13210*

CAMILLO PERACCHIA (81), *Department of Physiology, University of Rochester, School of Medicine and Dentistry, Rochester, New York 14642*

BENNETT M. SHAPIRO (257), *Department of Biochemistry, University of Washington, Seattle, Washington 98195*

ix

Kenjiro Wake (303), *Department of Anatomy, Faculty of Medicine, Tokyo Medical and Dental University, Yushima, Bunkyoku, Tokyo 113, Japan*

Fred D. Warner (1), *Department of Biology, Biological Research Laboratories, Syracuse University, Syracuse, New York 13210*

INTERNATIONAL

REVIEW OF CYTOLOGY

VOLUME 66

Dynein: The Mechanochemical Coupling Adenosine Triphosphatase of Microtubule-Based Sliding Filament Mechanisms

FRED D. WARNER AND DAVID R. MITCHELL

Department of Biology, Biological Research Laboratories, Syracuse University, Syracuse, New York

I. Introduction

External manifestations of force have always been an aspect of organismal affairs. However, realization that the generation of that force extended to nearly every level of subcellular organization and function has occurred only during the last 10–15 years. Motion both of and within living organisms fascinates biologists, probably because it represents the single macromolecular expression of biological activity that can be easily observed as it occurs. Cytoplasm flows, chromosomes separate, pseudopodia extend, muscle contracts, cilia beat. Motile systems can be appreciated for their own sake, as well as being readily accessible to experimental manipulation.

Biological motion dependent upon ATP hydrolysis is often generated by two distinct groups of proteins: actin and its mechanochemical coupling ATPase myosin, and tubulin and its ATPase dynein. In most instances, motion associated with either of these two systems is based upon a sliding filament mechanism.

1

F-actin and microtubules are the filamentous components used by the ATP phos-
phohydrolases myosin and dynein to generate movement by causing linear displacement or sliding of the filaments. The motion of the filaments is coupled to
otherwise stationary structures to produce, for example, contraction of the muscle sarcomere or separation of sister chromatids. Similarly, sliding filaments are
used to transduce bending moments that propel surrounding viscous environ-
ments as occurs along ciliated epithelia or among free-swimming protozoa or
spermatozoa.

The contraction of striated muscle has been studied for many years; however,
it was not until 1954 that H. E. Huxley (Huxley and Hanson, 1954) and A. F.
Huxley (Huxley and Niedergerke, 1954) proposed that contraction resulted from
overlapping arrays of actin and myosin filaments, actively sliding over one
another while maintaining approximately constant lengths. Evidence to support
the theory was derived from electron microscopic analysis of contracted versus
relaxed sarcomeres (Huxley and Hanson, 1955; Huxley, 1957). It was suggested
that sliding was produced by a cyclic crossbridge mechanism mediated by the
myosin ATPase (Huxley and Hanson, 1955), and it was shown that the myosin
active site (S1) fragment possessed the ability to react with or attach to F-actin
(Huxley, 1963).

By analogy, a sliding filament mechanism was suggested as an explanation for
the undulatory beating of cilia and flagella (Afzelius, 1959), but it was not until
1968 that P. Satir (Satir, 1965, 1968) was able to demonstrate that the
mechanism of propagated bending was in fact based on a sliding filament
mechanism. This was followed by demonstration that the dynein ATPase was
likewise involved in force production (Gibbons and Gibbons, 1973), presumably
in the form of a cyclic crossbridge mechanism. More recently it has been shown
that dynein possesses the ability to react with a site on the B-subfiber of doublet
microtubules (Warner, 1978).

In a sense, ciliary or flagellar motion represents the most experimentally
satisfying example of an energy dependent sliding filament mechanism. Sliding
can be seen as it occurs with the use of dark field or phase contrast illumination.
In 1971, I. R. Gibbons and K. Summers (Summers and Gibbons, 1971) added
the proteolytic enzyme trypsin to isolated sea urchin sperm flagella reactivated
with $MgATP^{2-}$. Unexpectedly, they saw the nine doublet microtubules of the
organelle actively slide or telescope to several times the original length of the
intact organelle. This retrospectively simple demonstration overcame numerous
obstacles in our study of microtubule-mediated motion and provided us with a
direct means for studying the molecular basis of the sliding mechanism, particularly as that mechanism is coupled to the transducing activity of the dynein
ATPase (Sale and Satir, 1977; Warner and Mitchell, 1978; Zanetti et al., 1979).

A major distinction between the activity of skeletal muscle and that of microtubule sliding in cilia or flagella resides in the sustained activation of the
sliding mechanism. Muscle contraction (sliding) is under regulation by the re-

peated cycling of 10^{-8} to 10^{-6} M free Ca^{2+} ion, which directly affects the myosin crossbridge; whereas the regulation of microtubule sliding appears to be removed from or not directly related to the primary force generating mechanism (Gibbons and Gibbons, 1972; Walter and Satir, 1979; Zanetti *et al.*, 1979). Microtubule sliding is either turned on, in which case a cilium can beat under steady state conditions for long time periods; or sliding is turned off and the organelle is quiescent. This on–off switch may be related to possible Ca^{2+} modulation of active beat parameters.

The only requirements for sustained microtubule sliding *in vitro* are a regulated pH and a steady state concentration of $MgATP^{2-}$. In the case of model detergent-extracted cilia, we presume this means that the dynein arms are undergoing constant cycling behavior when in an appropriate position to interact with a neighboring microtubule. The study of active sliding has permitted important advances in our understanding of the organization and functional enzymatic expression of the dynein ATPase. In this article, we describe recent developments in this area, and in addition, review overall aspects of dynein chemistry and ciliary activity, as well as other sliding microtubule systems. More detailed reviews of some aspects of these systems can be found in volumes edited by Sleigh (1974) and Goldman *et al.* (1976).

II. Sliding Filament Mechanisms

The utilization of an active sliding filament mechanism by the microtubule arrays in cilia, flagella, and spermtails as the basis for their motion is now firmly established (Satir, 1968; Summers and Gibbons, 1971; Warner and Satir, 1974). Active sliding also appears to be a likely mechanism for mitotic-based motion of chromosomes (Cande and Wolniak, 1978; McIntosh and Landis, 1971), as well as motion of other microtubule-constructed organelles (McIntosh, 1973; Mooseker and Tilney, 1973). This is not to say that microtubules do not perform other or related functions: their properties of assembly and disassembly may also be involved in the mitotic sequence (Margolis *et al.*, 1978); they apparently are involved in fast axoplasmic transport (Lasek and Hoffman, 1976); and they may serve an important cytoskeletal function as seen, for example, in nonmammalian red blood cells (Cohen, 1978). The extent to which other microtubule-based cytoplasmic or organelle motions are also founded in a sliding filament type dynein–microtubule interaction is not clear.

A. MODEL SYSTEM ANALYSIS

Isolated cilia or flagella that have had their limiting membranes solubilized or perturbed by nonionic detergents such as Triton X-100 will reactivate and beat normally upon addition of a hydrolyzable substrate for the dynein ATPase. Such "model" cilia have provided the basis for numerous *in vitro* studies of motion, particularly as that motion is related to the formation and propagation of active bending moments (see Brokaw and Gibbons, 1976 for review). However, these studies do not isolate the sliding behavior of the axonemal microtubules, but rather they represent a composite analysis of the major factors involved in normal motion including the transduction of sliding forces into propagated bending.

Current studies of microtubule sliding in cilia and flagella are therefore indebted to Summers and Gibbons' (1971) demonstration that active sliding could be uncoupled from rhythmic beating, thus permitting independent study of the sliding mechanism. Mild tryptic digestion of Triton X-100 demembranated flagella that would otherwise beat causes the nine doublet microtubules to actively slide or telescope apart when reactivated by $MgATP^{2-}$. Unlike sea urchin sperm flagella, demembranated *Tetrahymena* cilia spontaneously slide apart or disintegrate under minimal physiological conditions that include only a regulated pH and the addition of $MgATP^{2-}$ (Warner and Mitchell, 1978). This behavior is illustrated in Figs. 1 and 2. We have not been able to determine why cilia from

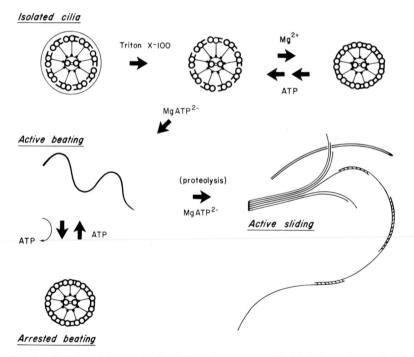

Isolated cilia

Triton X-100

Mg^{2+}

ATP

$MgATP^{2-}$

Active beating

(proteolysis)

$MgATP^{2-}$

Active sliding

ATP   ATP

Arrested beating

FIG. 1.   Depiction of the structural and dynamic responses of isolated, demembranated cilia or flagella to the addition of Mg^{2+} and ATP. The addition of Mg^{2+} alone or the withdrawal of ATP can result in uniform dynein arm attachment to the B-subfiber sites. The sliding response is spontaneous with *Tetrahymena* cilia, but typically requires proteolysis by trypsin when cilia from other organisms are used.

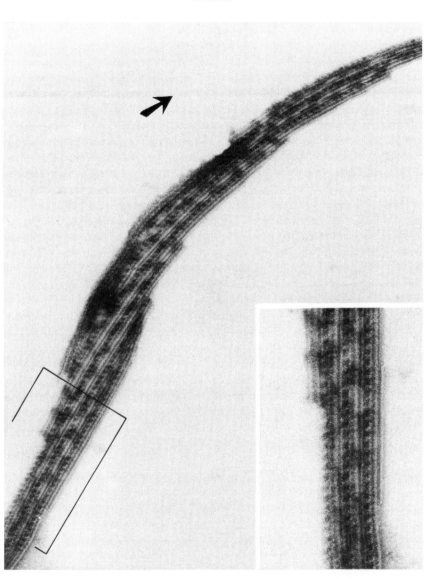

FIG. 2. A partially disintegrated *Tetrahymena* cilium after exposure to $MgATP^{2-}$. Sliding has occurred in the direction of the arrow in response to the activity of dynein crossbridges positioned between adjacent doublet microtubules (inset). $\times 71,000$. Inset, $\times 114,000$. (From Warner and Mitchell, 1978.)

this species disintegrate without externally applied proteolysis as required by cilia or flagella from other organisms (Lindeman and Gibbons, 1975; Summers and Gibbons, 1971; Walter and Satir, 1979; Witman *et al.*, 1968). There is no obvious damage to either the radial spokes or nexin links in *Tetrahymena* cilia, and in general, preservation of these structures is superior to that obtained with other organisms. However, we have noted that cilia isolated from early or middle log phase cell cultures will often reactivate and beat when exposed to MgATP^{2-}, but cilia isolated from late log phase or early stationary phase cultures always slide apart when reactivated. If *Tetrahymena* cilia that normally slide apart are additionally subjected to trypsin proteolysis, the extent of disintegration (decrease in optical absorbance at 350 nm) can only be increased about 20%, and proteolysis itself contributes to about half of that increase. Therefore, use of *Tetrahymena* cilia has considerable advantages in that we can study sliding and its related substructure free from structural damage otherwise caused by the use of trypsin. Obviously, however, there must exist some extremely sensitive component, structural or physiological, that is perturbed during growth and aging of the organism and thereby uncouples the sliding from the bending mechanism. This also questions the suggested significance of the spokes and nexin links as being the structures that, when digested by trypsin, result in the sliding reaction (Brokaw and Simonick, 1977; Summers and Gibbons, 1973).

It has become increasingly apparent to those of us working with reactivated model cilia that the sliding mechanism is insensitive to a variety of conditions that otherwise affect overall ciliary motion, suggesting that many aspects of control of ciliary motion may reside with the bending mechanism. Sliding seems either to be turned-on or turned-off, although the dynein substrate concentration has a rate limiting effect in that, within the range of 0.01–1 mM ATP, it is related linearly to the beat frequency (Brokaw, 1975).

The sliding reaction can be monitored quantitatively as changes in optical absorbance at 350 nm (as in Fig. 6) and the conformation of dynein crossbridges responsible for this activity can be studied at relatively high resolution by negative contrast electron microscopy (Sale and Satir, 1977; Warner and Mitchell, 1978). Similarly, crossbridge frequencies and distributions relative to doublet positions in disintegrating or reactivated cilia can be analyzed by thin section electron microscopy (Warner, 1978), and quantitative data thus obtained can be related to the physiological parameters of reactivation. These methods of analysis provide major tools in our continuing study of the sliding mechanism.

B. FORCE TRANSDUCTION: SLIDING AND BENDING

In order for sliding of microtubules to be converted into the rhythmic propagated bending characteristic of cilia and flagella, there needs to be some structural component(s) that has the capability to resist sliding shear and thus cause a bend to form. Active propagation of that bend similarly requires that the components resisting sliding have a transient aspect that must precede the leading edge of the bend in order for the bend to be propagated. Three components in the cilium could function in this way: the dynein arms, the nexin links, and the radial spokes. A case can be made for any or all of these structures being directly involved in the bending mechanism; however, evidence for that involvement exists only in the case of the radial spokes.

By using several nonmotile flagellar mutants of the alga *Chlamydomonas,* Allen and Borisy (1974) and Witman *et al.* (1978) have been able to show that mutants deficient in spoke structure or spoke–central sheath interactions are unable to form and propagate bends even though the sliding mechanism can be reactivated in these flagella. The direct implication of spoke involvement in the bending conversion is clear; the mechanism of that involvement remains uncertain.

Earlier, we had demonstrated that the radial spokes in mussel gill cilia undergo apparent cycles of attachment–stress–detachment during the course of normal ciliary beating (Warner and Satir, 1974). This cycle apparently occurs because the sliding displacement of the doublet microtubules relative to the central pair microtubules (spoke attachment site) exceeds the capability of the attached spokes to stretch or otherwise accommodate that displacement. This observation implied that the spokes were involved either in the formation, the propagation, or the regulation of the active bend. More direct evidence concerning spoke function remains to be obtained. However, it should be emphasized that motile arrays of microtubules, including some unusual sperm flagella that lack the elaborate radial spoke–central sheath system, still manage to initiate and propagate bends (Langford and Inoué, 1979; Nakamura and Kamiya, 1978; Schrevel and Besse, 1975). The organization of 9 + 2 cilia and flagella probably represents the greatest elaboration of structure and regulated function among microtubule-based systems, just as striated muscle represents the same for actin-based systems. Neither array precludes the essential motile features from being present in less or differently organized systems.

The geometrical features of a bend can be utilized to calculate the relative displacements of structures within the axoneme relative to values of bending considered normal for the organelle. This method was devised by Satir (1968) and elaborated upon by Warner and Satir (1974), and permits us to calculate sliding displacement (active or passive) for each pair of doublet microtubules during the formation of a normal bend. The data show that in the demonstrated absence of microtubule contraction (Satir, 1968; Warner and Satir, 1974), sliding displacements, be they in an active (effective stroke, d 3–4) or passive (recovery stroke, d 7–8) doublet pair position, do not exceed 30–40 nm in a beating cilium (Warner, 1976a). This emphasizes the importance of structures that constrain sliding, since, in their absence, sliding will continue for the length of the microtubules or 10–20 μm.

Considerable variability occurs in the beat frequency of cilia and flagella, both within and between species, although an average frequency is in the range of 30 Hz. Because of this high cycling frequency and presumably high sliding velocity, bends are difficult to preserve and study by methods other than high speed cinematography. Similarly, it is difficult to assign a particular physiological effect, which may, for example, manifest itself as a change in bend amplitude, to the bending mechanism per se when the physiological event may have also and primarily influenced active sliding. This places serious restrictions on potential analyses of bending as being a discrete, albeit an interacting, mechanism, separate from the generation of sliding shear by dynein–microtubule interactions. These restrictions become particularly severe when trying to understand how active and passive sliding might be under Ca^{2+}-modulated regulatory control that is potentially mediated by structures such as the radial spokes or central microtubule apparatus.

C. MECHANISMS OF CONTROL

1. *Regulation of Beat Parameters*

Complex regulatory mechanisms are required to transform the sliding forces responsible for ciliary motion into the bending waves typical of cilia and flagella. Since the sliding forces generated between adjacent doublets always have the same polarity such that doublet n pushes doublet $n + 1$ toward the tip of the axoneme (Sale and Satir, 1977), if unregulated, doublet pairs on opposite sides of the axoneme (doublets 3-4 versus 7-8) would produce bending in opposing directions at the same time. Obviously, opposing doublets cannot be active simultaneously in order for a net force and normal motion to result. While the dynein arms on one side of the axoneme are producing active sliding forces, those on the other side must be quiescent to allow passive sliding. Alternate regions of active sliding and passive response must propagate along the axoneme in synchrony with the shear resistance necessary to transform sliding into bending. The central pair–radial spoke complex may be involved in regulating alternate-side force generation. It possesses the necessary asymmetry (Linck, 1979), and the two central tubules apparently maintain a plane that is always perpendicular to the active bend plane (Gibbons, 1961; Satir, 1965; Tamm and Horridge, 1970). Omoto and Kung (1979) have extended these observations to show that the central pair may rotate 360 degrees counterclockwise during each complete beat cycle. The central pair itself is twisted to accommodate the three-dimensional dexioplectic beat pattern characteristic of protozoan cilia. The active or passive nature of central pair rotation has not been determined, so postulation of a cause-effect relationship must await further study.

Additional regulatory mechanisms must overlie those that produce bending in order to accommodate the effects of external stimuli on beat frequency, waveform, and effective stroke orientation. Considerable variation exists concerning these mechanisms from one class (or phylum) of organisms to another. Until unifying principles have been identified, each type of regulation must be considered separately.

Ciliate protozoa perform a complex avoidance response upon anterior mechanical stimulation, during which the organism reverses direction, turns, and resumes forward swimming. This is accomplished through changes in the power stroke orientation and beat frequency of the cilia, mediated through changes in intraciliary calcium levels (for reviews, see Naitoh and Eckert, 1974, and Machemer, 1977). Anterior stimulation causes membrane depolarization, which in turn opens membrane potential-sensitive calcium channels in the ciliary membrane and allows a transient influx of calcium from the external medium (Machemer, 1976; Eckert, 1972). The increased Ca^{2+} concentrations affect both beat frequency and orientation. The effective stroke orientation rotates progressively counterclockwise as the Ca^{2+} concentration is raised above 10^{-7} M (Naitoh and Kaneko, 1972), whereas the beat frequency reaches a minimum at about 5×10^{-7} M Ca^{2+} and then increases again. Two separate mechanisms appear to be involved, since intracellular injection of Ca^{2+} and EGTA give results that suggest that reorientation has a lower threshold for Ca^{2+} than do the frequency changes (Brehm and Eckert, 1978). Reorientation is seen in the absence of active beating in both live cells poisoned with Ni^{2+} (Naitoh, 1966) and in demembranated cell models (Naitoh, 1969; Naitoh and Kaneko, 1972, 1973), provided both Ca^{2+} and ATP are present. This strongly implicates an unidentified Ca^{2+}-sensitive ATPase in ciliary control.

When cells are given a posterior mechanical stimulation, the membrane potential becomes hyperpolarized, beat frequency increases, the effective stroke rotates clockwise, and the cell swims forward at enhanced speed, thus avoiding the stimulus. These responses to hyperpolarizing potential are not modulated by changes in the intracellular calcium concentration (Brehm and Eckert, 1978) and are retained in *Paramecium* mutants lacking the membrane potential-sensitive calcium channels involved in reversal (Dunlap and Eckert, 1977; Oertel *et al.*, 1977; Ogura and Takahashi, 1976). No clues exist concerning the nature of the mechanisms involved.

Calcium is also involved in mediating phototactic responses in flagellated algae such as *Chlamydomonas*. Studies employing photostimulation of live cells (Schmidt and Eckert, 1976), injections of EDTA, Mg^{2+}, and Ca^{2+} into live cells (Nichols and Rikmenspoel, 1978a,b), and reactivation of the intact flagellar apparatus (Hyams and Borisy, 1978) or isolated axonemes (Besson *et al.*, 1978) have all shown that raising the Ca^{2+} concentration above 10^{-6} M causes an abrupt change from a highly asymmetric, ciliary waveform to symmetrical flagellar-type undulations. Calcium has almost the exact opposite effect on the motility of reactivated sea urchin spermatozoa (Brokaw *et al.*, 1974; Brokaw and

Gibbons, 1976). Increasing the calcium concentration from 10^{-9} to 10^{-6} M causes increasing beat asymmetry, and higher concentrations inhibit motility entirely. Although sea urchin sperm are not known to exhibit chemotactic responses, Ca^{2+}-mediated asymmetry has been implicated in the chemotaxis of spermatozoa from other organisms (Brokaw, 1974). Nothing is known of the mechanism except that Ca^{2+} effects on beat symmetry can be eliminated by brief trypsin digestion without affecting other beat parameters (Brokaw and Gibbons, 1976).

Yet another response to Ca^{2+} is seen in metazoa (reviewed by Aiello, 1974) and is typified by the ciliated epithelium of fresh-water mussel gills (Murakami and Takahashi, 1975; Tsuchiya, 1977; Walter and Satir, 1978). Calcium levels above 10^{-7} M cause ciliary arrest, with all organelles arrested at the same stage of their beat cycle. Since Ca^{2+} has no direct effect on microtubule sliding (see following), it must interact with a higher level regulatory mechanism. One last intriguing case of Ca^{2+}-mediated regulation occurs in trypanosome flagella (Holwill and McGregor, 1975). Bending waves of reactivated trypanosome models travel from tip to base of the organelle during normal swimming, but progress from base to tip when the Ca^{2+} concentration is raised above 10^{-6} M.

The elucidation of mechanisms for each of these calcium effects remains a challenging problem in the study of motility. The solutions may lead to a better understanding of the role of calcium, not only in the regulation of microtubule systems, but also as a more general secondary messenger in the transduction of cell surface stimuli.

2. *Regulation of Microtubule Sliding*

Inasmuch as free Ca^{2+} seems to modulate only active beat parameters, what controls activation of the sliding mechanism per se? We presume that the dynein arms do not always exist in an "on" mode but must respond to some external stimulus, physiological or mechanical, to activate crossbridge cycling. It now appears from several studies that free Ca^{2+} has little or no effect on the activation of sliding (Gibbons and Gibbons, 1972; Walter and Satir, 1978, 1979; Zanetti *et al.*, 1979), nor does it affect the direction or polarity of sliding (Mitchell and Warner, 1978b), which is always toward the ciliary tip (Sale and Satir, 1977). Calcium in the range of 10^{-3} M does partially inhibit active sliding (see Fig. 7 and Section IV, A), but only by being in competition with free Mg^{2+}.

Mictrotubule sliding *in vitro* is activated solely by the provision of $MgATP^{2-}$. Normal ciliary or flagellar beating can occur in very low levels of ATP (10–20 μM) (Brokaw, 1975). The sliding disintegration reaction illustrated in Fig. 7 and studied in our laboratory is similarly responsive to low levels of ATP. In fact, there appears to be a threshold effect such that no sliding occurs at ATP concentrations below 10^{-6} M and sliding at maximum velocity occurs at about 10^{-5} M ATP. While the substrate concentration clearly can have a rate limiting effect on enzymatic activity and beat frequency (Brokaw, 1975), it appears unlikely that this "control" could occur by the cell manipulating the concentration of ATP^{4-}. In contrast, the concentration of free Mg^{2+} can be used to control the disintegration reaction (substrate concentration) over a wide range of cation concentrations (10^{-5} to 10^{-3} M). It is not known if the cilium possesses a mechanism to regulate its internal concentration of free Mg^{2+}, but Mg^{2+} is not generally viewed as a regulatory cation in view of its high concentration and widespread role as an enzyme substrate cofactor, although in the case of cilia or flagella, that level of control may be all that is needed. Determining the mechanism by which dynein is activated or deactivated will be an important step in our understanding of ciliary and flagellar motion.

III. Molecular Characteristics of Dynein

A. Protein Chemistry of the Molecule

1. Extraction Procedures

Studies of dynein biochemistry began with the observation that ATPase activity could be separated from the bulk of the axonemal protein after the ciliary membrane was removed by treatment with glycerol or detergents (Gibbons, 1963). Two dynein extraction techniques have been extensively used; unfortunately, this produced results that vary somewhat from organism to organism. This has led to considerable confusion in the dynein nomenclature, and only further comparative biochemistry will clarify the situation. The two procedures are Weber-Edsall (high ionic strength) extraction, and dialysis against low ionic strength media (1 mM buffer + 0.1 mM EDTA); clearly, ionic interactions are the prime forces stabilizing the axonemal protein structure.

In most organisms, limited exposure to 0.5–0.6 M KCl releases a single dynein species into solution. This species has variably been referred to as dynein 1, dynein A, or 30S dynein: the latter designation referring to the apparent sedimentation coefficient of the aggregate. Such crude KCl extracts contain about 60% of the ATPase activity and 20% of the protein of the axoneme, and generally produce a single peak of activity on sucrose gradients coinciding with a prominent protein peak. The second dynein species, dynein 2, dynein B, or 14S dynein, does not usually appear in KCl extracts, but there are exceptions (Brokaw and Benedict, 1971; Piperno and Luck, 1979).

Extraction by low-ionic-strength dialysis removes both species of dynein, which usually run on sucrose gradients as separate peaks of activity with sedimentation coefficients of 20–30S and 11–14S. In a few organisms, both species run as a single peak at 12–14S, and it is often possible to convert the larger, faster moving aggregates into more slowly sedimenting forms by expo-sure to urea or SDS (Hoshino, 1974), sonication (Hoshino, 1975), extremes of pH (Gibbons, 1963), and other harsh treatments (Gibbons and Fronk, 1979). Such a transition is often accompanied by either activation or a complete loss of activity. Dialysis removes more ATPase activity (95%) and more protein (60%) than does KCl extraction. Electron microscopy has shown that KCl extraction removes primarily the outer rows of dynein arms, and the dynein aggregates in KCl extracts will rebind to the axoneme to restore the arms (see Section V, B). Dialysis removes both rows of dynein arms and the central pair complex (Gib-bons, 1963), and in the case of *Chlamydomonas,* several nucleotide kinases (Watanabe and Flavin, 1976). Only the 20–30S dynein 1 portion of dialysis extracts rebinds to the extracted outer doublets, restoring the outer rows of arms. This location of dynein 1 has been confirmed by immunological studies using a ferritin-conjugated antibody against a dynein 1 fragment (Ogawa *et al.,* 1977b); no equivalent localization of dynein 2 is available, although the biochemical analysis of *Chlamydomonas* mutants may soon yield the answer (Huang *et al.,* 1979).

2. *Polypeptide Composition*

Tetrahymena 30S and 14S dyneins consist of multipeptide complexes with estimated molecular weights (MWs) of 5.4×10^6 and 6×10^5, respectively (Gibbons, 1965b). It should be noted that these estimates may be too large, since other evidence suggests that the 30S form may have MW closer to 1×10^6 and a sedimentation coefficient closer to 24S (Gibbons and Fronk, 1979; Mitchell and Warner, 1979). Both contain several high-molecular-weight proteins, as well as minor components of smaller size. The larger proteins have estimated MWs in the 250,000–560,000 range when run on SDS–polyacrylamide gels (Mabuchi and Shimizu, 1974; Hoshino, 1975). Although the true MWs remain in doubt due to the lack of standard proteins in this very high weight range, it is apparent that the major band of 30S dynein has a higher MW than the major band of 14S dynein (Mabuchi and Shimizu, 1974); the two bands appear in an approximate 2:1 ratio in whole axonemes. Both 30S and 14S dyneins contain not one but several high-molecular-weight bands, some of which may be unrelated contami-nants.

Hoshino (1974, 1975, 1977) has attempted further fractionation of *Tet-rahymena* 30S dynein by sucrose gradient sedimentation after treatment with urea, SDS, and proteolytic digestion. He has succeeded in generating three active fragments sedimenting at 24S, 18S, and 14S, of which the larger two retain the ability to rebind to extracted axonemes. Further electrophoretic or other biochem-ical analysis may reveal the relationships between these fragments, which could be used as an approach to understanding the molecular architecture of *Tet-rahymena* dyneins.

Echinoderms (several sea urchin species and one starfish) have been used extensively in studying dynein biochemistry, both because of their availability and because of their usefulness in correlating biochemical and enzymological data with motility parameters (e.g., beat frequency). Earlier studies (Mohri *et al.*, 1969; Ogawa and Mohri, 1972) revealed only a single ATPase component, and further purification by column chromatography was for the most part unsuccessful. Gibbons *et al.* (1976) describe a multistep extraction procedure that successively removes four different high-molecular-weight peptides. The most slowly migrating band was extracted by brief exposure to 0.5 *M* KCl, possessed ATPase activity, and was labeled dynein 1. One of the three remaining proteins copurified with ATPase activity on Sepharose and hydroxyapatite columns, and was labeled dynein 2. The other two bands did not appear to be associated with any enzymological activity. Dynein 1 sedimented at 13S, with a MW of 800,000 based on Stoke's radius and S value, and 520,000 MW based on Hedrick-Smith gels. Dynein 2, by the same two methods, has an estimated MW of 690,000 and 720,000, respectively. SDS gels revealed high-molecular-weight (HMW) bands for dyneins 1 and 2 of 330 ± 40K and 325 ± 40K.

More recently, Gibbons and Fronk (1979) have extracted dynein 1 from *Tripneustes gratilla* in a form with latent ATPase activity. This latent activity dynein 1 (LAD1) can be activated by several treatments, all of which decrease the sedimentation coefficient from 21S to 10–14S. The activated form has lost the ability to rebind to the extracted axonemes. The LAD1 complex has an estimated MW of 1,250,000 and produces bands at 330,000, 126,000, 95,000, and 77,000 MW on SDS gels. This large aggregate probably represents the intact arm structure (see Section V, B).

Extraction of starfish spermatozoa has likewise produced two dyneins, a 12S and a 20S form, differing in their HMW peptide components. The two forms respond differently to variations in pH, ionic strength, divalent cations, and sulfhydryl reagents (Mabuchi *et al.*, 1976).

Dyneins of *Chlamydomonas*, when extracted by dialysis, sediment at 12S and 18S (Watanabe and Flavin, 1976). Similarly, two activity peaks are eluted when the extract is placed on a DEAE-Sephadex column. Piperno and Luck (1979) also find 12S and 18E forms in 0.5 *M* NaCl extracts of *Chlamydomonas* flagella. The 12S dynein is composed of two peptides, 315,000 MW and 19,000 MW, in a 1:1 ratio. The more complex 18S dynein contains three HMW peptides (330K–300K), and 10 low-MW peptides (86K–15K) of unknown stoichiometry. One of the HMW peptides of 18S dynein coincides with the ATPase activity peak in the sucrose gradient, whereas the other two have a slightly different distribution. Separation of the crude extract on a hydroxyapatite column resolved a third ATPase component not revealed by the gradient. No HMW peptides correlated with this fraction, which may be identical to the membrane-derived Ca^{2+}-ATPase discovered earlier (Watanabe and Flavin, 1976).

Similar analyses, using ^{35}S-labeled proteins and two-dimensional gel electrophoresis, have been performed on paralyzed flagella mutants lacking inner and outer dynein arms (Huang *et al.,* 1979). This study confirmed that both the 12S and 18S dyneins are outer arm components, as previously suggested by Fay and Witman (1977). Only two of the thirteen polypeptides that copurify with 18S dynein are present in the outer-arm defective mutants, whereas neither of the two 12S peptides are found. Analysis of low-molecular-weight peptides suggests that the third, minor ATPase found in the extracts of wild-type axonemes may be missing in the inner-arm defective mutants. If further experimentation confirms the location of both major dyneins in the outer arms, it will represent a major difference between *Chlamydomonas* and the *Tetrahymena* and sea urchin systems.

3. *Proteolytic Digestion*

Proteolytic digestion, while a highly illuminating probe of myosin structure, has not proved as useful in elucidating dynein architecture. Ogawa (1973) and Ogawa and Mohri (1975) subjected sea urchin dynein 1 to tryptic digestion and purified the resulting fragment on a hydroxyapatite column. This fragment (F1A) had MW of 400,000 by the Hedrick-Smith gel method and 370,000 from sedimentation equilibrium, and produced two bands on SDS gels at 190,000 and 135,000 MW. Antiserum produced against the fragment inhibited the ATPase activity of dynein 1 and F1A, and precipitated dynein 1 from a crude dynein extract, while having no effect upon dynein 2 (Ogawa and Mohri, 1975; Ogawa and Gibbons, 1976). A chymotryptic fragment of dynein 1, F1A', was found to be very similar to F1A (Ogawa, 1978). The ATPase activity of F1A' is inhibited by anti-F1A, and SDS gels of F1A' produce two bands that comigrate with the two F1A peptides. Ogawa and Gibbons (1976) produced a tryptic fragment of dynein 2, F2A, which also retained its enzymatic activity. Hedrick-Smith gels gave an estimated MW of 360,000, slightly smaller than F1A. F2A is insensitive to anti-F1A antiserum.

Tryptic digestion of *Tetrahymena* 30S and 14S dyneins does not yield fragments directly comparable to F1A and F2A of sea urchins (Hoshino, 1977). Mild digestion of 30S dynein yields ATPase peaks at 24S and 12S, whereas prolonged digestion gives a single 12S peak. SDS gels of this fragment reveal four major peptides with many minor contaminants. Tryptic digestion of 14S dynein produces a 12S protein peak devoid of enzymatic activity, suggesting that the active site is susceptible to even mild proteolysis. Thus there appear to be major differences between the susceptibilities of *Tetrahymena* and sea urchin dyneins. However, some of these differences may be due to the relative complexities of the dynein aggregates first subjected to proteolysis, since the dynein 1 of Ogawa and Mohri (1975) may be only a portion of LADI, and Gibbons and Fronk (1979) suggest that LADI may be equivalent to 30S dynein. The same holds true for dynein 2, a single ATPase-containing HMW peptide, as opposed to the more complex 14S dynein.

4. *Modification of Dynein*

Following in the footsteps of the muscle biochemists, several workers have examined the effects of sulfhydryl reagents upon dynein. N-Ethylmaleimide (NEM) has been the most widely used. Treatment of isolated 30S dynein with this mild reagent for short periods of time or at low concentration leads to activation of the ATPase activity (Blum and Hayes, 1974; Shimizu and Kimura, 1974). Further reaction leads to complete inactivation of both 14S and 30S ATPase activity. Conditions that cause activation of 30S dynein also lower its ability to recombine with extracted axonemes. This appears to be true regardless of the activating agent since the same is true of activation by NEM, heat, acetone, Triton X-100, and several other sulfhydryl reagents and harsh treatments (Blum and Hayes, 1977, 1978; Gibbons and Gibbons, 1979; Shimizu *et al*, 1977).

Treatment of whole, demembranated axonemes leads to similar activation under mild conditions followed by inactivation with prolonged exposure, although the latter is not seen with every reagent tested (Blum and Hayes, 1976, 1977, 1978). NEM treatment renders sea urchin flagella nonmotile, while only inhibiting a portion of the ATPase activity (Cosson and Gibbons, 1978). Homogenizing the axonemes lowers the activity still further. In fact, there is no difference between the activity of homogenized axonemes from NEM-treated and control sperm, suggesting that NEM only effects the ''motility-coupled'' ATPase activity. Since NEM induces rigor wave formation in beating sea urchin sperm, it may interfere with the dynein crossbridge cycle during a stage when the arm is bridged to the B-tubule. NEM is likewise unable to inhibit the ATPase activity of broken *Tetrahymena* cilia or to prevent Mg^{2+}-induced bridging (Mitchell and Warner, 1978a), although very low concentrations (10^{-5} M) completely inhibit sliding disintegration in $MgATP^{2-}$ (unpublished observations).

The modification of dynein by NEM can be partially prevented by ATP but not by other nucleotides (Shimizu and Kimura, 1977), suggesting that one sulfhydryl (SH) group may be located near the active site. Neither the stoichiometry of NEM addition nor the nature of the residues susceptible to modification is known, although progressive activation followed by inactivation requires a minimum of at least two sites.

5. *Immunological Analysis*

Antisera have been prepared against dynein 1, a tryptic fragment of dynein 1 (F1A), and several components of 14S and 30S dyneins of *Tetrahymena*. They have been used as probes of dynein subunit structure, intraaxonemal location, the interrelationships between the two axonemal dyneins, and the roles of dyneins in motility. Ogawa and Mohri (1975) prepared a rabbit antiserum against F1A from the sea urchin *Anthocidaris crassispina*. This antiserum formed a single precipitin line with both intact dynein 1 and F1A from three sea urchin species. The ATPase activities of dynein 1 and F1A from all three species were inhibited by 80 and 100%, respectively. Anti-F1A precipitates two proteins from dialysis extracts of *A. crassispina* flagella, revealed as one HMW dynein band and tubulin on SDS gels. This same antiserum gave 80% inhibition of dynein 1 from *Tripneustes gratilla*, whereas dynein 2 from this species was only inhibited by 5% (Ogawa and Gibbons, 1976). This shows that a great deal of sequence homology exists between dyneins from different species, whereas little homology was detected between dyneins 1 and 2 from the same species.

The beat frequency of reactivated flagella from several sea urchin species is also progressively inhibited by anti-F1A. Motility can be completely inhibited (less than 1% of control frequency), although 45–60% of the axonemal ATPase activity remains (Gibbons *et al.*, 1976; Okuno *et al.*, 1976). Under modified conditions (Ogawa *et al.*, 1977b), 80% of the axonemal activity is inhibited; the remainder is attributed to dynein 2, which has been estimated to account for 15% of the activity of sea urchin axonemes (Gibbons *et al.*, 1976). Since anti-F1A also inhibits the sliding disintegration of trypsin-treated axonemes (Masuda *et al.*, 1978), it is obvious that dynein 1 is directly involved in the sliding mechanism.

An antiserum against intact dynein 1 from *Strongylocentrotus purpuratus* spermatozoa had no effect on the ATPase activity of isolated dynein 1 or dynein 2 (*S. purpuratus*) or F1A (*A. crassispina*), although 70% of the dynein 1 formed a precipitate with the antibody (Ogawa *et al.*, 1977a). Both beat frequency and ATPase activity of whole sperm were inhibited by anti-dynein 1. Neither anti-dynein 1 nor anti-F1A reacts with an active fragment of *S. purpuratus* dynein 1, suggesting that the effects of these antisera on motility and enzymatic activity are not due to direct interaction with active-site determinants.

Nishino and Watanabe (1977) have used immunological methods to examine the relationship between *Tetrahymena* 14S and 30S dyneins. Of the four HMW bands seen on SDS gels of *Tetrahymena* axonemes, the first and fourth (in order of molecular weight) are 30S dynein components, whereas the closely spaced second and third bands are derived from 14S dynein. Each dynein also produces a unique set of low-molecular-weight bands, but considerable antigenic cross-reactivity between the low- and high-molecular-weight bands may indicate that the former are fragments of the latter, as suggested by Hoshino (1975).

The larger 30S HMW peptide induced antigens that produced three precipitin lines against denatured 30S dynein in Ouchterlony immunodiffusion tests. These antigens detected two of these three components (A and B) in 14S dynein as well, as seen by continuous precipitin lines (lines of identity). Antisera against the two HMW bands of 14S dynein formed two lines of identity between 14S and 30S dynein, representing components A and C, respectively. Antibodies against the lower 30S HMW band reacted only with 30S dynein, forming a single line.

All of this data suggests a very interesting subunit composition. Since antibodies against the 30S upper HMW band detect three different components, A, B, and C, this band must contain three different comigrating proteins. Because the three precipitin lines form lines of identity between 30S and 14S dynein, 14S dynein must contain three proteins with similar antigenic determinants. One antiserum against the upper 30S HMW band detected component C in 30S dynein but not in 14S dynein, showing that this component has some dissimilar antigenic determinants between the two dyneins. Based upon stoichiometric considerations, Gibbons and Fronk (1979) suggest that the major HMW band of LAD1 also consists of 2–3 comigrating polypeptides.

It is evident that considerable work is necessary before the present confusion over dynein biochemistry subsides. The dyneins are a class of proteins of extremely high molecular weight and occur as large aggregates together with several lower molecular weight proteins to form functional dynein arms. The larger dynein aggregates are able to interact with specific sites on both the A- and B-subfibers of outer doublet microtubules, as well as to cleave the terminal phosphate of $MgATP^{2-}$. Several fragments of these aggregates have already been produced, and a detailed correlation between their retained functional capabilities and their protein composition could yield much useful information. The recent application of high resolution two-dimensional gel electrophoresis to the study of *Chlamydomonas* dyneins (Huang *et al.*, 1979; Piperno *et al.*, 1979) has revealed a far greater complexity than had previously been suspected. At the same time, this technique coupled with genetic and functional analyses promises to provide a means of dissecting the motile apparatus one gene product at a time and of unraveling many of its complexities.

B. ENZYMATIC BEHAVIOR OF DYNEIN

Many of the enzymological properties of dyneins are dependent upon the relationship between the active-site peptide and the rest of the axoneme. The situation is similar to that encountered in the actin-myosin system, the F1 ATPases of chloroplasts and mitochondria, and multienzyme complexes in general. The enzymatic moiety is a portion of a larger macromolecular aggregate (the dynein arm), which, in turn, functions as an integral part of an even more complex suprastructure, the axoneme. The specific activity, pH dependence, divalent cation dependence, ionic strength dependence, and K_m for $MgATP^{2-}$ are all affected by the interactions of the active subunit with axonemal components. Since very little is known about these interactions at present, even less is understood regarding their effects on enzyme activity.

Since two different ATPase-containing peptides can be distinguished in most ciliary and flagellar axonemes by biochemical or immunological techniques, many studies of dynein have used some isolated, "purified" form of enzyme rather than intact axonemes. It is difficult to speak of the purification of dyneins in the same sense one would speak of the purification of a monofunctional cytosolic enzyme, since a functional unit cannot be strictly defined. Instead, one must find catalytically active peptides or complexes and then examine their interactions with other components in ever larger complexes until one reaches an understanding of the mechanochemistry of the entire organelle.

1. *Dynein in Situ*

Hoffmann-Berling (1955) used glycerinated "models" of trypanosomes and grasshopper spermatozoa to show a direct relationship between ATP concentration and beat frequency. This relationship has been confirmed and extended by many others since (Brokaw, 1961, 1975; Holwill, 1969; Gibbons and Gibbons, 1972). Brokaw and Benedict (1968) demonstrated a correlation between beat frequency and rate of ATP hydrolysis. They also found that the enzymatic activity could be functionally divided into motility-dependent and motility-independent fractions by shearing flagellar axonemes to render them nonmotile and recording the resultant decrease in specific activity. Such broken axonemes retain only 35–40% of the ATPase activity of unbroken, motile sperm axonemes.

Correlations between the motility-dependent fraction of the enzymatic activity and motility parameters such as beat frequency have generally used one of two variables. When substrate ($MgATP^{2-}$) concentrations are varied, there is a linear relationship between beat frequency and motility-coupled ATPase activity. If the viscosity of the medium is increased by adding polyethylene glycol (PEG), both beat frequency and motility-coupled activity decrease, but the relationship is not linear (Gibbons and Gibbons, 1972). However, since the activity of both motile and nonmotile sperm is decreased by PEG, the nonlinearity may be due to direct effects of PEG on activity, in addition to its effects on viscous drag. It is as yet unclear whether the motility-dependent and -independent fractions represent the activities of separate enzyme species (dyneins 1 and 2) or merely an alteration in kinetics taking place on uncoupling from motility.

Gibbons and Gibbons (1976) have shown that extraction of dynein 1 with 0.5 *M* KCl lowers the beat frequency and that rebinding of the extracted dynein restores beat frequency. In addition, the pH-activation profile and KCl-dependence of dynein 1 more nearly mimics that of motile sperm than that of nonmotile broken axonemes (Gibbons and Fronk, 1972), suggesting that dynein 1 may be the primary force-generating enzyme for the sliding tubule mechanism; a similar conclusion has been drawn from studying the effects of anti-dynein 1 antisera on microtubule sliding (Masuda *et al.,* 1978). Unlike the dynein, broken axonemes gave nonlinear double-reciprocal plots under optimal conditions for motility, suggesting the presence of two components of differing substrate affinity (Gibbons and Gibbons, 1972).

2. *Isolated Dynein*

Recent work has shown that the activity of extracted dyneins is dependent upon the extraction method and subsequent treatment. The specific activity of dynein 1 and 30S dynein is quite low after extraction by the KCl method. Subsequent rough treatment such as exposure to acetone, heat, sulfhydryl reagents (Blum and Hayes, 1977), Triton X-100, EDTA, or aging (Gibbons and Fronk, 1979) can cause a several-fold activation and reduce the ability of the dynein to rebind to the axoneme. When the more native form [which Gibbons and Fronk (1979) call latent activity dynein, LAD] rebinds to the axoneme, its activity is enhanced several fold (Blum and Hayes, 1977; Gibbons and Fronk, 1979).

Both dynein 1 and dynein 2 (A and B; 30S and 14S) require a divalent cation, since the true substrate is a metal–ATP complex and not free ATP (Hayashi, 1974). Mg^{2+} gives the highest activity and is generally regarded as being the physiological cation, although Ca^{2+}, Mn^{2+}, Co^{2+}, Ni^{2+}, and Fe^{2+} also support some activity (Gibbons, 1966). Both dyneins are highly specific for ATP; other nucleoside triphosphates are hydrolyzed at less than one-tenth the rate of ATP, although contaminating adenylate kinases have often resulted in an apparent ADPase activity that can be separated by further purification (Watanabe and Flavin, 1976).

Gibbons and Fronk (1972) have shown that dynein 1 has a simple pH activity curve and is activated by increasing KCl concentrations; dynein 2 is inhibited by KCl (Ogawa and Gibbons, 1976). Likewise, 30S dynein is activated by KCl, whereas 14S dynein is inhibited (Gibbons, 1966). Free ATP^{4-} is a competitive inhibitor of dynein 1 Mg-ATPase (Hayashi, 1974) although free Mg^{2+} has no effect on the enzyme. Most dynein preparations are inhibited by free Ca^{2+} when $CaATP^{2-}$ is utilized as substrate, and inhibition by free Mn^{2+} has also been reported (Ogawa and Mohri, 1972). *Tetrahymena* 30S dynein is only inhibited by Ca^{2+}, whereas 14S dynein is inhibited by free Mg^{2+} as well (Hoshino, 1974). No K_i has been reported for either ATP or a divalent cation [disregarding the incorrect kinetic analysis of Hayashi (1974)], probably because of the many difficulties encountered in accurately determining ATP–cation equilibria (Storer and Cornish-Bowden, 1976). This same difficulty may contribute to the variability in reported K_m values, which are generally between 20 and 100 μM.

Several authors have noted the ability of microtubules or soluble doublet fractions to stimulate the activity of isolated dynein preparations. In addition to the activation that occurs upon rebinding of dynein to outer doublets, crude ciliary tubulin (Otokawa, 1972; Ogawa and Mohri, 1972) and purified brain tubulin (Hoshino, 1976) will also stimulate dynein ATPase activity. Hoshino (1976) reports activation of *Tetrahymena* 30S dynein, 14S dynein, and a 12S tryptic fragment of 30S dynein by purified porcine brain tubulin (free of accessory proteins), suggesting that dynein arms in the axoneme interact directly with tubulin dimers of the B-subfiber. Hoshino's kinetic analysis of this activation suggests that free dynein binds tubulin more readily than the dynein–substrate complex does, a result strikingly similar to the effects of substrate binding on the myosin–actin interaction of striated muscle (Lymn and Taylor, 1971). A similar conclusion was reached from studying the effects of ATP and ATP analogs on the attachment of dynein arms to the B-subfiber (Takahashi and Tonomura, 1978; Warner, 1978). Although a detailed kinetic analysis such as Lymn and Taylor present has not yet been used on the dynein–tubulin system, one can tentatively conclude that substrate binding to the dynein active site releases dynein arms from the B-subfiber. This step is supported indirectly by studies using the nonhydrolyzable imido analog of ATP, AMP-PNP, which relaxes rigor-wave sperm (Penningroth and Witman, 1978), and by the direct observation that ATP releases dynein from B-subfiber in the presence of vanadate, a potent inhibitor of dynein ATPase (unpublished observations). Extending the speculation, ATP hydrolysis may create an activated enzyme–product complex, which releases mechanical energy upon reattachment to the B-subfiber.

A report recently appeared in the literature describing unique kinetics associated with solubilized dynein from *Mytilus edulis* gills (Nakamura and Masuyama, 1977). Although the steady-state rates of liberation of ADP and P_i were equal, there was an initial burst of P_i not accompanied by an equal burst of ADP. The rate of P_i liberation required 1–2 minutes to reach a steady-state velocity, although ADP was liberated at a constant rate from time zero. The P_i burst was not affected by the addition of 0.2 M KCl, which stimulated the steady-state velocity, while replacement of Mg^{2+} with Ca^{2+} inhibited the burst and had little effect on the steady-state velocity. The interpretation of this otherwise interesting data is clouded by the unfortunate use of a crude dialysis extract, which may well have contained more than one dynein ATPase as well as other axonemal proteins. Although the authors outline some plausible mechanisms for an initial P_i burst, it is more difficult to explain the 1–2 minute lag prior to attainment of steady-state P_i release. Obviously, only further experimentation can lead to a clearer answer, but the data suggest the formation of an easily detected phosphorylated intermediate.

Very little work has been devoted to the examination of dynein ATPase inhibition kinetics. As mentioned above, free ATP^{4-} is a competitive inhibitor, as is its nonhydrolyzable analog AMP–PNP (unpublished observation). CTP (Nagata and Flavin, 1978), and ADP and GDP (Gibbons and Gibbons, 1972) are also (presumably competitive) inhibitors, whereas free Ca^{2+} and free Mn^{2+} inhibit by an unknown mechanism. Vanadate, an uncompetitive dynein ATPase inhibitor, was identified as a contaminant of Sigma equine muscle ATP preparations (Cantley *et al.*, 1977). Both 30S and 14S dyneins from *Tetrahymena* are sensitive to vanadate, the Mg-ATPase activity being far more sensitive than the Ca-ATPase activity (Kobayashi *et al.*, 1978; Nagata and Flavin, 1978). Increasing the KCl concentration increases the apparent K_i, which in the absence of KCl is in the micromolar range. The mechanism of vanadate inhibition is at present unknown; vanadate at higher concentrations also inhibits Na,K-ATPase, alkaline phosphatase, and actomyosin ATPase (Kobayashi *et al.*, 1978). It inhibits microtubule sliding in sea urchin flagella (Sale and Gibbons, 1979), and in the mitotic apparatus of mammalian cells (Cande and Wolniak, 1978), although it has no effect on the ATP-induced dissociation of the dynein arm–B-subfiber complex in cilia and flagella (Gibbons and Gibbons, 1978; Mitchell and Warner, 1979; Okuno, 1979; Sale and Gibbons, 1979). This suggests that vanadate interferes with the active-site mechanism without preventing substrate binding at the active site.

C. STRUCTURAL CONFORMATION OF DYNEIN ARMS

The dynein arms occupy two rows along the A subfiber of the outer doublet microtubules. Individual arms are spaced at 24 nm within each row (Amos *et al.*, 1976; Warner and Mitchell, 1978) and probably have a staggered or offset periodicity between rows that reflects their respective positions on the helical lattice of tubulin subunits (Warner, 1976a). The offset positioning may add to the asynchrony of arm crossbridge cycling that is necessary for cumulative sliding to occur. Structural and chemical solubility differences have long been noted between the outer and inner rows of arms (Allen, 1968; Gibbons and Gibbons, 1973), but both rows must be regarded as being functionally equal since KCl extraction of the outer rows of arms results only in a reduction in the beat frequency while other motile parameters remain unaffected (Gibbons and Gibbons, 1973).

By using both *Unio* and *Tetrahymena* cilia, we have been able to identify the major subunits present in both the 14S and 30S dynein fractions in the electron microscope (Warner *et al.*, 1977). Both inner and outer rows of arms can be extracted from *Tetrahymena* cilia by brief exposure to 0.5 *M* KCl. When the extract is run on a linear 0–30% sucrose density gradient, three protein peaks are found corresponding to sedimentation coefficients of 30, 14, and 8S. Each of the peaks possesses substantial Mg^{2+}-activated ATPase activity. The 30S peak will recombine as functional dynein arms with the extracted axonemes (see Section V,B). When 14S dynein is examined in the electron microscope after negative contrasting with uranyl acetate, a relatively homogeneous population of 9.3 nm spherical subunits is observed (Fig. 3). The size of these subunits corresponds to a MW of about 350,000. Although at this level of resolution the native dynein

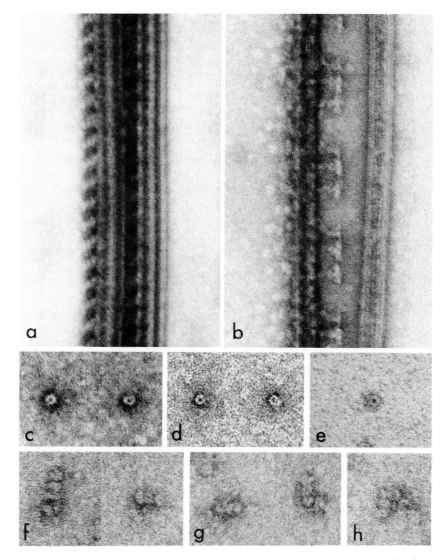

FIG. 3.   Electron micrographs of negatively contrasted dynein. Figure (a) shows the three large subunits of intact dynein arms, whereas Fig. (b) shows the arm subunits after partial dissociation. Figures (c–e) show single subunit 14S dynein from *Unio* (c, d) and *Tetrahymena* (e). Figures (f, h) shows 30S dynein from *Tetrahymena,* which consists in part of 3-subunit aggregates, similar to intact arms. (a, b), ×225,000; (c–h), ×320,000. (From Warner *et al.,* 1977, 1978; and unpublished results.)

molecule appears to possess internal symmetry, the images suggest that the molecule is built up from smaller structural subunits. In general, however, the enzyme has an appearance that is similar to many oligomeric enzymes when viewed by these methods. Most importantly, the molecule clearly lacks the highly asymmetric structure associated with myosin, even though a high-molecular-weight tryptic fragment of dynein can be obtained (Ogawa and Mohri, 1975), suggesting the presence of some internal molecular asymmetry.

The 30S dynein fraction consists of a relatively homogeneous population of subunit aggregates, in which two to four of the 9 nm subunits are visible (Fig. 3). Unfortunately, the spacial relationships between the subunits is not well resolved: the overall population bears only a general resemblance to the structures we see as intact arms on whole cilia.

High resolution images of negatively contrasted *Tetrahymena* cilia disintegrated in $MgATP^{2-}$ show extensive segments of microtubules where the dynein arms can be seen to comprise three large spherical subunits having a center-to-center spacing of about 8 nm (Fig. 3) (Warner and Mitchell, 1978). These subunits resemble in size and shape the 9.3 nm subunits identified as the major subunit of dynein 1 (Warner *et al.*, 1977).

For the subunits to have assembled into a 3-subunit chain implies high specificity and polarity of their respective bonding sites and suggests that there must be some minor differences between the major subunits. Furthermore, at least one of the three subunits has a binding site for A-subfiber tubulin and at least one has an active, but transient binding site for B-subfiber tubulin, suggesting that the subunits as a functional polymer are highly polarized or asymmetric.

As noted, Gibbons and Fronk (1979) characterized whole arm (21S) dynein from sea urchin sperm flagella, based on ATPase activity and SDS gel electrophoresis. They found that the 21S particle consisted of 2–3 high-molecular-weight subunits (330,000) and three medium weight subunits (126,000, 95,000, and 77,000). Recombination of 21S dynein with the extracted flagella restored the ATPase activity and beat frequency (Gibbons and Gibbons, 1979). It appears probable that the three large morphological subunits of *in situ* dynein arms and isolated 30S dynein from *Tetrahymena* cilia correspond to the three high-molecular-weight subunits studied by Gibbons and Fronk. Furthermore, their analysis emphasizes that functional dynein arms are a complex polymer of at least six subunits whose structure, assembly, and cooperative function remains to be determined.

With regard to intact or *in situ* dynein arms, we presently recognize three distinct conformational states, although we cannot yet equate these states with their physiological position in the crossbridge cycle. Analyses of arm structure have come predominantly from electron microscopic studies of $MgATP^{2-}$ disintegrated *Tetrahymena* cilia (Sale and Satir, 1977; Warner and Mitchell, 1978). The technique affords relatively high resolution images of the arms, but it is difficult to control the physiological environment of those arms with respect to free Mg^{2+}, $MgATP^{2-}$, $ADP \cdot P_i$, and free ATP^{4-}.

Because of a general inability to structurally define the inner rows of arms and a similar inability to distinguish them from the outer rows of arms in negatively contrasted preparations, the remarks and observations that follow are limited to the outer rows of dynein arms. Setting aside the dynein substrate factors, we can recognize an apparently relaxed or unactivated conformational state of the arms, since it appears in cilia preparations that have not been exposed to exogenous $MgATP^{2-}$. However, we cannot yet rule out the possibility that dynein bound $ADP \cdot P_i$ is carried through the isolation procedure. Nevertheless, the arms can be seen in an extended configuration, uniformly tilted at an angle of 32° from the perpendicular and toward the base of the cilium (Warner and Mitchell, 1978). Importantly, this orientation is opposite to the direction of force generation in these cilia (Sale and Satir, 1977). The base-tilted arms are generally not seen to be attached to an adjacent B-subfiber. In $MgATP^{2-}$ reactivated cilia, the extended arm configuration is also present (as in Fig. 3a) but sometimes appears to be attached to the B-subfibers (Warner and Mitchell 1978). There is considerable ambiguity in these images, however, and it must be remembered that the interaction of only a few dynein arms with the B-subfiber could result in a morphological condition that appeared as if all arms were attached. However, both extended attached and unattached configurations are consistent with mechanical states of their crossbridging cycling (see Section V,A).

In $MgATP^{2-}$-disintegrated *Tetrahymena* cilia, we see a second structural conformation that we at first attributed to the inner rows of arms (Warner and Mitchell, 1978) but now think may be related to a conformational change in the arms. This arm conformation only appears between two juxtaposed doublets, never along free doublets, implying that the conformation may be dependent upon physical interaction with the B-subfiber. The arms have the characteristic base-directed tilt, but the terminal or distal region or subunit of the arm appears to be oriented or tilted toward the tip of the cilium (Fig. 4a). The images are not as clear as those of the extended arms, but they bear considerable resemblance to the unique orientation of the doublet number 5 arms that typifies the doublet 5–6 bridge of certain cilia and flagella (Linck, 1979; Warner, 1976b). The arms along doublet number 5 in both *Unio* and *Pecten* gill cilia have their terminal subunit tilted toward the cilium tip (Fig. 4b) and it is apparently attached to the B-subfiber, although it is not yet clear if doublets 5 and 6 are capable of relative sliding displacement. The tilted subunit conformation of the arms is likewise consistent with their mechanical cycling (Section V,A). Satir (1979) has observed an apparently unattached but potentially similar arm conformation that he terms "flattened" and suggests may be the true inactive state of the arms.

Obviously, clarification and accurate description of the various structural and physiological states of the dynein arms is absolutely essential before we can

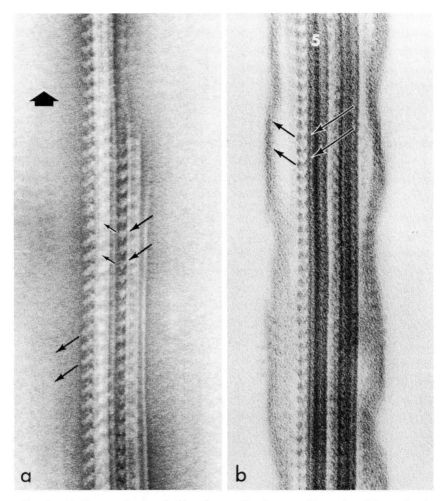

FIG. 4. (a) Two overlapping doublets from a *Tetrahymena* cilium after disintegration in MgATP^{2-}. The dynein arms on subfiber A of the free doublet are extended and tilt towards the cilium base. Those arms on the adjacent doublet are likewise tilted toward the base where they attach to the A-subfiber, but in the region of the adjacent B-subfiber, the terminal segment of the arm appears to be oriented toward the cilium tip (arrows). ×145,000. (b) This micrograph shows the dynein arms along doublet 5 of a *Unio* cilium where the base–tip orientation is very clear (arrows). The tip directed subunits are normally attached to the B-subfiber of doublet 6 (missing from the section plane) in this species. ×145,000.

understand their cycling behavior. To this end, important advances have been made with regard to arm behavior under various substrate conditions and these are described in the ensuing section.

IV. Force Generation by Dynein

The rhythmic beating of cilia, flagella, and spermtails results from an energy-dependent sliding filament mechanism. Active sliding involves the ATPase dynein, which when mechanically coupled, utilizes $MgATP^{2-}$ as its primary substrate. The enzyme's activity is directly related to the formation of transient crossbridges between doublet microtubules in order to produce relative displacement or sliding of the tubules. The major problem associated with the sliding filament mechanism in cilia and flagella has been the inability to demonstrate or preserve physical interaction of the dynein arms with B-subfiber sites. However, we have recently been able to show that dynein arms of both *Tetrahymena* and *Unio* cilia will attach to their adjacent B-subfibers solely in response to the addition of a divalent cation such as Mg^{2+} (Fig. 5) (Warner, 1978; Mitchell and Warner, 1978b; Zanetti *et al.*, 1979).

Although our approach to the analysis of dynein's interaction with microtubules is rather different, by comparison, myosin decoration of F-actin does not manifest a similar dependence on a divalent cation (Tonomura, 1973). This permits us to question if the Mg^{2+}-dependent attachment of dynein to the B subfiber is physiological, i.e., is it a manifestation of part of the crossbridge cycle, and if so, is it the physiological equivalent of the stable myosin–F-actin complex.

A. Crossbridge Cycling and Enzymatic Behavior

Dynein presumably interacts with subfiber B tubulin as part of its force-generating cycle, but unlike the actin stimulation of myosin ATPase, there appears to be no major stimulation of isolated dynein ATPase by tubulin. Hoshino (1976) has reported that the ATPase activity of 30S dynein from *Tetrahymena* cilia was stimulated by about 35% in the presence of solubilized 6S brain tubulin. This value agrees with our own observation of 30% enhancement of both 30S and 14S *Tetrahymena* dynein activity when measured in the presence of either intact ciliary doublet microtubules or solubilized 6S subfiber B tubulin.

The sensitivity of dynein arms to both Mg^{2+} and ATP can be monitored by both spectrophotometry (A_{350}) and electron microscopy (Fig. 6). The addition of divalent cations to *Tetrahymena* cilia suspended in 40 mM HEPES buffer and 0.1 mM EDTA causes an immediate 5% increase in absorbance. This increase is accompanied by uniform attachment of the dynein arms to the B-subfibers. Since

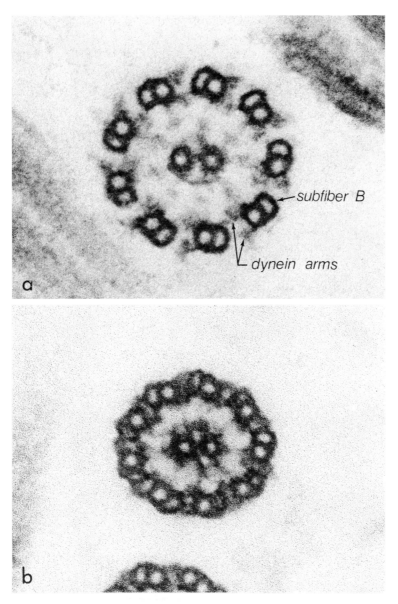

FIG. 5.   Two demembranated cilia from *Tetrahymena*, treated identically except that the cilium in (b) was fixed in the presence of 6 m*M* MgSO$_4$ and all dynein arms have become attached to their adjacent B-subfibers. ×240,000. (From Zanetti *et al.*, 1979.)

numerous factors can influence turbidity changes at 350 nm, the extent to which the bridging of the dynein arms is related to the increase in turbidity is uncertain, but the two methods (EM and A_{350nm}) nevertheless demonstrate responses induced solely by the addition of Mg^{2+} ion. The subsequent provision of 0.1 mM ATP to these cilia causes an immediate 30–40% decrease in turbidity, accompanied by active sliding disintegration of the cilia (Fig. 6).

Attachment of the dynein arms to the B-subfiber is in equilibrium with the free Mg^{2+} concentration (Table I; Zanetti *et al.*, 1978). Is this Mg^{2+}-dependent

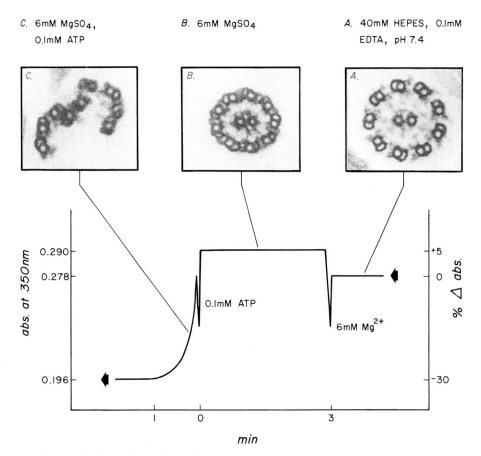

FIG. 6. The figure depicts the morphological condition of *Tetrahymena* cilia as associated with their optical absorbance properties (A_{350nm}) in response to Mg^{2+} and ATP. The addition of Mg^{2+} to demembranated cilia (A) causes a 5% increase in turbidity that is accompanied by uniform attachment of the dynein arms to the B subfibers (B). When ATP is added to these cilia, there is an immediate 35% decrease in turbidity accompanied by spontaneous sliding disintegration of the cilia (C). (From Zanetti *et al.*, 1979; and unpublished results.)

TABLE I

Dynein Crossbridge Response to Changes in Divalent Cation Concentration

Cation concentration (mM)	Percentage of bridging[a]	
	Mg^{2+}	Ca^{2+}
0.05	7.1 ± 3.7	9.4 ± 5.3
0.10	9.3 ± 3.7	13.8 ± 3.8
0.50	31.4 ± 2.8	37.9 ± 5.8
1.00	48.6 ± 9.8	68.1 ± 2.8
1.50	73.2 ± 11.0	88.1 ± 1.2
2.00	88.1 ± 4.1	98.9 ± 3.4
3.00	100	100
6.00	100	100
10.00	100	100

[a] The percentage of bridging was determined from 60 cilia at each concentration of divalent cation. The values represent the mean and standard deviation of two separate counts. (Data from Zanetti et al., 1979.)

reaction a physiological manifestation of part of the dynein crossbridge cycle or is the cation simply backing up the cycle to a state that is stable for the attached crossbridge? Several features of the reaction argue that it may have functional significance. Most notably, the reaction occurs at Mg^{2+} concentrations (1–3 mM) considered physiological for this cation, and it is sensitive to physiological concentrations of added ATP. At low Mg^{2+} concentrations (0.1–2 mM), the attached arms are unequally but symmetrically distributed with respect to the cross-sectional axoneme profile. The addition of 0.1 mM ATP to *Unio* cilia suspended in 2 mM Mg^{2+} (88% arm attachment) reduces the frequency of arm attachment to 48%, and with a similar but unequal distribution of the remaining attached arms (Warner, 1978). Those doublet pairs positioned perpendicular to the bend plane where relative sliding is necessarily minimal (Warner, 1976) show a greater percentage of bridged arms than those doublet pairs lying parallel to the bend plane where sliding is maximal. We have suggested (Mitchell and Warner, 1978b) that this distribution may reflect the normal distribution of sliding within the organelle if the unattached arms represent those arms undergoing a higher cycling frequency and hence have a lower probability of being captured (fixed) in the attached state. Preliminary results with the nonhydrolyzable ATP analog AMP-PNP indicate that nucleotide mediated release of the attached crossbridge is not dependent upon nucleotide hydrolysis. If ATP releases the crossbridge, how do its hydrolysis and the Mg^{2+} requirement for the crossbridge reaction fit into the crossbridge cycle?

Does the Mg^{2+}-dependent attached crossbridge exist in a state that is resistant to sliding? The muscle term ''rigor'' has been applied to sperm flagella whose active bend form was arrested by dilution of the ATP supply (Gibbons and Gibbons, 1974; Gibbons, 1975). Muscle rigor occurs because the myosin crossbridge is unable to detach owing to ATP depletion, ADP–ATP exchange being necessary for relaxation of the otherwise stable attached crossbridge. Are the Mg^{2+}-induced dynein bridges the equivalent of myosin rigor crossbridges?

Dynein crossbridges cannot be preserved for electron microscopy after dilution of $MgATP^{2-}$, even though the bend form is preserved. Mg^{2+} must be present in order to preserve the attached crossbridges as it was in Gibbons' (1975) original description of bridged arms in ''rigor'' sperm flagella (I. R. Gibbons, personal communication). Bridged dynein arms can be only partially relaxed by ATP or EDTA, but dilution of the Mg^{2+} concentration to 10^{-5} M results in a crossbridge frequency of near zero (Warner, 1978; Zanetti et al., 1979; Mitchell and Warner, 1979). However, only ATP (Gibbons and Gibbons, 1973) or AMP–PNP (Penningroth and Witman, 1978) relaxes rigor sperm flagella. Similarly, if high concentrations of Mg^{2+} (10 mM) are used to maintain dynein bridging at 100%, the subsequent addition of 0.1 mM ATP results in normal motion (sliding) but relaxation of only 10–15% of the attached arms (Table II).

Another phenomenon has been reported that seems to conflict with our interpretation of the attached dynein crossbridge. Chelation of free Mg^{2+} by excess EDTA does not relax the arrested or rigor bend forms of *Lytechinus* sperm flagella (Penningroth and Witman, 1978). Similarly, the rigor bend forms of *Tripneustes* sperm flagella can be maintained after dilution of free Mg^{2+}, as well

TABLE II

DYNEIN RESPONSE TO SELECTED DIVALENT CATIONS[a]

	Bridging frequencies (%)			ATPase activity (%)	Disintegration (%)
Cation	Control	ATP	ADP		
None	0	0	0	2	3
Mg^{2+}	88	75(13)	69(19)	100	100
Ca^{2+}	99	27(72)	28(71)	66	58
Co^{2+}	92	69(23)	84(18)	80	2

[a] Standard conditions: demembranated *Tetrahymena* cilia suspended in 40 mM HEPES buffer at ph 7.4. Cation and nucleotide concentrations were 2 and 0.2 mM, respectively. The numbers in parentheses are the percentage of reduction of bridging frequency under the stated conditions. The comparable reduction in frequency by ADP probably reflects conversion of ADP to ATP by an adenylate kinase. The standard error in the bridging frequency counts is ±4.8% (nine bridged arms per cross section = 100% bridging). (From Zanetti et al., 1979; and unpublished results.)

as by dilution of ATP (Gibbons and Gibbons, 1978). Based on the premise that the arms rather than the spokes or nexin links are the structures responsible for maintaining the arrested bend forms (Gibbons and Gibbons, 1974), both observations imply that free Mg^{2+} is not necessary to maintain attached dynein arms. Our data on the actual dynein crossbridges and their sensitivity to ATP imply exactly the opposite: free Mg^{2+} is absolutely essential to maintain the attached crossbridge. Clarification of the conflict should firmly establish either the dissimilarity or similarity of this aspect of dynein behavior to the myosin rigor crossbridge.

We are presently studying the kinetics of dynein crossbridge release by nucleotides and find that the sensitivity of the crossbridges to ATP saturates at an ATP concentration of about 10^{-5} M. Increasing the ATP concentration to as high as 2 mM does not cause additional release of crossbridges beyond the level that can be accounted for by complexing of free Mg^{2+} by ATP^{4-}, which simply shifts the Mg^{2+}-dependent bridging equilibrium to a lower value. Similarly, we find that when Ca^{2+} is substituted for Mg^{2+}, the sensitivity of the resulting crossbridges to ATP increases by several times (Table II).

The effects of divalent cations such as Mg^{2+} and Ca^{2+} have generally been seen only as components of the enzyme substrate and have not been observed to have any direct regulatory effect on either active beating or sliding (Gibbons and Gibbons, 1972; Zanetti et al., 1979). Several other cations promote enzymatic activity, but only Mn^{2+} can substitute for Mg^{2+} and still result in normal motion (Gibbons and Gibbons, 1972). The addition of 10^{-3} M Ca^{2+} to Mg^{2+}-activated cilia partially inhibits the sliding disintegration reaction, but probably only because Ca^{2+} competes for the relevant Mg^{2+}-binding sites in the reaction (Fig. 7; Gibbons and Gibbons, 1972; Zanetti et al., 1979) rather than exerting a direct regulatory influence on sliding. In contrast, as discussed previously, Ca^{2+} concentrations in the range of 10^{-7} to 10^{-5} M do have a regulatory influence on ciliary or flagellar beating. Fluxes in the free cation concentration cause either ciliary arrest (Satir, 1975; Tsuchiya, 1977) or reversal of the direction or type of beat (Eckert, 1977; Hyams and Borisy, 1978; Nichols and Rikmenspoel, 1978b).

B. Dynein–Microtubule Recombination

The dynein arms of isolated cilia generally retain a stable, noncovalent attachment to the A-subfiber, but prolonged exposure to 0.5 M KCl or chelating agents such as EDTA causes the arms to dissociate from the A-subfiber. In Tetrahymena, these extraction procedures release 30S dynein into solution, and it has been demonstrated in several species, including Tetrahymena, that incubation of the extracted axonemes with (30S) dynein and 2–5 mM $MgSO_4$ restores the arms to their native positions on the A-subfiber (Gibbons, 1963, 1965b; Haimo and Rosenbaum, 1976; Mabuchi et al., 1976; Mitchell and Warner,

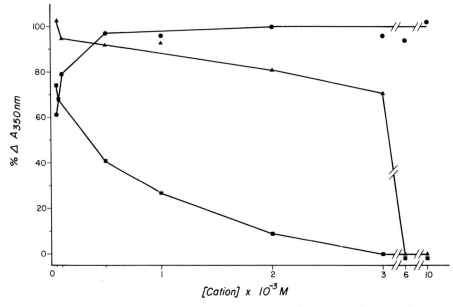

FIG. 7. The graph depicts the sliding disintegration response (ΔA_{350nm}) of isolated *Tetrahymena* cilia after exposure to 0.1 m*M* ATP and variable concentrations of Mg^{2+} (●); and 2 m*M* Mg^{2+} plus variable concentrations Ca^{2+} (▲), and Co^{2+} (■). Both Ca^{2+} and Co^{2+} inhibit $MgATP^{2-}$ activated sliding, probably by being in competition for the relevant Mg^{2+} binding sites. (Data from Zanetti *et al.*, 1979.)

1978b; Shimizu, 1975). This reaction apparently reflects a simple noncovalent reassembly phenomenon, unrelated to the force-generating behavior of the arms.

Studies of the rebinding of dynein to extracted axonemes have used several techniques for monitoring the extent of the recombination. Axonemal suspensions scatter light, and the turbidity of the solution is dependent in part on the mass of the light-scattering particles (Gibbons, 1965a). Dynein itself is too small to contribute substantially to the turbidity (A_{350nm}), but as it binds to the axonemes, their mass apparently increases and so does the turbidity of the suspension. The axonemes can also be removed from the solution by centrifugation, and the pellet and supernatant can be tested for protein content or ATPase activity. In addition, the pelleted axonemes can be fixed and examined by electron microscopy to determine what structures have reappeared. Early studies using these techniques demonstrated that dynein will rebind in the presence of Mg^{2+} (Gibbons, 1963); maximal binding occurs at about 5 m*M* $MgCl_2$ or $CaCl_2$ and is inhibited by ATP in a dose-dependent fashion (Shimizu, 1975). Electron microscopy has shown that rebinding is accompanied by the reappearance of the outer row of arms on the A-subfibers (Gibbons, 1963); the unequivocal recombination of inner arms has not been reported. Only the larger (20–30S) dynein aggregates recombine. These particles are probably intact dynein arms, and treatments that disrupt the aggregate generally destroy its recombining ability (Gibbons and Gibbons, 1979; Hoshino, 1974), although Hoshino (1975) has produced fragments of 30S dynein sedimenting at 24S and 18S that retain their binding properties.

Dynein arms generate sliding forces between adjacent doublet tubules, hence the distal end of the arm must interact with the B-subfiber of the adjacent doublet. Arms forming bridges between the A- and B-subfibers of intact axonemes are seen when Mg^{2+} or another divalent cation is present during fixation for electron microscopy (Warner, 1978). Hence, dynein arms possess tubule binding sites at either end; the proximal end binds to A-subfiber sites, while the distal end bridges to sites on the B-subfiber. The B-subfiber bridging frequency increases with the cation concentration, saturating at about 3 mM Mg^{2+} (Zanetti et al., 1979), and is markedly decreased when 0.1 mM ATP is also present in the fixation solutions (Warner, 1978).

The conditions that promote bridging to the B-subfiber are very similar to those that support dynein recombination with extracted axonemes. In fact, when Tetrahymena 30S dynein and extracted axonemes are recombined and fixed in the presence of 6 mM $MgSO_4$, the arms are attached at both ends, forming bridges between adjacent doublets (Fig. 8). Thus, it may be possible that B-subfiber attachment is the primary event in recombination, and that A-subfiber binding occurs only after proper repositioning of the arms through the Mg^{2+}-dependent B-subfiber association.

In order to clearly distinguish between interactions with the A-subfiber and the B-subfiber, the binding of isolated dynein to each site must be examined separately. Takahashi and Tonomura (1978) rebound 30S dynein to the B-subfiber of trypsin-treated Tetrahymena axonemes. Trypsin treatment of demembranated axonemes followed by exposure to $MgATP^{2-}$ causes sliding disintegration, separating the axonemes into individual doublets. This exposes the dynein binding sites on the B-subfiber but unfortunately also disrupts arms along the A-subfiber to some extent. In the presence of 2.5 mM Mg^{2+}, Takahashi and Tonomura showed 30S dynein binds to the B-subfiber of such doublets, forming a row of arms of similar dimensions and spacing as those on the A-subfiber. Arms on the B-subfiber tilt toward the base of the axoneme but at a slightly greater angle than A-subfiber arms. When 30S dynein was added to dialysis-extracted doublets, arms reappeared on both the A- and B-subfibers; addition of 1 mM ATP removed only the B-subfiber arms, which rebound after hydrolysis of the nucleotide.

Demembranated Tetrahymena axonemes as isolated by the procedures used in our laboratory (Warner et al., 1977), will undergo $MgATP^{2-}$-induced disintegration without prior proteolysis. About 85% of the axonemes slide apart, as esti-

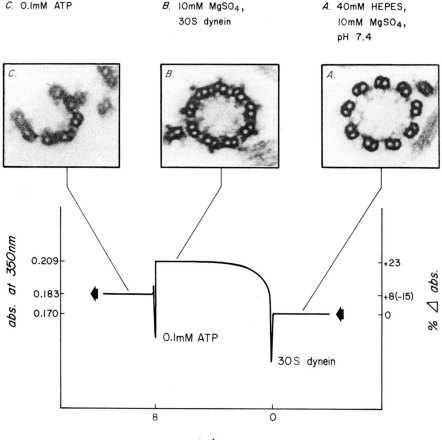

C. 0.1mM ATP   *B.* 10mM MgSO$_4$,   *A.* 40mM HEPES,
                                      30S dynein   10mM MgSO$_4$,
                                                                            pH 7.4

FIG. 8. The figure depicts the morphological condition of *Tetrahymena* cilia as associated with their optical absorbance properties (A_{350nm}) after extraction by dialysis to remove the dynein arms (A) followed by the subsequent addition of 30S dynein (B) and 0.1 m*M* ATP (C). The dynein is restored to its native attachment sites on both the A- and B-subfibers accompanied by a 23% increase in turbidity. The subsequent addition of ATP causes active disintegration of the cilia accompanied by a 15% decrease in turbidity. (Data from Mitchell and Warner, 1979.)

mated from observation by electron and phase-contrast microscopy. Dynein isolated by 0.6 *M* KCl extraction will bind to the exposed B-subfibers in the presence of a divalent cation, producing decorated doublets with an arrowhead appearance (Fig. 9) (Mitchell and Warner, 1979). Dynein binding also can be monitored by the increase in turbidity of a suspension of doublet tubules (Fig. 8); the increase in turbidity (A_{350nm}) is proportional to the amount of protein bound to the tubules, and the amount of added dynein that will bind is a saturable function of the divalent cation concentration (Fig. 10); Ca^{2+} promotes B-subfiber

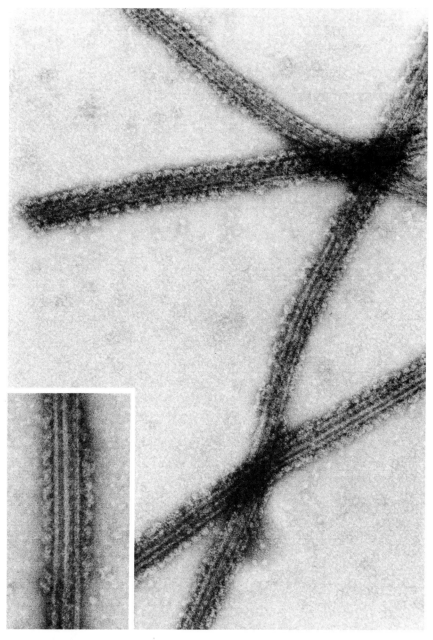

Fig. 9.   Isolated doublets from *Tetrahymena* cilia after recombination with 30S dynein in the presence of 6 m*M* MgSO$_4$. Dynein has decorated the exposed B-subfiber binding sites with a periodicity (24 nm) and polarity (base-directed) similar to the native position of dynein on the A-subfibers. ×118,000. Inset, ×180,000. (From Mitchell and Warner, 1979.)

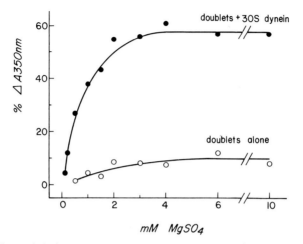

FIG. 10.   The graph depicts the recombination of 30S dynein (40 μg/m1) with the B-subfibers of isolated doublet microtubules (70 μg/m1) from *Tetrahymena* cilia as measured by ΔA_{350nm} in response to increasing concentrations of MgSO$_4$ (●). The control curve (○) shows the response of isolated doublets in the absence of added 30S dynein. (Data from Mitchell and Warner, 1979.)

dynein binding more efficiently than Mg^{2+}. This cation dependence appears to be the same phenomenon as the cation-dependent bridging of dynein arms in intact axonemes (Zanetti *et al.*, 1979).

When ATP is added to recombined doublets suspended in 6 mM Mg^{2+}, the turbidity of the suspension decreases to as little as 40% of its initial value. A portion of this decrease is seen when no dynein has been added and is therefore due to residual sliding of the few remaining intact axonemes. The remainder, by subtraction, is due to the release of arms from the B-subfiber. The turbidity drop due to dynein release reaches a maximum at about 10 μM ATP. The binding of dynein to the B-subfiber and its release by ATP are uneffected by 1-100 μM *meta*-vanadate, concentrations that completely inhibit microtubule sliding (Gibbons *et al.*, 1978).

Recombination of 30S dynein differs in two important aspects with regard to A- and B-subfiber sites. After recombination with the A-subfiber, the arms are stable in the presence of ATP or after dilution of free Mg^{2+} to a concentration that does not promote recombination ($<10^{-4}M$). In contrast, recombination with the B-subfiber is sensitive to both factors: the B-subfiber–dynein complex is inherently unstable in the absence of a divalent cation, and the addition of as little as 5 μM ATP will completely dissociate the complex in the presence of 10^{-3} M Mg^{2+}.

Similarly, the orientation of the recombined 30S dynein differs on the A- and B-subfibers. A-subfiber decoration manifests base-directed arms, whereas decoration of the B-subfiber also occurs with an arm orientation toward the cilium base (Fig. 9). But this is in effect opposite to the arms' orientation when attached only to the A-subfiber: when attached only to the B-subfiber, the arm has in effect become tip-directed with respect to the A-subfiber. Thus, the A- and B-subfiber binding sites on the dynein arm are likely to have opposite polarities, assuming that decoration of the B-subfiber is done by that dynein subunit that normally interacts with the B-subfiber rather than by the subunit that interacts with the A-subfiber. The sensitivity of the B-subfiber decoration to nucleotides argues that this is the case, although we need to be certain that the base-directed arms on the B-subfiber do not simply represent the preferred orientation of the arms in the absence of attachment to the A-subfiber.

If the terminal segment or subunit of the arm is tilted toward the cilium tip when the arms are attached to both the A- and B-subfibers, it would suggest that the simple addition of Mg^{2+} induces a major conformational change, as well as an enhanced protein–protein interaction. Such a marked change is by no means impossible (consider the effects of Ca^{2+} on the vorticellid spasmoneme; Amos *et al.*, 1976b), but the relationship of such a change to any postulated crossbridge cycle remains unclear.

C. Dynein Activity in Other Systems

In virtually every instance where microtubule arrays have been studied, neighboring tubules are seen to be joined in a periodic manner by links or crossbridges. The crossbridges generally fall into at least one of three functional categories: (1) active force transducing crossbridges; (2) rigid or inextensible bridges that probably serve an anchoring or stabilizing function; and (3) elastic or extensible bridges that serve to hold tubules together but must give in order to accommodate sliding. For general reviews of these components, refer to studies by Amos *et al.* (1976a), McIntosh (1974), Wellings and Tucker (1979), and Warner (1976a,b).

Of these three crossbridge categories, only the first is known to be mechanochemically active, and definitive evidence for and characterization of that activity exists only in the case of the dynein arms of cilia and flagella. However, if microtubule-based sliding filament mechanisms are to work elsewhere, we must presume that they interact with force-producing crossbridges that may be dynein or a dynein-like enzyme. Compelling evidence for sliding filament mechanisms and a dynein-like ATPase has been produced for only two other systems, the mitotic apparatus and the motile axostyle of certain protozoa. McIntosh (1973) has presented morphological data that support a sliding mechanism for axostyle motility and both Mooseker and Tilney (1973) and Bloodgood (1975) have been able to isolate the axostyle and demonstrate *in vitro*

ATPase-dependent motion. In both instances, the enzymatic activity shared the major features of dynein ATPase.

Studies of motion during the mitotic sequence have a long history (for review see McIntosh, 1979) and the best morphological evidence for both chromosome separation and spindle elongation supports either a sliding filament type mechanism (McIntosh and Landis, 1971; McDonald *et al.*, 1977) or a combined sliding, microtubule assembly–disassembly mechanism (Margolis *et al.*, 1978). Until recently, however, there have been few biochemical studies that supported a sliding mechanism. Fortunately, Cande and Wolniak (1978) have been able to show that vanadate, the uncompetitive inhibitor of dynein ATPase discussed earlier, completely blocks ATP-dependent anaphase motion of model PtK_1 cells. The blockage is reversible upon addition of norepinephrine, which apparently reduces vanadate to an inactive V^{4+} oxidation state (Cantley *et al.*, 1977; Gibbons *et al.*, 1978). Accepting the specificity of vanadate inhibition for dynein rather than myosin ATPases, the study provides strong evidence for dynein participation in mitosis. Similarly, Sakai *et al.* (1976) have shown that elongation of isolated sea urchin spindles is blocked by addition of dynein F1A antiserum but not by myosin antiserum. Based on immunofluorescent localization, Mohri *et al.* (1976) have characterized distribution of dynein antiserum during the course of mitosis. The antiserum reacts not only with the cytoplasm but is present in the isolated spindle as well.

Since actin and myosin appear to be ubiquitous and interacting components of eukaryotic cytoplasm, does the same hold true for dynein and microtubules? While the widespread occurrence of cytoplasmic microtubules is well documented, the presence of cytoplasmic dynein is supported only indirectly by the use of immunofluorescence probes (Kobayashi *et al.*, 1978; Mohri *et al.*, 1976). Isolation and characterization of chemically identifiable cytoplasmic dynein has yet to be accomplished although brief descriptions of a dynein-like protein contaminant of brain tubulin preparations have appeared (Burns and Pollard, 1974; Gaskin *et al.*, 1974). The recent study of Haimo *et al.* (1979) demonstrating the capability of cytoplasmic microtubules to interact with flagellar dynein is most encouraging.

V. Concluding Remarks

Based upon limited observations of different structural states of the dynein arms, a mechanical crossbridge cycle can be constructed that, as it must, parallels the known morphological features of the myosin crossbridge cycle. Unfortunately, we cannot yet relate the chemical features of dynein activity to the structural features. Two recent and similar models of dynein mechanochemistry have been suggested by Satir (1979) and Sale and Gibbons (1979), both of which incorporate and essentially parallel the main features of myosin mechanochemistry as originally proposed by Tonomura *et al.* (1969) and Lymn and Taylor (1971). However, our finding that the *in vitro* formation of a stable dynein–B-subfiber complex is dependent upon a divalent metal cation with a dose-dependence similar to *in situ* dynein–B-subfiber interaction, confirms our earlier observations (Warner, 1978; Zanetti *et al.,* 1979) and illustrates a major difference in dynein behavior when compared with the stable myosin–F-actin complex. The difference acquires greater significance when considering the possibility that Mg^{2+} alone may be responsible for a major conformational change in the dynein arm as well as modulating protein–protein interactions.

Progress in dynein mechanochemistry is likely to depend upon clarification of the nature of the dynein molecule as a motility-coupled ATPase. Dynein is a complex, multisubunit aggregate capable of forming transient crossbridges in an enzymatic cycle resembling the cycle of the myosin crossbridge. We must presume that the many subunits of the dynein arm (at least 6?) interact in a concerted way to accomplish several related functions. The subunits must provide binding sites for both A- and B-subfiber tubulin, they must provide a structural or anchoring element against which work can be done, they must be capable of undergoing at least one major conformational change, they must react with a divalent cation in order to react with B-subfiber tubulin, and last, most of these activities must be sequentially integrated with active site processing of $MgATP^{2-}$. Additional complexity occurs when considering potential differences between the inner and outer rows of dynein arms, functional coupling of dynein activation with a Ca^{2+}-regulatory mechanism, and adenylate kinase processing of the enzymatic reaction product.

Major advances in our knowledge of dynein behavior are likely to come from continued biochemical analyses of the active dynein complex, coupled with kinetic analyses of the reaction sequence. The enormous potential for molecular studies provided by the flagellar mutants of *Chlamydomonas* should have a major impact on the field. In several instances, we are restricted in our studies by still having to sort out many of the phenomenological and structural details of ciliary and flagellar motion, although it is these kinds of details that have provided much of the excitement, insight, and encouragement for workers in the field.

ACKNOWLEDGMENTS

Much of the work described in this paper has been carried out over a period of several years in the laboratory of Dr. F. D. Warner. In many instances, these studies were significantly advanced by participation of graduate students David R. Mitchell and Nina C. Zanetti, as well as by the capable technical assistance of Carol R. Perkins. The research has been supported by grants GB41552 from the National Science Foundation and GM20690 from the National Institutes of Health.

REFERENCES

Afzelius B. A. (1959). *J. Biophys. Biochem. Cytol.* **5**, 269–278.
Aiello, E. (1974). *In* "Cilia and Flagella" (M. L. Sleigh, ed.), pp. 353–376. Academic Press, New York.
Allen, C., and Borisy, G. G. (1974). *J. Cell Biol.* **63**, 5a (Abstr).
Allen, R. D. (1968). *J. Cell Biol.* **37**, 825–831.
Amos, L. R., Linck, R. W., and Klug, A. (1976a). *In* "Cell Motility" (R. D. Goldman, T. D. Pollard and J. L. Rosenbaum, ed.), pp. 847–867. Cold Spring Harbor Laboratory Press, Cold Spring Harbor, New York.
Amos, W. B., Routledge, L. M., Weis-Fogh, T., and Yew, F. F. (1976b). *Symp. Soc. Exp. Biol.* **30**, 273–301.
Besson, M., Fay, R. B., and Witman, G. B. (1978). *J. Cell Biol.* **79**, 306a (Abstr.).
Bloodgood, R. A. (1975). *Cytobios* **14**, 101–120.
Blum, J. J., and Hayes, A. (1974). *Biochemistry* **13**, 4290–4298.
Blum, J. J., and Hayes, A. (1976). *J. Supramol. Struct.* **5**, 15–25.
Blum, J. J., and Hayes, A. (1977). *J. Supramol. Struct.* **6**, 155–167.
Blum, J. J., and Hayes, A. (1978). *J. Supramol. Struct.* **8**, 153–171.
Brehm, P., and Eckert, R. (1978). *J. Physiol.* **283**, 557–568.
Brokaw, C. J. (1961). *Exp. Cell Res.* **22**, 151–162.
Brokaw, C. J. (1974). *J. Cell. Physiol.* **83**, 151–158.
Brokaw, C. J. (1975). *J. Exp. Biol.* **62**, 701–719.
Brokaw, C. J., and Benedict, B. (1968). *Arch. Biochem. Biophys.* **125**, 770–778.
Brokaw, C. J., and Benedict, B. (1971). *Arch. Biochem. Biophys.* **142**, 91–100.
Brokaw, C. J., and Gibbons, I. R. (1976). *In* "Swimming and Flying in Nature" Vol. 1 (T.Y.-T. Yu, C. J. Brokaw and C. Brennen, eds.), pp. 89–126. Plenum, New York.
Brokaw, C. J., and Simonick, T. F. (1977). *J. Cell Biol.* **75**, 650–665.
Brokaw, C. J., Josslin, R., and Bobrow, L. (1974). *Biochem. Biophys. Res. Commun.* **58**, 795–800.
Burns, R. G., and Pollard, T. D. (1974). *FEBS Lett.* **40**, 274–280.
Cande, W. Z., and Wolniak, S. M. (1978). *J. Cell Biol.* **79**, 573–580.
Cantley, L. C., Josephson, L., Warner, R., Yanagisawa, M., Lechene, C., and Guidotti, G. (1977). *J. Biol. Chem.* **252**, 7421–7423.
Cohen, W. D. (1978). *J. Cell Biol.* **78**, 260–273.
Cosson, M. P., and Gibbons, I. R. (1978). *J. Cell Biol.* **79**, 286a (Abstr.).
Dunlop, K., and Eckert, R. (1977). *J. Cell Biol.* **70**, 245a (Abstr.).
Eckert, R. (1972). *Science* **176**, 473–481.
Eckert, R. (1977). *Nature (London)* **268**, 104–105.
Fay, R. B., and Witman, G. B. (1977). *J. Cell Biol.* **75**, 286a (Abstr.).
Gaskin, F., Kramer, S. B., Cantor, C. R., Adelstein, R., and Shelanski, M. L. (1974). *FEBS Lett.* **40**, 281–286.
Gibbons, B. H., and Gibbons, I. R. (1972). *J. Cell Biol.* **54**, 75–97.
Gibbons, B. H., and Gibbons, I. R. (1973). *J. Cell Sci.* **13**, 337–357.
Gibbons, B. H., and Gibbons, I. R. (1974). *J. Cell Biol.* **63**, 970–985.
Gibbons, B. H., and Gibbons, I. R. (1976). *J. Biochem. Biophys. Res. Commun.* **73**, 1–6.
Gibbons, B. H., and Gibbons, I. R. (1978). *J. Cell Biol.* **79**, 285a (Abstr.).
Gibbons, B. H., and Gibbons, I. R. (1979). *J. Biol. Chem.* **254**, 197–201.
Gibbons, B. H., and Ogawa, K., and Gibbons, I. R. (1976). *J. Cell Biol.* **71**, 823–831.
Gibbons, I. R. (1961). *J. Biophys. Biochem. Cytol.* **11**, 179–205.
Gibbons, I. R. (1963). *Proc. Natl. Acad. Sci. U.S.A.* **50**, 1002–1010.
Gibbons, I. R. (1965a). *J. Cell Biol.* **26**, 707–712.
Gibbons, I. R. (1965b). *Arch. Biol. (Liege)* **76**, 317–352.

Gibbons, I. R. (1966). *J. Biol. Chem.* **241**, 5590-5596.

Gibbons, I. R. (1975). *In* "Molecules and Cell Movement" (S. Inoué and R. E. Stephens, eds.), pp. 207-231. Raven, New York.

Gibbons, I. R., and Fronk, F. (1972). *J. Cell Biol.* **54**, 365-381.

Gibbons, I. R., and Fronk, F. (1979). *J. Biol. Chem.* **254**, 5017-5029.

Gibbons, I. R., Cosson, M. P., Evans, J. A., Gibbons, B. H., Houck, B., Martinson, K. H., Sale, W., and Tang, W.-J. Y. (1978). *Proc. Natl. Acad. Sci. U.S.A.* **75**, 2220-2224.

Goldman, R. D., Pollard, T. D., and Rosenbaum, J. L. (1976). "Cell Motility," Vols. A, B, C. Cold Spring Harbor Laboratory Press, Cold Spring Harbor, New York.

Haimo, L., and Rosenbaum, J. L. (1976). *J. Cell Biol.* **70**, 216a (Abstr.).

Haimo, L., Telzer, B. R., and Rosenbaum, J. L. (1979). *Proc. Natl. Acad. Sci. U.S.A.* **76**, 5759-5763.

Hayashi, M. (1974). *Arch. Biochem. Biophys.* **165**, 288-296.

Hoffman-Berling, H. (1955). *Biochim. Biophys. Acta* **16**, 146-154.

Holwill, M. E. J. (1969). *J. Exp. Biol.* **50**, 203-222.

Holwill, M. E. J., and McGregor, J. L. (1975). *Nature (London)* **255**, 157-158.

Hoshino, M. (1974). *Biochim. Biophys. Acta* **351**, 142-154.

Hoshino, M. (1975). *Biochim. Biophys. Acta* **403**, 544-553.

Hoshino, M. (1976). *Biochim. Biophys. Acta* **462**, 49-62.

Hoshino, M. (1977). *Biochim. Biophys. Acta* **492**, 70-82.

Huang, B., Piperno, G., and Luck, D. J. L. (1979). *J. Biol. Chem.* **254**, 3091-3099.

Huxley, A. F., and Niedergerke, R. (1954). *Nature (London)* **173**, 971-973.

Huxley, H. E. (1957). *J. Biophys. Biochem. Cytol.* **3**, 631-648.

Huxley, H. E. (1963). *J. Mol. Biol.* **7**, 281-308.

Huxley, H. E. (1968). *J. Mol. Biol.* **37**, 507-520.

Huxley, H. E., and Hanson, J. (1954). *Nature (London)* **173**, 973-976.

Huxley, H. E., and Hanson, J. (1955). *Symp. Soc. Exp. Biol.* **9**, 228.

Hyams, J. S., and Borisy, G. G. (1978). *J. Cell Sci.* **33**, 235-253.

Kobayashi, T., Martensen, T., Nath, J., and Flavin, M. (1978). *Biochem. Biophys. Res. Commun.* **81**, 1313-1318.

Langford, G. M., and Inoué, S. (1979). *J. Cell Biol.* **80**, 521-538.

Lasek, R. J., and Hoffman, P. N. (1976). *In* "Cell Motility" (R. D. Goldman, T. D. Pollard and J. L. Rosenbaum, eds.), pp. 1021-1050. Cold Spring Harbor Laboratory Press, Cold Spring Harbor, New York.

Linck, R. W. (1979). *In* "The Spermatozoan" (D. W. Fawcett and J. M. Bedford, eds.), pp. 99-115. Urban & Sharzenberg, Baltimore.

Lindeman, C. B., and Gibbons, I. R. (1975). *J. Cell Biol.* **65**, 147-162.

Lymn, R. W., and Taylor, E. W. (1971). *Biochemistry* **10**, 4617-4624.

Mabuchi, I., and Shimizu, T. (1974). *J. Biochem. (Tokyo)* **76**, 991-999.

Mabuchi, I., Shimizu, T., and Mabuchi, Y. (1976). *Arch. Biochem. Biophys.* **176**, 564-576.

Machemer, H. (1976). *J. Exp. Biol.* **65**, 427-448.

Machemer, H. (1977). *In* "Fortschritte der Zoologie," Vol. 24, pp. 195-210. Fischer, Stuttgart.

Margolis, R. L., Wilson, L., and Kiefer, B. I. (1978). *Nature (London)* **272**, 450-452.

Masuda, H., Ogawa, K., and Miki-Nomura, T. (1978). *Exp. Cell Res.* **115**, 435-439.

McDonald, K., Pickett-Heaps, J. D., and McIntosh, J. R. (1977). *J. Cell Biol.* **74**, 377-388.

McIntosh, J. R. (1973). *J. Cell Biol.* **56**, 324-339.

McIntosh, J. R. (1974). *J. Cell Biol.* **61**, 166-187.

McIntosh, J. R. (1979). *In* "Microtubules" (J. S. Hyams and K. Roberts, eds.), pp. 381-441. Academic Press, New York.

McIntosh, J. R., and Landis, S. C. (1971). *J. Cell Biol.* **49**, 468-497.

Mitchell, D. R., and Warner, F. D. (1978a). *J. Cell Biol.* **79**, 293a (Abstr.).

Mitchell, D. R., and Warner, F. D. (1978b). *In* "Cell Reproduction" (E. R. Dirksen, D. M. Prescott and C. F. Fox, eds.), pp. 631–637. Academic Press, New York.

Mitchell, D. R., and Warner, F. D. (1979). *J. Cell Biol.* **83,** 177a (Abstr).

Mohri, H., Hasegawa, S., Yamamoto, M., and Murakami, S. (1969). *Sci. Pap. Coll. Gen. Educ., Univ. Tokyo* **19,** 195–217.

Mohri, H., Mohri, T., Mabuchi, I., Yazaki, I., Sakai, H., and Ogawa, K. (1976). *Dev. Growth Differ.* **18,** 391–398.

Mooseker, M. S., and Tilney, L. G. (1973). *J. Cell Biol.* **56,** 13–26.

Murakami, M., and Takahashi, K. (1975). *Nature (London)* **257,** 48–49.

Nagata, Y., and Flavin, M. (1978). *Biochim. Biophys. Acta* **523,** 228–235.

Naitoh, Y. (1966). *Science* **154,** 660–662.

Naitoh, Y. (1969). *J. Gen. Physiol.* **53,** 517–529.

Naitoh, Y., and Kaneko, H. (1972). *Science* **176,** 523–524.

Naitoh, Y., and Kaneko, H. (1973). *J. Exp. Biol.* **58,** 657–676.

Naitoh, Y., and Eckert, R. (1974). *In* "Cilia and Flagella" (M. L. Sleigh, ed.), pp. 305–352. Academic Press, New York.

Nakamura, K.-I., and Masuyama, E. (1977). *Biochim. Biophys. Acta* **481,** 660–666.

Nakamura, S., and Kamiya, R. (1978). *Cell Struct. Funct.* **3,** 141–144.

Nichols, K. M., and Rikmenspoel, R. (1978a). *Exp. Cell Res.* **116,** 333–340.

Nichols, K. M., and Rikmenspoel, R. (1978b). *J. Cell Sci.* **29,** 233–247.

Nishino, Y., and Watanabe, Y. (1977). *Biochim. Biophys. Acta* **490,** 132–143.

Oertel, D., Schein, S., and Kung, C. (1977). *Nature (London)* **268,** 120–124.

Ogawa, K. (1973). *Biochim. Biophys. Acta* **293,** 514–525.

Ogawa, K. (1978). *Zool. Mag. (Tokyo)* **87.**

Ogawa, K., and Gibbons, I. R. (1976). *J. Biol. Chem.* **251,** 5793–5801.

Ogawa, K., and Mohri, H. (1972). *Biochim. Biophys. Acta* **256,** 142–155.

Ogawa, K., and Mohri, H. (1975). *J. Biol. Chem.* **250,** 6476–6483.

Ogawa, K., Okuno, M., and Mohri, H. (1975). *J. Biochem. (Tokyo)* **78,** 729–737.

Ogawa, K. Asai, D. J., and Brokaw, C. J. (1977a). *J. Cell Biol.* **73,** 182–192.

Ogawa, K., Mohri, T., and Mohri, H. (1977b). *Proc. Natl. Acad. Sci. U.S.A.* **74,** 5006–5010.

Ogura, A., and Takahashi, K. (1976). *Nature (London)* **264,** 170–171.

Okuno, M. (1979). *Biophys. J.* **25,** 209a (Abstr.).

Okuno, M., Ogawa, K., and Mohri, H. (1976). *Biochem. Biophys. Res. Commun.* **68,** 901–906.

Omoto, C. K., and Kung, C. (1979). *Nature (London)* **279,** 532–534.

Otokawa, M. (1972). *Biochim. Biophys. Acta* **275,** 464–466.

Penningroth, S. M., and Witman, G. B. (1978). *J. Cell Biol.* **79,** 827–833.

Piperno, G., and Luck, D. J. L. (1979). *J. Biol. Chem.* **254,** 3084–3090.

Sakai, H., Mabuchi, I., Shimoda, S., Kuriyama, R., Ogawa, K., and Mohri, H. (1976). *Dev. Growth Differ.* **18,** 211–219.

Sale, W. S., and Gibbons, I. R. (1979). *J. Cell Biol.* **82,** 291–298.

Sale, W. S., and Satir, P. (1977). *Proc. Natl. Acad. Sci. U.S.A.* **74,** 2045–2049.

Satir, P. (1965). *J. Cell Biol.* **26,** 805–834.

Satir, P. (1968). *J. Cell Biol.* **39,** 77–94.

Satir, P. (1975). *Science* **190,** 586–588.

Satir, P. (1979). *In* "The Spermatozoan" (D. W. Fawcett and J. M. Bedford, eds.). Urban & Sharzenberg, Baltimore.

Schmidt, J. A., and Eckert, R. (1976). *Nature (London)* **262,** 713–715.

Schrevel, J., and Besse, C. (1975). *J. Cell Biol.* **66,** 492–507.

Shimizu, T. (1975). *J. Biochem. (Tokyo)* **78,** 41–49.

Shimizu, T., and Kimura, I. (1974). *J. Biochem. (Tokyo)* **76,** 1001–1008.

Shimizu, T., and Kimura, I. (1977). *J. Biochem. (Tokyo)* **82,** 165–173.

Shimizu, T., Kaji, K., and Kimura, I. (1977). *J. Biochem. (Tokyo)* **82,** 1145–1153.

Sleigh, M. L. (1974). "Cilia and Flagella." Academic Press, New York.

Storer, A. C., and Cornish-Bowden, A. (1976). *Biochem. J.* **159,** 1–5.

Summers, K. E., and Gibbons, I. R. (1971) *Proc. Natl. Acad. Sci. U.S.A.* **68,** 3092–3096.

Summers, K. E., and Gibbons, I. R. (1973). *J. Cell Biol.* **58,** 618–629.

Takahashi, M., and Tonomura, Y. (1978). *J. Biochem. (Tokyo)* **84,** 1339–1355.

Tamm, S. L., and Horridge, G. A. (1970). *Proc. R. Soc. London Ser. B.* **175,** 219–233.

Tonomura, Y. (1973). "Muscle Proteins, Muscle Contraction and Cation Transport." Univ. of Tokyo Press, Tokyo.

Tonomura, Y., Nakamura, H., Kinoshita, N., Onishi, J., and Shigekawa, M. (1969). *J. Biochem. (Tokyo)* **66,** 599–618.

Tsuchiya, Y. (1977). *Comp. Biochem. Physiol.* **56A,** 353–361.

Walter, M. F., and Satir, P. (1978). *J. Cell Biol.* **79,** 110–120.

Walter, M. F., and Satir, P. (1979). *Nature (London)* **278,** 69–70.

Warner, F. D. (1976a) *In* "Cell Motility" (R. D. Goldman, T. D. Pollard and J. L. Rosenbaum, eds.), pp. 891–914. Cold Spring Harbor Laboratory Press, Cold Spring Harbor, New York.

Warner, F. D. (1976b). *J. Cell Sci.* **20,** 101–114.

Warner, F. D. (1978). *J. Cell Biol.* **77,** R19–R26.

Warner, F. D., and Mitchell, D. R. (1978). *J. Cell Biol.* **76,** 261–277.

Warner, F. D. and Satir, P. (1974). *J. Cell Biol.* **63,** 35–63.

Warner, F. D., Mitchell, D. R., and Perkins, C. R. (1977). *J. Mol. Biol.* **114,** 367–384.

Watanabe, T., and Flavin, M. (1976). *J. Biol. Chem.* **251,** 182–192.

Wellings, J. V., and Tucker, J. B. (1979). *Cell Tissue Res.* **197,** 313–323.

Witman, G. B., Plummer, J., and Sander, G. (1978). *J. Cell Biol.* **76,** 729–747.

Zanetti, N. C., Mitchell, D. R., and Warner, F. D. (1979). *J. Cell Biol.* **80,** 573–588.

INTERNATIONAL REVIEW OF CYTOLOGY, VOL. 66

Structure and Function of Phycobilisomes: Light Harvesting Pigment Complexes in Red and Blue-Green Algae

ELISABETH GANTT

Radiation Biology Laboratory, Smithsonian Institution, Rockville, Maryland

I. Introduction

Phycobilisomes are specialized aggregated structures composed of phycobili-proteins, which are photosynthetic accessory pigments in red and blue-green algae. Phycobiliproteins, which can account for as much as 24% of the dry weight of blue-green algal cells (Myers and Kratz, 1955) and 40–60% of the total soluble protein (Bennett and Bogorad, 1973; Gantt and Lipschultz, 1974), are the major light harvesters in these organisms. Chlorophyll *a*, which absorbs light primarily in the blue region and the red region of the visible spectrum, leaves a large absorption gap. This is filled in by the phycobiliproteins, which have an absorption range of 500–660 nm. These pigments work in conjunction with chlorophyll *a* to optimize light harvesting for photosynthesis, particularly under light limiting conditions.

Blue-green algae, for which a reclassifcation as cyanobacteria has been proposed (Stanier and Cohen-Bazire, 1977), have an origin whose fossil record dates back at least 3 billion years (Muir and Grant, 1976; Knoll and Barghoorn, 1977). It can be reasonably assumed that the phycobiliproteins have also existed for a very long time; however, they did not make the transition as a major light harvesting pigment in land plants. The only phycobiliprotein in higher plants is phytochrome, which occurs in very small amounts and serves as a photoregulatory receptor.

In cryptophyte algae (flagellated unicells of indefinite taxonomic position), phycobiliproteins also serve as major photosynthetic accessory pigments, but phycobilisomes are absent (summarized in Gantt, 1979). Instead, these pigments occur within the intrathylakoidal spaces, which, as in the phycobilisomes, provide for close packing of the pigments and enhanced energy transfer efficiency.

The intense coloration and fluorescence properties of the phycobiliproteins have long been of interest to protein chemists; phycobiliproteins occur as inherently labeled proteins. Several recent reviews concerning phycobiliproteins (Bogorad, 1975; Glazer, 1977; Gysi and Chapman, 1980; O'Carra and O'Eocha, 1976; Rüdiger, 1975; Troxler, 1977) and a summary (Gantt, 1977) detail their characteristics, including their chromophore structure, subunit structure, and amino acid sequences. The abbreviations used by Bogorad (1975) for phycobiliproteins will be followed in this article. The three major classes of phycobiliproteins are the red-colored phycoerythrins (PEs), and the blue-colored phycocyanins (PCs) and allophycocyanins (APCs). Sometimes these pigments have a prefix to designate their algal origin, i.e., C- for cyanophycean; R- for rhodophycean; and B-, b- for bangiophycean, which is a subgroup of the latter. The emphasis of this article will be on the phycobilisome characteristics, their relationship to the photosystems in the thylakoid membrane, and major problems to be addressed in future investigations. Several significant findings have been made in this area since it was last reviewed (Gantt, 1975).

II. Morphological Studies on Phycobilisomes

A. Location and Morphology

In red algal chloroplasts or in procaryotic blue-green algae, the phycobilisomes are located on the external (stroma) surface of the thylakoid membrane pair, i.e., on the side facing away from its own membrane pair. In appearance they resemble oversized ribosomes on rough endoplasmic reticulum. They are not encased in a membrane; this would have been initially advantageous in

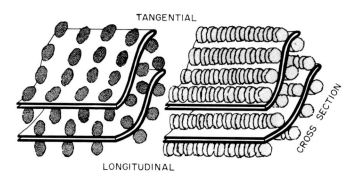

TANGENTIAL

LONGITUDINAL

CROSS SECTION

FIG. 1.   Schematic comparison of phycobilisome arrangements on thylakoid lamellae. Globular-shaped types are on the left and disc-shaped types on the right. The planes of section pass across the phycobilisome rows (cross section), parallel with the lamellae (tangential section), and along the rows (longitudinal section).

isolating them in an intact state. The phycobilisomes occur in both short and long extensive rows with a highly regular spacing, suggesting an orderly crystalline array.

1. *Shape*

The shape of phycobilisomes varies with the species. In some species, phycobilisomes are disc-shaped, but they range through intermediates to globular-shaped. To determine the shape, various sectional views are required. Three sectional views of phycobilisome shapes and arrangements are diagramed in Fig. 1. In cross-sectional view, all phycobilisomes appear rounded, but in longitudinal or tangential (to the membrane) sections, the different shapes can be distinguished.

Among the algae thus far examined, the disc-shaped phycobilisomes predominate in most species. In longitudinal and tangential sections, the disc-shaped types are usually observed as regularly spaced short dense lines (Figs. 2–4). Less common is the globular type, thus far found only in red algae that have high concentrations of R-PE, B-PE, and b-PE. In the red algae *Porphyridium cruentum*[1] (Fig. 5) and *Griffithsia pacifica,* where PE constitutes 84–89% of the total phycobilisome, the phycobilisomes have a prolate or block shape.

Lichtlé and Giraud (1970) reported a cord type of phycobilisome arrangement in *Batrachospermum*. In these, the phycobilisomes were not individually distinguishable, giving the appearance of solid cords. It is possible that this is not a distinctive type but may consist of densely packed disc-shaped phycobilisomes or normally spaced phycobilisomes that are obscured by interferring electron dense

---

[1]Also designated as *Porphyridium purpureum.*

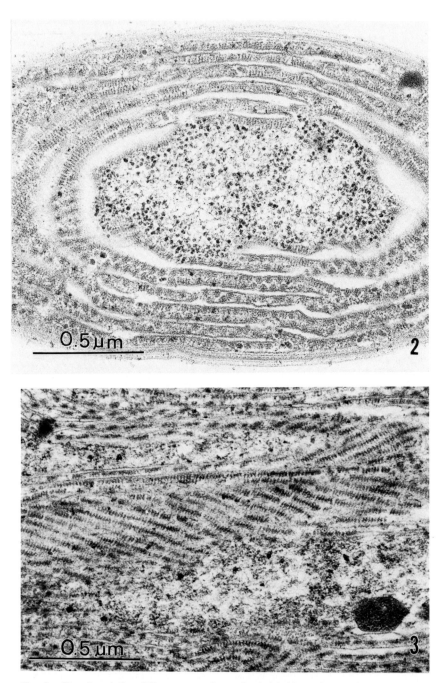

Fig. 2. Disc-shaped phycobilisomes extend over the thylakoid membranes of the thermophilic blue-green alga *Synechococcus lividus*. (From Edwards and Gantt, 1971.)

Fig. 3. Regularly spaced rows of disc-shaped phycobilisomes mostly in tangential section extend over the thylakoid lamellae of *Oscillatoria splendida*. (From Lichtlé and Thomas, 1976.)

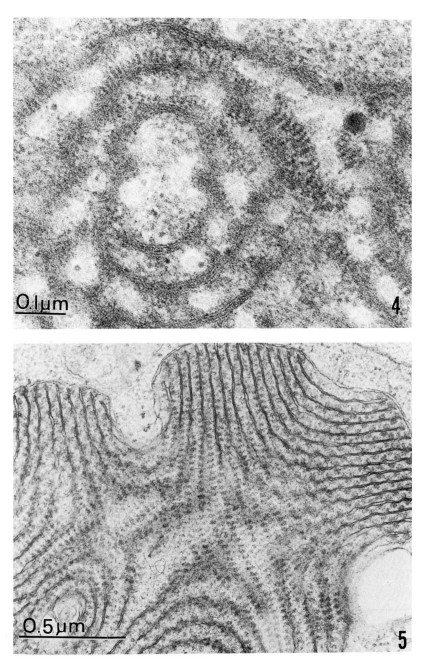

Fig. 4. Distinct disc-shaped phycobilisomes on thylakoid membranes in *Nostoc muscorum*. Glycogen storage bodies are interspersed between phycobilisome rows. (From Wildman and Bowen, 1974.)

Fig. 5. Prolate-shaped phycobilisomes in the red alga *Porphyridium cruentum* in several sectional planes. Short projections can be seen on the margins of the phycobilisomes in cross section on the right. (From Gantt and Conti, 1966b.)

material between them. Actually, individual phycobilisomes frequently are not well defined in electron micrographs of algae containing the nonglobular type. It is often difficult to ascertain from sections where one phycobilisome begins and ends and whether only one or more than one disc constitutes one phycobilisome.

Perhaps the most primitive type of phycobilisome exists in *Gloeobacter* (Fig. 6), a blue-green alga in which internal thylakoids are absent but which has all three phycobiliproteins (C-PE, C-PC, APC) (Rippka *et al.*, 1974). Definitive phycobilisomes are not visible; instead, there is an unusual electron-dense layer on the plasma membrane. This electron-dense layer, 80 nm wide, has a faint regularity and is likely to be comprised of stack-like phycobilisomes with APC

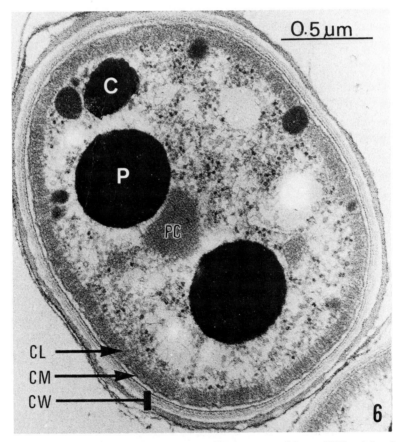

FIG. 6. Section of *Gloeobacter violaceus* showing the outer wall layers (CW) and the plasma membrane (CM), which contains a cortical electron-dense layer (CL) with the appearance of bristles that are presumptive phycobilisomes. Inclusions: cyanophycean (C) and polyphosphate (P) granules, and paracrystalline (PC) bodies. (From Rippka *et al.*, 1974.)

occurring near the plasma membrane and with C-PC and C-PE extending into the cytoplasm.

2. *Size*

A summary of species (Table I) reveals a range of sizes. These size determinations were made primarily from electron micrographs of embedded and sectioned cells. The broad face view, obtained in cross sections (Fig. 1), has a diameter of 20–58 nm. In thickness, the disc-shaped types vary from 6 to 10 nm, whereas the globular-shaped phycobilisome and those with an intermediate shape range from 15 to 33 nm.

Negatively stained phycobilisomes, whether attached to thylakoid vesicles or free, probably yield values which are closer to the *in vivo* state. A comparison of the values obtained form sections and from negatively stained preparations reveals that those from sections are about one-third less. This reduction in size is attributable to the dehydration in absolute alcohol or acetone required for embedment in resins. Such a comparison, however, has been made for few algae: *P. cruentum* (Gantt and Conti, 1966b; Gantt and Lipschultz, 1972); *Rhodella violacea* (Wehrmeyer and Schneider, 1975; Mörschel *et al.*, 1977); and LLP 7409 (Bryant *et al.*, 1979). Accurate size and shape determinations are essential for calculation of phycobilisome volume and for estimates of the phycobiliprotein content and arrangement.

3. *Effect of Light Quality and Intensity*

In at least two red algae, phycobilisome size or shape does not seem to be affected by light intensity or wavelength (Guerin-Dumartrait *et al.*, 1973; Waaland *et al.*, 1974). Phycobilisomes isolated from *P. cruentum,* grown either at high or low light intensities (2000 vs 350 fc), had not only the same size but also the same relative proportions of constitutive phycobiliproteins (experiments from our laboratory). In *G. pacifica,* the phycobilisome number per lamellar area increased when the PE to chlorophyll ratio increased under low light intensities (Waaland *et al.*, 1974). Lichtlé (1978) obtained similar results with *Rhodochorton purpureum.* Because the number but not the phycobilisome size was increased, there is probably a finite phycobilisome size.

Chromatic adaptation, an elaboration of the phycobiliprotein that absorbs the predominate wavelength, occurs in many blue-green algae (Bennett and Bogorad, 1973; Fujita and Hattori, 1960; Tandeau de Marsac, 1977; Bryant *et al.*, 1979). Wildman and Bowen (1972, 1974) found disc-shaped phycobilisomes in numerous blue-green algal species. They surmised that variations of the phycobiliprotein ratios did not change the phycobilisome shape, however some of their data suggest that the thickness of the phycobilisomes may be changed with light intensity. Edwards *et al.* (1968) observed phycobilisomes with smaller diameters in some preparations, but it is not clear that this was an intensity effect.

TABLE I

PHYCOBILISOME DIMENSIONS AND SPACING IN CYANOPHYTES AND RHODOPHYTES

Species	Shape[a]	Average dimensions[b] (nm)		Average spacing (nm) (Center-to-center)		Phycobiliprotein pigments	References
		Diameter	Thickness	Parallel rows	Within rows		
Cyanophytes[e]							
Anacystis nidulans	D	35	10	65		PC, APC	Cosner (1978); Wildman and Bowen (1974)
Chroococcus minutus	D	35				PC, APC	Thomas (1972)
Cyanophora paradoxa	D	35				PC, APC	Bourdu and Lefort (1967)
Glaucocystis nostochinearum	D–G	35[c]	15			PC, APC	Bourdu and Lefort (1967)
		30					Lefort (1965)
Fremyella diplosiphon	D	35		45		PE, PC, APC	Gantt and Conti (1969)
LPP-7409	D	58 × 28	7	80		PE, PC, APC	Bryant et al. (1979)
		66 × 38[d]	16[d]				
Nostoc	D	40				PE, PC, APC	Gray et al. (1973)
Oscillatoria brevis	D	32 × 23	6	77		PE, PC, APC	Lichtlé and Thomas (1976)
Oscillatoria limosa	D	24 × 18	6		10	PE, PC, APC	Lichtlé and Thomas (1976)
Oscillatoria splendida	D	25 × 20	8	50		PC, APC	Lichtlé and Thomas (1976)

Synechococcus lividus	D	35	6	45	7	PC, APC	Edwards and Gantt (1971)
Synechococcus 6312	D	41 × 23	8	55		PC, APC	Bryant *et al.* (1979)
Anabaena variabilis, A. cylindrica, Aphanizomenon flos-aquae, Arthrospira jenneri, Calothrix, Microcoleus vaginatus, Nostoc muscorum, Symploca muscorum, Tolypothrix distorta	D	35–38	10			Variable pigment content	Wildman and Bowen (1974)
Rhodophytes							
Antithamnion glanduliferum	D–G	37	18	40	27	PE, PC, APC	Lichtlé and Thomas (1976)
Batrachospermum virgatum	C	33		65		PE, PC, APC	Lichtlé and Giraud (1970)
Griffithsia pacifica	G	45 × 27 / 63 × 38[d]	38[d]			PE, PC, APC	Waaland *et al.* (1974); Gantt and Lipschultz (1980)
Griffithsia floculosa	G	37				PE, PC, APC	Peyriere (1968)
Hypoglossum woodwardii	G	27				PE, PC, APC	Lichtlé, Ph.D. Thesis (1978)
Polysiphonia elongata	G	33		63		PE, PC, APC	Lichtlé, Ph.D. Thesis (1978)
Porphyra leucosticta	G	20		40		PE, PC, APC	Sheath *et al.* (1977)
Porphyridium aerugineum	D	30		40		PC, APC	Gantt *et al.* (1968)

(*continued*)

TABLE I (*continued*)

Species	Shape[a]	Average dimensions[b] (nm)		Average spacing (nm) (Center-to-center)		Phycobiliprotein pigments	References
		Diameter	Thickness	Parallel rows	Within rows		
Porphyridium cruentum	G	36		45		PE, PC, APC	Gantt and Conti (1965, 1966b)
				65[c]			Guerin-Dumartrait *et al.* (1970)
		47 × 32[d]	33[d]				Gantt and Lipschultz (1972)
Rhodella violacea	D	33	7			PE, PC, APC	Wehrmeyer and Schneider (1975)
		53 × 33[d]					Mörschel *et al.* (1977)
Rhodochorton purpureum	D–G	37	21	67	31	PE, PC, APC	Lichtlé (1973)
Rhodothamniella floridula	D–G	37	17	77	24	PE, PC, APC	Lichtlé (1973)

[a] G, globular; D, disc-shaped; D–G, intermediate G and D; C, continuous cords in longitudinal view, sometimes with appearance of very close disc-shaped packing.

[b] From sectioned material (prepared by dehydration and embedding) unless otherwise noted.

[c] PBS dimension from freeze–fracture images.

[d] PBS dimensions from isolated and negatively stained preparations.

[e] For alternate nomenclature see Rippka *et al.* (1979).

Self-shading of cells, which normally occurs in vigorously growing cultures, unless rigorously controlled can cause complications, because changes in size could result from variations in light intensity. Recently Bryant *et al.* (1979) observed that phycobilisome size decreased in *Synechocystis* 6701 when PE decreased as a function of chromatic adaptation. Green light grown phycobilisomes had a sedimentation velocity value of $45S_{20w}$, whereas those in red light had one of $38S_{20w}$.

B. Developing Plastids

Phycobilisomes have been shown to exist in developing chloroplasts of vegetative cells of red algae. In the apical zone of *Polysiphonia* (Lichtlé and Giraud, 1969) proplastids with only one or two lamellae have fully formed phycobilisomes with a diameter of 30–35 nm. With maturation of the chloroplasts the thylakoid number and the phycobilisomes increased. In proplastids of *Nitophyllum* (Honsell *et al.*, 1978) phycobilisomes appear with the same spacing on the thylakoids as in mature chloroplasts. Although it has been postulated that thylakoids in red algal proplastids arise from the inner chloroplast limiting membranes as in green plants, identifiable phycobilisomes have not been found on this membrane.

III. Phycobilisome Composition

A. Cytological Evidence

Localization of phycobiliproteins within cells has long been of interest to plant cell biologists, and phycobilisome-type structures in association with the thylakoids were anticipated before they were actually discovered. This assumption was prompted by the efficient energy transfer between phycobiliproteins and chlorophyll (Arnold and Oppenheimer, 1950; Tomita and Rabinowitch, 1962). The high content of phycobiliproteins per cell (Myers and Kratz, 1955) was another reason for assuming a close and compact association.

The initial demonstration for the location of phycobiliproteins in the phycobilisomes came from whole cell extractions on *P. cruentum* (Gantt and Conti, 1966a,b). Cells extracted to remove chlorophyll and carotenoids were pink in color, which by spectral analysis was identifiable as being due to phycobiliproteins. These cells retained the regularly arrayed granules that were later identified as phycobilisomes. Removal of the phycobiliproteins from broken cells also resulted in removal of the phycobilisomes. This was further confirmed with a blue-green alga *Gloeocapsa* (Cohen-Bazire and Lefort-Tran, 1970) and by the isolation of phycobilisomes.

B. Isolation of Phycobilisomes

1. *Isolation Conditions*

Intact phycobilisomes were first isolated by use of glutaraldehyde as a fixative (Gantt and Conti, 1966b; Cohen-Bazire and Lefort-Tran, 1970). This kept the constituent phycobiliproteins from dissociating. Whereas they appeared to remain functionally active by transferring energy to chlorophyll (Clement-Metral and Lefort-Tran, 1971), further analysis of their composition was not possible until fixatives were replaced by a high ionic strength buffer for phycobilisome isolation in *P. cruentum* (Gantt and Lipschultz, 1972). The procedure was subsequently modified for some blue-green algae and another red alga (Gray *et al.*, 1973; Koller *et al.*, 1977; Tandeau de Marsac and Cohen-Bazire, 1977) and eventually for numerous species (Gantt *et al.*, 1979).

For isolation of phycobilisomes, cells are suspended in 0.75 M phosphate buffer (pH 6.8), and after disruption in a French press or by sonication, a nonionic detergent (Triton X-100) is added to release the phycobilisomes from the membrane. The phycobilisomes remain in the supernatant when cell fragments are removed by centrifugation. Subsequent centrifugation on a sucrose step gradient (0.25–2.0 M) concentrates the phycobilisomes in the 1 M sucrose layer in about 3 hours, leaving the loose pigments in the upper sucrose layers (Fig. 7).

For most algae, isolation at 20–23°C yields the best preserved phycobilisomes (Gantt *et al.*, 1979). Phycobilisomes thus isolated are stable for hours or days when stored as a pellet or as a concentrated solution in 0.75 M phosphate (pH 6.8). At lower temperature (4–10°C), partial dissociation occurs sometimes with complete loss of allophycocyanin.

Alternate procedures have used sonication for breaking cells, with omission of the sucrose step gradient (Searle *et al.*, 1978), or have used a linear sucrose gradient with extended centrifugation times (15 hours) (Koller *et al.*, 1977). Whereas omission of a sucrose gradient results in intact phycobilisomes, they will inevitably be more contaminated with chlorophyll, free phycobiliproteins, other proteins, and other cell constituents.

2. *Characteristics of Isolated Phycobilisomes*

The main criteria by which phycobilisomes are characterized include their absorption and fluorescence spectra, the uniformity of structure in electron micrographs, and their position on the isolation gradient.

Phycobilisomes recovered after centrifugation from the 1 M sucrose layer of a sucrose–phosphate step gradient have an appearance similar to those attached to thylakoid membranes (Gantt and Conti, 1966b; Cohen-Bazire and Lefort-Tran, 1970). In *P. cruentum*, the uniformity in size and shape (Fig. 7) is particularly

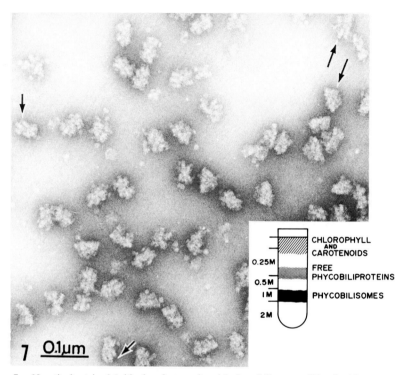

FIG. 7. Negatively stained (with phosphotungstic acid) phycobilisomes of *Porphyridium cruentum* recovered from the 1 *M* sucrose layer of an isolation gradient as diagramed in the inset. Short rows of aggregates are discernable on peripheries (arrows).

notable. Due to their prolate shape, they appear more regular than the disc-shaped type in *R. violacea* (Koller *et al.*, 1977). As pointed out by Koller *et al.* (1977), the thinness of phycobilisomes of *R. violacea* allows their substructure to be more readily seen than that of the thicker *P. cruentum* phycobilisomes. Definite substructure in the latter is only apparent on the edge of the phycobilisome, whereas in the former, it is visible over the entire phycobilisome (compare Fig. 7 with Fig. 15).

The absorption peaks in the visible region are dependent on the phycobiliprotein composition of the cells and major peaks from 498 to 567 nm for PE, from 610 to 630 nm for PC, with overlaps of APC peaking at 650 to 653 nm (Figs. 8 and 9). The sum of the phycobiliprotein absorption peaks (λ_{max}) divided by the absorption at 275 nm will be in a ratio of four or more when a phycobilisome preparation has little contamination. Sometimes small contaminants coisolate with the procedure given above; these can be removed by further purification

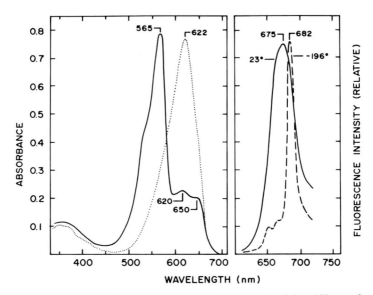

FIG. 8.   Absorption (left) and fluorescence emission (right) spectra of phycobilisomes from *Fremyella diplosiphon* that had been grown under green light (——) and red light (.....). Phycoerythrin (λ_{max} 565 nm) was formed only in green light, whereas phycocyanin (λ_{max} 620 nm) and allophycocyanin (λ_{max} 650 nm) were present under both conditions. Fluorescence emission maxima of the two preparations were indistinguishable at room (23°C) and liquid nitrogen (−196°C) temperatures.

steps (Gantt *et al.*, 1979). Chlorophyll contamination, which can be as high as 1 μg chlorophyll/100 μg protein, can be reduced to 1 μg chlorophyll/4 gm protein by an additional treatment with Triton X-100.

Fluorescence emission at about 675 nm (23°C) is the single most meaningful criterium for assessing the intactness of phycobilisomes. Whether phycobilisomes contain only APC and PC, or PE in addition (Fig. 8), the emission at room temperature occurs at about 670 to 675 nm and at 678 to 685 nm at liquid nitrogen temperature (Gantt *et al.*, 1979). Free phycobiliproteins in a phycobilisome preparation, or attached but energetically uncoupled phycobiliproteins, can be identified by their independent emission peaks (Fig. 9) at 578 to 660 nm. There is no convincing evidence that chlorophyll is a constituent of isolated phycobilisomes, although there is a 685 nm (−196°C) peak from chlorophyll (Cho and Govindjee, 1970) from photosystem II, which in membrane preparations cannot be separately distinguished from energetically uncoupled phycobilisomes. Reduction of chlorophyll contamination to one chlorophyll molecule per ten phycobilisomes, and exhaustive chlorophyll extraction has not diminished the 685 nm emission (Gantt *et al.*, 1976b; Katoh and Gantt, 1979).

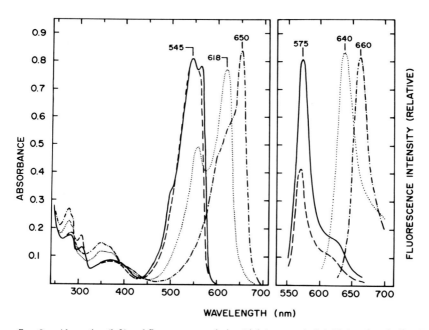

Fɪɢ. 9. Absorption (left) and fluorescence emission (right) spectra in 0.1 M phosphate buffer pH 6.8 of the major phycobiliprotein purified from *Porphyridium cruentum* phycobilisomes. Shown are B-phycoerythrin (———), b-phycoerythrin (---), R-phycocyanin (. . . .), and allophycocyanin (- · - ·). For the spectrum of allophycocyanin-B, also a component, see Fig. 17. (After Gantt and Lipschultz, 1974.)

C. PHYCOBILISOME STRUCTURE

1. Pigment Localization

Pigment analysis of *P. cruentum* phycobilisomes showed that the phycobili-proteins account for the major portion of protein content (Gantt and Lipschultz, 1974): the PEs (B- and b-) account for 84%, R-PC 11%, and APC about 5%. Ley *et al.* (1977), using independently determined extinction coefficients, obtained the same approximate composition with one significant exception, their APC value of 10%, of which <1% is attributed to APC-B, is higher. In *R. violacea,* the phycobilisomes contain 58% B-PE, 25% C-PC, and 17% APC (Koller *et al.,* 1977).

Arrangement of phycobiliprotein within the phycobilisomes has been partly inferred from controlled time-course dissociation studies. The release of free phycobiliproteins vs the undissociated portion was followed by absorption spectra (Fig. 9) and fluorescence emission (Gantt *et al.*, 1976a). The progressive

release of PE > R-PC > APC (Fig. 10), which was independent of pH (5.4–
8.0), led to the proposal that, in a *P. cruentum* phycobilisomes, APC exists as a
core surrounded by a hemispherical layer of R-PC, with PE being on the
periphery. Such a structure is consistent with maximum energy transfer, as well
as with immunochemical localization results.

2. *Immunochemical Evidence*

In order to assess whether APC was located on the "membrane side" of the
phycobilisome, immunochemical techniques were adapted for electron microscopic examination. Small thylakoid vesicles, which had the phycobilisomes on
the external surface, were prepared. On these, phycobilisomes could be dissociated as were the isolated phycobilisomes in Fig. 10. Phycobilisomes attached
to thylakoid vesicles remained intact under nondissociating conditions (0.75 M
phosphate), whereas, under dissociating conditions (0.03 M phosphate), there
was an enrichment of APC in the thylakoid vesicles (Gantt *et al.*, 1976b) relative
to PE, due to a greater loss of PE. Reaction of monospecific rabbit antisera
against PE and APC with intact and partially dissociated phycobilisomes confirmed that APC is indeed located on the side nearest the thylakoid (Gantt and
Lipschultz, 1977). In this procedure (diagrammed in Fig. 11), ferritin was
utilized as the electron-dense marker to localize the reacted antiserum. Further
removal of phycobiliproteins from such thylakoid vesicles, by exhaustive dilute

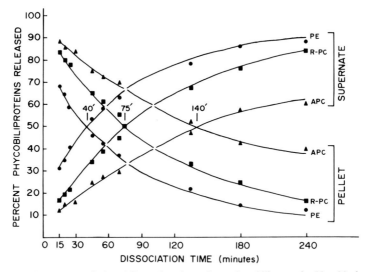

FIG. 10. Time course of phycobiliprotein release from phycobilisomes in 55 mM phosphate
buffer from the supernatant fractions and pellet fractions of dissociating phycobilisomes of *Porphyridium cruentum*. APC, allophycocyanin; PE, phycoerythrin; R-PC, R-phycoeyanin. (After Gantt
et al., 1976a.)

buffer rinses, resulted in the absence of any further reaction with anti-APC, which strongly suggests that APC is not a structural part of the thylakoid.

In isolated phycobilisomes of *Nostoc* species containing about equal amounts of the three major phycobiliprotein types, PE and APC reacted more readily than PC, suggesting they are on exposed surfaces. A comparison of antisera reaction times on intact and dissociated phycobilisomes showed a slower reaction only with anti-PC on intact phycobilisomes; this probably indicated that PC was inside the phycobilisome (Gantt *et al.*, 1976a).

3. *Phycobiliprotein Assembly Forms*

Stable *in vitro* forms of phycobiliproteins have been extensively studied by biochemical and hydrodynamic methods, and some by electron microscopy.

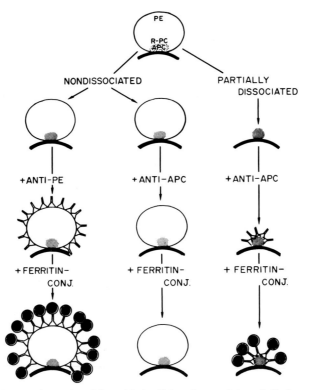

FIG. 11. Diagram of procedure followed in localizing phycoerythrin and allophycocyanin with monospecific antisera (rabbit) and ferritin-conjugates (goat antirabbit). In a phycobilisome, the pigment distribution is allophycocyanin (APC, stippled) near membrane; with R-phycocyanin adjacent (R-PC), and phycoerythrin (PE) on the periphery. In intact phycobilisomes, phycoerythrin became labeled (on left), but not allophycocyanin (middle). In partially dissociated phycobilisomes, allophycocyanin became labeled (on right). (After Gantt and Lipschultz, 1976b.)

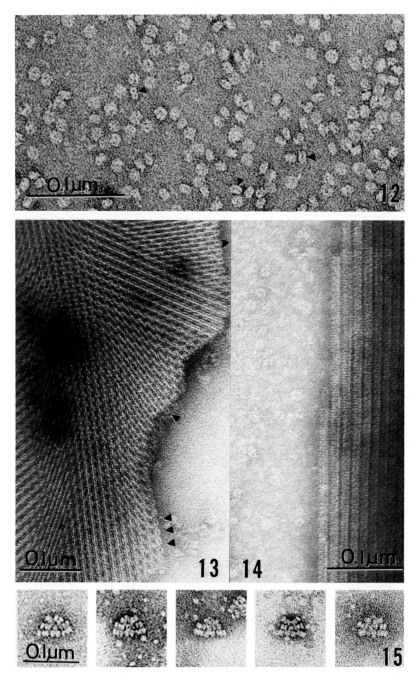

FIGS. 12-15.   Building blocks of phycobilisomes, negatively stained purified phycobiliproteins.

Generally they are aggregates of polypeptides α and β, with molecular weights (MWs) of 11,000 to 20,000, and each containing one to four chromophores (Bogorad, 1975; Glazer, 1977; Gysi and Chapman, 1979; O'Carra and O'Eocha, 1976; Rüdiger, 1975; Troxler, 1977). Sometimes they also contain larger polypeptides (γ) with MWs of approximately 29,000, e.g., in B-PE (Gantt and Lipschultz, 1974; Glazer and Hixson, 1977) and in APC-I (Zilinskas et al., 1978). The common assembly states are $(\alpha\beta)_3$ or a multiple thereof. Examples are C-PC $(\alpha\beta)_6$ (Berns, 1971; Glazer, 1977) and B-PE $(\alpha\beta)_6 + \gamma$. The more stable aggregates range from approximately 90,000 to 260,000 MW (Bogorad, 1975).

Electron micrographs of negatively stained preparations of isolated phycobiliproteins reveal that they are disc-shaped (Figs. 12–14) with axial ratios of 1:2 or 1:3. Particles of B-phycoerythrin (260,000 daltons) stained with phosphotungstic acid had a diameter of 10 nm and a height of 5.4 nm, but when stained with uranyl oxalate, the respective dimensions were 11.5 and 6.4 nm (Gantt, 1969). C-PC similarly has dimensions of approximately 12 × 6.0 nm (Berns and Edwards, 1965; Eiserling and Glazer, 1974).

At this time, most phycobiliprotein aggregates have not been individually visualized in solution by negative staining. Particularly intractable have been APC and b-PE. However, Bryant et al. (1976) have succeeded in obtaining the first dimensions of APC and confirmation for C-PC from examinations of small crystals. In cross fractures of a PC crystal (Fig. 13), discs of 12 nm diameter can be discerned. The appearance suggests that the crystal is composed of repeating units in agreement with the measurements of loose PC. On examination of phycoerythrocyanin, they found that it was indistinguishable from C-PC. APC microcrystals (Fig. 14) exhibited stacks with a repetition of 3.4 nm and a diameter of 12.5 nm. Each stack is assumed to be a trimeric aggregate $(\alpha\beta)_3$ with MW of approximately 105,000. In phycobilisomes they appear to exist as $(\alpha\beta)_6$. Such APC aggregates (196,000 daltons) have been recovered by Brown and Troxler (1977).

Examination of crude cell extracts and isolated phycobilisomes show convincing similarities. Short stacks of C-PC recovered from crude blue-green algal extracts (Eiserling and Glazer, 1974) resemble those from purified PC. Furthermore, when Kessel et al. (1973) compared fresh cell extracts with purified

---

FIG. 12. B-phycoerythrin disc-shaped molecules oriented mostly in broad face view. Arrows indicate side views.

FIG. 13. Fragment of a phycocyanin crystal with arrow heads indicating disc-like structures.

FIG. 14. Edge of an allophycocyanin microcrystal with 3.4 nm periodicity visible. (Figs. 13 and 14 from Bryant et al., 1976.)

FIG. 15. Selected and oriented electron micrographs of Rhodella violacea phycobilisomes. Six spoke-like structures radiate from a center that appears to consist of three ring-shaped aggregates. (From Mörschel et al., 1977.)

samples, they found that large aggregates ($18S_{20w}$) predominated in the fresh extracts, while smaller aggregates (6 and $11S_{20w}$) were more common in the purified samples. Their electron micrographs indicate that stacking of the disc occurs predominately on their broad face. Whereas this is also a common occurrence in purified B-PE, interaction along the short face is also possible (Gantt, 1969). Molecules with multifaceted binding sites would seem to allow greater versatility in the phycobilisome shape and perhaps allow for greater stability. It has been possible to obtain larger aggregates in 2 M NaCl in $vitro$ in C-PC derived from some halophilic blue-green algal strains by Kao et al. (1976). However, this was not universally applicable, and it appears that other factors are required for aggregation.

4. Concerning the Phycobiliprotein in Vivo State

We are only beginning to realize that the in $vivo$ state of phycobiliproteins is more complex than might be predicted by examination of those phycobiliproteins stable under commonly used experimental conditions. Observations on dissociating phycobilisomes reveal changes in absorption and fluorescence emission that cannot be entirely explained by the disaggregation of the commonly purified forms. For example, in $P.$ $cruentum$, major absorption changes at 545 and 573 nm are present in difference spectra of dissociating vs intact phycobilisomes (Gantt et al., 1976a) that suggest that b-PE and B-PE are spectrally much more similar in $vivo$ than when in solution, where b-PE can appear in various molecular sizes (55,000 to 110,000 daltons). Significant absorption changes have also been observed in examining concentrated vs dilute phycobiliprotein solutions of $Nostoc$ species (Gray and Gantt, 1975). Further, of the two C-PC forms (PC_{637} and PC_{623}) isolated from phycobilisomes of $Agmenellum$ $quadruplicatum$ (Gray et al., 1976), PC_{637} is probably closer to the in $vivo$ state than the more stable PC_{623}. This supposition is based on the fact that PC_{637} has a longer absorption and fluorescence emission (653 nm) state that is more akin to some observed PC forms of dissociating phycobilisomes (Gantt et al., 1979). With time, a gradual change from the 637 nm absorption peak to 623 nm occurs, as well as concomitant changes in their emission from 653 to 643 nm. Along with the spectral changes, there was also a decrease in the sedimentation rate, with that of the PC_{623} form being lower.

Of significance in assessing the in $vivo$ state is the question of whether or not the phycobiliproteins are able to recombine in $vitro$, and further, whether it is a functional recombination. Can the recombined complexes transfer energy as in a phycobilisome?

5. Are There Binding Proteins in Phycobilisomes?

In an examination of phycobilisomes from eight species of blue-green algae, Tandeau de Marsac and Cohen-Bazire (1977) found additional polypeptide bands on sodium dodecyl sulfate-polyacrylamide gel electrophoresis (SDS-PAGE). They suggested that the lower MW forms (30,000 to 70,000) may be involved in phycobiliprotein binding. Furthermore, in strains undergoing chromatic adaptation, cessation of PE synthesis was accompanied by a decrease of one of the bands, whereas with increased PC synthesis, two other bands increased. Yamanaka *et al.* (1978) and Bryant *et al.* (1979) have also found additional polypeptide bands on electrophoresis upon SDS treatment.

Direct evidence for the involvement of binding proteins in phycobilisomes has recently become available in our laboratory. Complexes consisting of PE and PC have been obtained from phycobilisomes, of the red alga *P. sordidum,* and the blue-green alga *Nostoc.* It is possible to dissociate these complexes into their individual phycobiliproteins. Upon mixing the PE and PC fractions, recombination occurs to a high degree: approximately 60% in *P. sordidum* and 80% in *Nostoc.* These complexes show the same efficient energy transfer, sedimentation characteristics, and pigment composition as the original. However, a comparison on SDS-PAGE showed the recombinable pigment fractions had additional polypeptide bands (25,000–30,000 daltons) that were lacking in the nonrecombinable fractions. Thus the additional polypeptides appear to be necessary for the binding.

It has been estimated that as much as 12–15% of the total stainable protein (Coomassie blue in SDS-PAGE) is accounted for by noncolored proteins (Tandeau de Marsac and Cohen-Bazire, 1977; Yamanaka *et al.,* 1978; Bryant *et al.,* 1979). These results appear different from those of Gantt and Lipschultz (1974), Gray and Gantt (1975), and Mörschel *et al.* (1977) who found that the major, if not entire, protein content phycobilisomes was accounted for by the phycobiliproteins. However, the discrepancies are due in part to different methods of analysis, particularly the SDS-PAGE systems. It must also be remembered that phycobiliproteins fade extensively (64% PE, 90% PC) during SDS-PAGE treatment and that a stoichiometric relationship for Coomassie blue staining of various phycobiliprotein fractions has not been established.

D. Phycobilisome Models

The first phycobilisome model was based strictly on morphological data from blue-green algae (Edwards and Gantt, 1971; Wildman and Bowen, 1972). It was of a disc-shaped phycobilisome (approximately 32 × 7 nm) with a hexagonal structure that could accommodate 12–14 PC molecules. This was somewhat oversimplified since it did not consider the presence and location of APC and did not include PE.

For efficient energy transfer, the most logical arrangement of the phycobiliproteins would be to have them in order of transfer events PE → PC → APC → chlorophyll. In this arrangement, APC would be nearest to the photosynthetic membrane followed by PC and then PE on the outside. The simplest type of phycobilisome would thus have APC near the membrane (Fig. 16A) and PC on the stroma side. Since images from electron micrographs reveal finger-like projections on the periphery, it is assumed that they consist of stacked discs of PC and possibly APC. It is possible that each stack within a cluster (phycobilisome) could function independently providing each contained a far-emitting APC form.

An elaboration on the basic structure is seen in Fig. 16B. Each finger-like projection consists of PE on the outside, with C-PC in the middle near an APC core. This model, an obvious intermediate between A and C, is supported by pigment and electron microscopy data from studies on *R. violacea* (Mörschel *et al.*, 1977). Mörschel *et al.* (1977) obtained the first electron microscopic evidence for a regular phycobilisome structure (Fig. 15). It usually consists of a core of three units with six projections radiating from it. Bryant *et al.* (1979), in their detailed studies of several blue-green algal phycobilisomes, have found that the same arrangement exists. This suggests that a homologous structure exists in the disc-shaped phycobilisomes of red and blue-green algal phycobilisomes. Recently, Koller *et al.* (1978) isolated a complex composed of three discs that have the same diameter and spacing as the projections. Each of these has a molecular weight of approximately 790,000 and pigment molar ratios of 2 B-PE to 1 PC. This is the same ratio as in *R. violacea* phycobilisomes, which in addition, also contain one molar ratio of APC (Mörschel *et al.* 1977). By analogy with *P. cruentum* (Gantt *et al.*, 1976a,b) and from inference of the most logical energy transfer sequence, it can be assumed that APC is in the center and B-PE on the outer side. Furthermore, the volume occupied by the phycobiliprotein stacks and core is in reasonable agreement with the phycobilisome volume expected from electron micrograph dimensions. A phycobilisome of *R. violacea* can thus be assumed to consist of six APC molecules (130,000 MW each), six C-PC molecules (260,000 MW each), and twelve B-PE molecules (280,000 MW each), totaling approximately 5.6–6.0 × 10^6 MW (Koller *et al.* 1978).

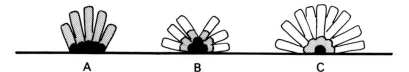

A   B   C

FIG. 16. Phycobilisome models showing probable phycobiliprotein arrangements. Consistent with maximum energy transfer, allophycocyanin (black) would be near the thylakoid membrane and serve to funnel excitation energy to chlorophyll. Adjacent to allophycocyanin is phycocyanin (stippled) and phycoerythrin (clear area) attached to phycocyanin. (A) This represents the basic type of phycobilisome consisting only of allophycocyanin and phycocyanin. (B) This is an elaboration on the basic type with addition of phycoerythrin as in *Rhodella violacea*, a disc-type phycobilisome (Mörschel *et al.*, 1977; Koller *et al.*, 1978). (C) This cutaway model shows the arrangement as proposed for *Porphyridium cruentum* (Gantt *et al.*, 1976a), a globular-type of phycobilisome.

Molecular weight estimates for two blue-green algal phycobilisomes were in a similar range 3.5–7.5×10^6 (Bryant *et al.*, 1979).

The phycobilisome structure for *P. cruentum* diagrammed in Fig. 16C is directly supported by kinetic studies of phycobiliprotein release, pigment localization with monospecific antisera, shape determinations from negatively stained electron micrographs and pigment analysis (Gantt and Lipschultz, 1974; Gantt *et al.*, 1976a,b). In calculating the volume of a prolate-shaped phycobilisome, the formula of an ellipsoid is used and the dimensions are: 47.5 nm diameter, 33 nm height, and 33 nm width. This yields an average volume of 2.8×10^4 nm^3. Since the most accurate values are available for B-PE (10×5 nm), it has been assumed to represent the standard unit as to size and shape. It has a calculated volume of approximately 39 nm^3. Thus a phycobilisome can contain approximately 72 molecules of that size. Since each such molecule has an apparent MW of 270,000, the calculated phycobilisome weight is approximately 19.4×10^6. From the percentage phycobiliprotein composition (84% PE, 11% R-PC, 5% APC) made on a per protein basis, it can be assumed that the pigments occupy similar volumes. This would result in 60 molecules of PE, 8 of R-PC, and 4 of APC (each molecule being 270,000 MW).

The phycobilisome of *P. cruentum* has a considerably larger volume than that of *R. violacea*, but the above calculations are only as valid as the assumptions made from as yet incomplete data. They will no doubt be modified as more precise data become available, which will require application by all laboratories of the same standard methods for electron microscopy, and use of low angle crystallography combined with biochemical and hydrodynamic measurements on stable phycobilisomes.

IV. Energy Transfer

A. Quantum Yield and Fluorescence Polarization

Quantum yield and fluorescence polarization determinations of phycobilisomes from *P. cruentum, Nostoc,* and *Fremyella* (with variable pigment compositions) have shown that phycobilisomes are energetically effective macromolecular aggregates. Energy migration to APC was indicated by the predominant APC fluorescence emission and was independent of phycobiliprotein excitation. High efficiency inside the phycobilisomes was evident from low polarized fluorescence (Grabowski and Gantt, 1978a,b).

Excitation of phycobilisomes (500–650 nm) resulted in degrees of fluorescence polarization between ±0.02, whereas in isolated phycobiliproteins, it was 2 to 12 times greater. In addition, 94–98% of the excitation energy of PE was transferred to PC and APC as determined from comparisons of fluorescence spectra of intact and dissociated phycobilisomes. Independent confirmation in *P. cruentum* was obtained by Searle (1978) and colleagues with time resolved picosecond measurements.

Fluorescence quantum yields of phycobilisomes (0.60–0.68) were very similar to those of purified APC (0.68). Phycobilisomes isolated from *F. diplosiphon* and *Nostoc*, which showed striking chromatic adaptation to different wavelengths, exhibited no significant alteration of their fluorescence quantum yields.

The mean transfer time, calculated on the basis of experimental results by Grabowski and Gantt (1978b) was 280 ± 40 psec for transfer of excitation from the PE to PC layer of the phycobilisome. Searle *et al.* (1978), at high excitation intensities, reported a fluorescence risetime between PC and APC of 120 psec.

B. RESONANCE TRANSFER AMONG PHYCOBILIPROTEINS

It is now generally accepted that the migration of excitation energy from phycobiliproteins to chlorophyll *a* molecules proceeds by inductive resonance (a case of very weak interaction). The efficiency of transfer, according to Förster (1960, 1965) is inversely proportional to the sixth power of the distance between the donor and acceptor chromophores. In the phycobiliproteins, it requires that the donor molecule fluorescence must have a sufficiently long lifetime to make the transition possible, and that this emission must overlap the absorption spectrum of the acceptor. Furthermore, the relative orientations of both donor and acceptor oscillators must permit strong interaction. Duysens (1952) proposed that energy migration in whole cells occurred from PE → PC → chlorophyll. Dale and Teale (1970) and Teale and Dale (1970), from characterization of several phycobiliproteins, had calculated that transfer from PE → PC was the favored migration route as compared with PE transfer to chlorophyll *a* directly.

For determinations of excitation energy transfer in phycobilisomes of *Nostoc* and *P. cruentum* phycobilisomes, the absolute quantum yields, fluorescence polarization spectra, and fluorescence lifetimes of isolated pigments were used (Grabowski and Gantt, 1978a,b). The most probable transfer path among *P. cruentum* phycobiliprotein chromophores is indicated by Förster's critical distances (in Å over arrows):

$$\text{b-PE} \overset{49}{\leftrightarrow} \text{b-PE} \overset{55}{\rightarrow} \text{B-PE} \overset{53}{\leftrightarrow} \text{B-PE} \overset{63}{\rightarrow} \text{R-PC} \overset{55}{\leftrightarrow} \text{R-PC} \overset{64}{\rightarrow} \text{APC} \overset{61}{\leftrightarrow} \text{APC}$$

This scheme indicates that, although some transfer occurs among chromophores of the same pigment (double-headed arrows), the most probable transfer path (direct arrows) is indicated by the increased distances. The scheme as shown is incomplete in that it does not take into account the various existing APC forms

(Ley *et al.,* 1977; Zilinskas *et al.,* 1978). Also, since the b-PE may be greatly altered *in vitro,* and thus have a crucially lower fluorescence lifetime, its position in the transfer scheme may be different *in vivo.*

C. Bridging Pigment from the Phycobilisome to the Photosynthetic Membrane

When phycobilisomes were first reported to have a fluorescence emission peak at 675 nm, it was attributed to APC (Gantt and Lipschultz, 1973) because it is the phycobiliprotein with the longest known emission. It is also the funnel through which energy migrates, because the 675 nm emission was found to decrease on dissociation of phycobilisomes from *P. cruentum.* In addition APC is also an efficient light harvesting pigment as indicated by the oxygen evolution in blue-green algal cells (Lemasson *et al.,* 1973). It was originally postulated that the longer wavelength fluorescence emission (675 vs 660 nm) was due to an aggregated state of APC in the phycobilisome when it served as the bridging pigment to chlorophyll. In 1975, Glazer and Bryant isolated from several blue-green algae an APC form that had two absorption peaks, 617 and 671 nm, but that, significantly, had an emission peak at approximately 680 nm, and thus was designated as APC-B and as the final emitter of phycobilisomes. Subsequently, Ley *et al.* (1977) also found an APC-B form in *P. cruentum.* They furthermore showed that crystals of the 660 nm emitting form of APC did not show a long fluorescence emission.

On the basis of its spectral characteristics, APC-B is the logical bridging pigment between phycobilisomes and chlorophyll. Although it is usually present in very small amounts, it is not expected to be limiting as long as there is one APC-B molecule per phycobilisome. Calculations indicate (Grabowski and Gantt, 1978b) that the de-excitation rate of PC, or APC, in phycobilisomes is much greater (at least 10^8 sec^{-1}) than the excitation rate (10^4 sec^{-1}).

There is yet another form which, on the basis of the fluorescence emission, must be considered as a bridging pigment. APC-I isolated from phycobilisomes of *Nostoc* (Zilinskas *et al.* 1978) has a fluorescence emission peak at 680 nm (23°C). Its absorption spectrum with a λ_{max} at 654 nm distinguishes it from APC-B. Four forms of APC have been recovered from *Nostoc* phycobilisomes by separation with calcium phosphate adsorption chromatography and isoelectric focusing. They are distinctive by their absorption spectra (Fig. 17) and their circular dichroism spectra (O. Canaani, personal communication) but have great similarity by immunoprecipitation reactions. It has been suggested that some of these forms, particularly APC-II and APC-III are directly related in that III reverts to II (Zilinskas *et al.,* 1978). Since APC-B and APC-I are distinctive far-emitting forms, the question arises as to why there would be two bridging pigments. In this event, it would increase the possibility of two distinct

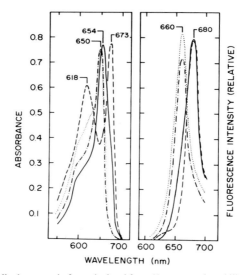

FIG. 17. Four allophycocyanin forms isolated from *Nostoc* sp. phycobilisomes. Absorption (left) and emission spectra (right) were made at room temperature in phosphate buffer pH 6.8. Allophycocyanin (APC) I (———) and allophycocyanin-B (- - -) with very distinct absorption spectra have overlapping emission peaks at 680 nm. The absorption peaks of APC II (....) and APC III (- · - ·) are superimposed at 650 nm; they have an emission peak at 660 nm. (After Zilinskas *et al.*, 1978; and courtesy of O. Canaani.)

phycobilisome populations, one population funnelling energy through APC-I, the other through APC-B. It is also possible that the two far-emitting APCs are part of one transfer sequence, an indication of which would be forthcoming from more refined emission spectra.

V. Phycobilisomes in Relation to the Photosynthetic Membrane

A. ARRAYS ON AND IN MEMBRANES

Since the observations of the regular arrays of phycobilisomes on thylakoid membranes, speculations have existed that certain structured membrane components may be the photoactive centers. In many species of red and blue-green algae, the parallel rows of phycobilisomes (Figs. 3 and 5) have a center-to-center spacing of 40–77 nm (Table I). The spacing is somewhat variable with the species.

In freeze–fractures across the phycobilisome rows and membranes, phycobilisomes can be seen (Fig. 18) and bear a strong resemblance to those in sections. However, when the membranes themselves are cleaved in certain planes, small

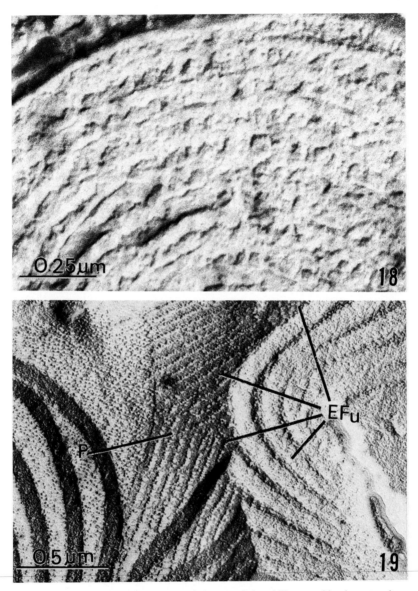

FIG. 18.   Freeze-etch replica showing regularly arrayed phycobilisomes with substructure in cross fracture of *Glaucocystis nostochinearum*. (From Bourdu and Lefort, 1967.)

FIG. 19.   Fracture through the plastid of *Porphyridium cruentum* showing parallel arrays of small particles (P) on the EFu fracture face. (From Neushul, 1971.)

particles can be observed in one of these, that may be underlying the phycobili-somes. In the EFu fracture face (terminology as in Staehelin *et al.*, 1978), parallel rows of small particles (approximately 10 nm) are revealed (Figs. 19) in many species that also possess parallel phycobilisome rows with comparable spacing (Bourdu and Lefort, 1967; Guerin-Dumartrait *et al.*, 1970; Lefort-Tran *et al.*, 1973; Lichtlé and Thomas, 1976; Neushul, 1971). It is not possible to obtain phycobilisomes and underlying small particles in the same fracture planes. The phycobilisomes are usually seen only in fractures across the mem-brane, and the 10 mm particles are revealed in fractures parallel to the mem-brane. In *P. cruentum*, Neushul (1970) observed that the spacing of the small particle rows were either 50 or 100 nm. This observation, and the fact that an interdigitation of alternate phycobilisome rows from opposing membranes is sometimes observed, led him to propose a model. This model provided for the possibility that phycobilisomes could transfer to one or both thylakoid mem-branes. Such a model, however, does not seem to have an obvious functional basis and, further, would suggest that the phycobilisomes would have two APC sites; this later suggestion is not supported by the present evidence. A model more consistent with available information is that proposed by Lefort-Tran *et al.* (1973), a modified form of which is shown in Fig. 20. It allows for the possibility that each phycobilisome is associated with one or more of the 10 nm particles, and is consistent with the parallel spacing of particles and phycobilisomes in algae where this occurs.

The most definitive study on correlating the spacing of the phycobilisome rows with small membrane particles was carried out by Lichtlé and Thomas (1976). They compared the phycobilisome number in tangential cuts of chemically fixed sections with the number of 10 nm particles on EFu fracture faces. They found ratios of 0.5–2.0 phycobilisomes per particle (Table II).

Whereas the positive structural correlations are supportive of a structural and functional association between the phycobilisomes and the 10 nm membrane particle, it must be pointed out that a regular structural array is not always observed. In *Spermothamnion* and *Griffithsia* (Staehelin *et al.*, 1978), both phycobilisomes and EFu particles were randomly arranged, yet regularly arrayed

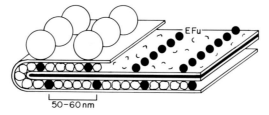

Fig. 20. Schematic representation of suggested arrangement of phycobilisomes (large, clear particles) on the thylakoid membrane and the parallel rows of small particles (black) in the cleaved membrane (EFu face). (Adapted after Lefort-Tran *et al.*, 1973.)

TABLE II

COMPARISON OF PHYCOBILISOME NUMBER PER AREA IN SECTIONS, AND NUMBER OF 10 NM PARTICLES APPEARING ON THE EFu FACE IN FREEZE-CLEAVED PHOTOSYNTHETIC MEMBRANES

Species	Phycobilisomes per μm^2	10 nm Particles per μm^2	Ratio of phycobilisomes to particles (μm^2)	References
Oscillatoria splendida	1400	1300	1	Lichtlé and Thomas (1976)
Oscillatoria brevis	1200	500	2	Lichtlé and Thomas (1976)
Oscillatoria limosa	1200	650	2	Lichtlé and Thomas (1976)
Antithamnion glanduliferum	230	600	0.5	Lichtlé and Thomas (1976)
Griffithsia pacifica (50 fc)	270	670	0.4	Waaland et al. (1974); Staehelin et al. (1978)
Griffithsia pacifica (300 fc)	165	770	0.2	Waaland et al. (1974); Staehelin et al. (1978)
Porphyridium cruentum	400[a]	2000	0.2	Neushul (1970)

[a] Determinations from Gantt's laboratory on independent culture.

phycobilisomes were observed in *Spermothamnion* in another study (Gantt and Conti, 1966a). In *Antithamnion* (Lichtlé and Thomas, 1976), the phycobilisomes, but not the EFu particles had a regular array. Perhaps culture conditions, particularly illumination, or specimen preparation may influence the regularity or randomness of these thylakoid structures.

B. Association with the Photosystems

1. *Energy Transfer from Phycobiliprotein to the Reaction Centers*

The phycobiliproteins serve as antennae pigments that funnel the excitation energy absorbed primarily to photosystem II (PSII) reaction centers from which it can then be passed to photosystem I (PSI) reaction centers. In *P. cruentum* (Ley and Butler, 1977a,b), with the PSI reaction centers closed, 95% of the energy absorbed by PE was transferred to PSII as indicated by the fluorescence yield at 695 nm ($-196°C$). Under physiological conditions, the energy is distributed to both photosystems, as in *Anacystis nidulans* where 40% of the quanta absorbed by PC are contributed to PSI and 60% to PSII (Wang *et al.*, 1977). However, 84% of the quanta absorbed by chlorophyll go to PSI and only 16% to PSII. These results suggests that since the chlorophyll contribution in PSII is low the size of the reaction center II must be small. There is general agreement that this is a common state in many red and blue-green algae (Mimuro and Fujita, 1977; Ried *et al.*, 1977; Tel-Or and Malkin, 1977; Wang *et al.*, 1977).

2. *Photosystem II Particles*

It can be logically assumed that a PSII structural particle would be in close contact with the phycobilisome. A prime candidate for this is the 10 nm intrathylakoidal particle, as suggested initially by Lefort-Tran *et al.* (1973). Staehelin *et al.* (1978) pointed out in a recent summary that there is no direct proof for this as yet, but the possibility is likely. The circumstantial evidence for this includes (a) similar membrane spacing of phycobilisomes and small particles, (b) the expected small size of the PSII particle from the small chlorophyll number, (c) the direct energy transfer from phycobiliproteins to PSII chlorophyll, and (d) positive correlations of increased particle density with increased PSII activity (Wollman, 1979). One might expect a constant ratio of PSII particles and phycobilisomes if a direct relationship exists. The values expressed in Table II indicate that there appear to be anywhere from 0.5 to 4 particles per phycobilisome. In some algae such as in *Oscillatoria splendida* a 1:1 relationship seems to exist. But before acceptance of any constant ratio a more direct correlation would be desirable. It is of particular interest to point out the work of Stevens and Myers (1976). In pigment mutants of *A. nidulans* that varied in both chlorophyll and PC content, they determined that, in the parent and in a PC-poor mutant, the PSII reaction centers per cell and the number of calculated phycobilisomes were essentially constant. However, in a PC-rich mutant, the PSII reaction centers per cell decreased to about half that of the wild type and the phycobilisomes per reaction center increased to 3.5. Interestingly, this mutant had a higher turnover rate of the reaction centers. Although comparisons of these results with those from structural determinations suggest that similar relationships exist between phycobilisomes and PSII reaction centers, and between phycobilisomes and purported PSII particles direct proof is still lacking.

C. *In Vivo* FORMATION OF PHYCOBILISOMES

At this time, nothing is known about the site of phycobiliprotein synthesis. Several studies indicate that the biosynthesis and breakdown of the chromophores and the apoprotein are coordinated. Troxler (1972) showed that when PC-deficient cells (grown in darkness) of *Cyanidium caldarium* were placed in light, phycobiliprotein synthesis resumed. In this case, there were no detectable quantities of either free polypeptide subunits or chromophores. Furthermore, the results of Bennett and Bogorad (1973) from chromatically adapting *F. diplosiphon* cultures and those of Lau *et al.* (1977) from PC-deficient *A. nidulans* cells show that the disappearance and reappearance of phycobilin absorption and of apoprotein follow similar time courses.

For investigating the phycobiliprotein synthesis and phycobilisome formation, it is possible to use cells that have diminished amounts of phycobiliproteins. Phycobiliprotein synthesis is greatly decreased or absent when cells are grown without carbon dioxide (Miller and Holt, 1977) or in media deficient in minerals such as nitrate (Fujita and Hattori, 1960; Allen and Smith, 1969; Lau *et al.*, 1977) and phosphate (Ihlenfeldt and Gibson, 1975). Upon readdition of these components, phycobiliprotein synthesis resumes. In phycobiliprotein-deficient cells, the phycobilisomes also decrease in number (Wehrmeyer and Schneider, 1975; Miller and Holt, 1977) and increase after addition of the critical components.

Results in our laboratory have shown that when *P. cruentum* cells are grown in a nitrate-deficient medium their phycobiliprotein content decreases to <1 $\mu g/$ cell, and the photosynthetic membranes are completely devoid of phycobilisomes. In 24 hours after readdition of nitrate and resumption of a small amount of phycobiliprotein synthesis, phycobilisomes are observable in a few small regions of the chloroplast, usually on peripheral thylakoids (Fig. 21). In 36–48 hours cells become fully pigmented with extensive phycobilisomes. It is interesting to note that neither the phycobilisome size nor the phycobiliprotein ratio varies between cells from 24- or 48-hour recovering cultures. The peripheral and restricted location, and the invariant pigment composition suggest that the phycobiliprotein synthesis and phycobilisome formation are probably coordinated.

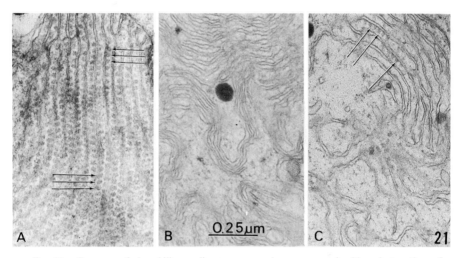

FIG. 21.   Sequence of phycobilisome disappearance and reappearance in chloroplast sections of *Porphyridium cruentum:* (A) with normal phycobiliprotein content, phycobilisomes (arrows) are closely packed; (B) deficient in phycobiliproteins and with thylakoid membranes devoid of phycobilisomes; and (C) small groupings of phycobilisomes 24 hours after resumption of phycobiliprotein synthesis.

VI. Problems and Future Directions

At this point the structure and function of phycobilisomes has only begun to be studied. Many areas remain open for exploration. Now that phycobilisome isolation has become routine, further characterization of the phycobilisome constituents is feasible. Once conditions are found whereby the less stable forms of phycobiliprotein and the binding proteins can be isolated it should be possible to accomplish the reconstitution of phycobilisomes *in vitro.*

A major area for future investigations relates to where phycobiliprotein synthesis occurs and how it is regulated. Action spectra (Diakoff and Scheibe, 1973; Haury and Bogorad, 1977) and photoinduction of phycobiliprotein synthesis (Ohki and Fujita, 1978) are expected to provide important leads. Phycobiliprotein regulation bears directly on the *in vivo* formation of phycobilisomes and raises such questions as: Do phycobilisomes form on specific predetermined membrane sites? Does the site involve a binding protein? Answers to some of these questions will become available by using recently developed conditions (Katoh and Gantt, 1979) with which functional phycobilisome attachment can be assayed. Furthermore, is the adaptive strategy to different light conditions the same in red algae and blue-green algae? Chromatic adaptation in *P. cruentum,* involving the chlorophyll forms, and the effect on the energy distribution from phycobilisomes to the photosystems, was recently reported by Ley and Butler (1979), this differs from the more familiar adaptation involving the phycobiliprotein ratios. Recently Gendel *et al.* (1979) in examining phycoerythrin synthesis in chromatically adapting *Fremyella* provided suggestive evidence that such adaptation involves gene regulation at the transcriptional level.

Further questions relate to the association of phycobilisomes to the photosystems. The available evidence suggests that there is one phycobilisome type that is functionally associated primarily with PSII. Is this association with the chlorophyll pool or directly with the reaction center? The possibility that there is more than one type of phycobilisome with separate associations to photosystems I and II should remain under consideration because it cannot be ruled out by the existing data. Characterization of photosystem particles on blue-green thylakoid membranes (Newman and Sherman, 1978; Reiman and Thornber, 1979) is an important initial step in investigating this area.

ACKNOWLEDGMENTS

Research cited from this laboratory was supported by DOE Contract No. EY-76-S-05-4310. The help of C. A. Lipschultz and research support are gratefully acknowledged.

REFERENCES

Allen, M. M., and Smith, A. J. (1969). *Arch. Microbiol.* **69,** 114.

Arnold, W., and Oppenheimer, J. R. (1950). *J. Gen. Physiol.* **33,** 423.

Bennett, A., and Bogorad, L. (1973). *J. Cell Biol.* **58,** 419.

Berns, D. S. (1971). *In* "Biological Macromolecules" (S. N. Timasheff and G. Fasman, eds.), p. 105. Dekker, New York.

Berns, D. S., and Edwards, M. R. (1965). *Arch. Biochem. Biophys.* **110,** 511.

Bogorad, L. (1975). *Annu. Rev. Plant.* **26,** 369.

Bourdu, R., and Lefort, M. (1967). *C. R. Acad. Sci. (Paris)* **265,** 37.

Brown, A. S., and Troxler, R. F. (1977). *Biochem. J.* **163,** 571.

Bryant, D. A., Glazer, A. N., and Eiserling, F. A. (1976). *Arch. Microbiol.* **110,** 61.

Bryant, D. A., Guglielmi, G., Tandeu de Marsac, N., Castets, A. M., and Cohen-Bazire, G. (1979). *Arch. Microbiol.* **123,** 113.

Cho, F., and Govindjee (1970). *Biochim. Biophys. Acta* **216,** 151.

Clement-Metral, J. D., and Lefort-Tran, M. (1971). *FEBS Lett.* **12,** 225.

Cohen-Bazire, G., and Lefort-Tran, M. (1970). *Arch. Microbiol.* **71,** 245.

Cosner, J. (1978). *J. Bacteriol* **135,** 1137.

Dale, R. E., and Teale, F. W. J. (1970). *Photochem. Photobiol.* **12,** 99.

Diakoff, S., and Scheibe, J. (1973). *Plant Physiol.* **51,** 382.

Duysens, L. N. M. (1952). Ph.D. Thesis, University of Utrecht, Netherlands.

Edwards, M. R., and Gantt, E. (1971). *J. Cell Biol.* **50,** 896.

Edwards, M. R., Berns, D. S., Ghiorse, W. C., and Holt, S. C. (1968). *J. Phycol.* **4,** 283.

Eiserling, F. A., and Glazer, A. N. (1974). *J. Ultrastruct Res.* **47**, 16.
Förster, Th. W. (1960). *In* "Comparative Effects of Radiation" (M. Burton, J. S. Kirby-Smith and J. L. Magee, eds.), p. 300. Wiley, New York.
Förster, Th. W. (1965). *In* "Modern Quantum Chemistry" (O. Sinanoglu, ed.), p. 93. Academic Press, New York.
Fujita, Y., and Hattori, A. (1960). *Plant Cell Physiol.* **1**, 281.
Gantt, E. (1969). *Plant Physiol.* **44**, 1629.
Gantt, E. (1975). *BioScience* **25**, 781.
Gantt, E. (1977). *Photochem. Photobiol.* **26**, 685.
Gantt, E. (1979). *In* "Biochemistry and Physiology of Protozoa" (M. Levandowsky and S. A. Hutner, eds.), 2nd Ed., Vol. 1, p. 121. Academic Press, New York.
Gantt, E., and Conti, S. F. (1965). *J. Cell Biol.* **26**, 365.
Gantt, E., and Conti, S. F. (1966a). *J. Cell Biol.* **29**, 423.
Gantt, E., and Conti, S. F. (1966b). Brookhaven Symp. in Biology, No. 19, p. 393.
Gantt, E., and Conti, S. F. (1969). *J. Bacteriol.* **97**, 1486.
Gantt, E., and Lipschultz, C. A. (1972). *J. Cell Biol.* **54**, 313.
Gantt, E., and Lipschultz, C. A. (1973). *Biochim. Biophys. Acta* **292**, 858.
Gantt, E., and Lipschultz, C. A. (1974). *Biochem.* **13**, 2960.
Gantt, E., and Lipschultz, C. A. (1977). *J. Phycol.* **13**, 185.
Gantt, E., and Lipschultz, C. A. (1980). *J. Phycol.* *16* (in press).
Gantt, E., Edwards, M. R., and Conti, S. F. (1968). *J. Phycol.* **4**, 65.
Gantt, E., Lipschultz, C. A., and Zilinskas, B. (1976a). *Biochim. Biophys. Acta* **430**, 375.
Gantt, E., Lipschultz, C. A., and Zilinskas, B. A. (1976b). Brookhaven Symp. in *Biology,* No. 28, p. 347.
Gantt, E., Lipschultz, C. A., Grabowski, J., and Zimmerman, B. K. (1979). *Plant Physiol.* **63**, 615.
Gendel, S., Ohad, I., and Bogorad, L. (1979). *Plant Physiol.* **64**, 786.
Glazer, A. N. (1977). *Mol. Cell Biochem.* **18**, 125.
Glazer, A. N., and Bryant, D. A. (1975). *Arch. Microbiol.* **104**, 15.
Glazer, A. N., and Hixson, C. S. (1977). *J. Biol. Chem.* **52**, 32.
Grabowski, J., and Gantt, E. (1978a). *Photochem. Photobiol.* **28**, 39.
Grabowski, J., and Gantt, E. (1978b). *Photochem. Photobiol.* **28**, 47.
Gray, B. H., and Gantt, E. (1975). *Photochem. Photobiol.* **21**, 121.
Gray, B. H., Lipschultz, C. A., and Gantt, E. (1973). *J. Bacteriol.* **116**, 471.
Gray, B. H., Cosner, J., and Gantt, E. (1976). *Photochem. Photobiol.* **24**, 299.
Guerin-Dumartrait, E., Sarda, C., and Lacourly, A. (1970). *C. R. Acad. Sci. (Paris)* **270**, 1977.
Guerin-Dumartrait, E., Hoarau, J., Leclerc, J.-C., and Sarda, C. (1973). *Phycologia* **12**, 119.
Gysi, J. R., and Chapman, D. J. (1980). *In* "CRC Handbook on Solar Resources" (A. Mitsui, C. C. Black, and O. Zaborsky, eds.) (in press).
Haury, J., and Bogorad, L. (1977). *Plant Physiol.* **60**, 835.
Honsell, E., Avanzini, A., and Ghirardelli, L. A. (1978). *J. Submicrosc. Cytol.* **10**, 227.
Ihlenfeldt, M. J. A., and Gibson, J. (1975). *Arch. Microbiol.* **102**, 22.
Kao, O. H. W., Edwards, M. R., MacColl, R., and Berns, D. (1976). *Experientia* Suppl. **26**, 291.
Katoh, T., and Gantt, E. (1979). *Biochim. Biophys Acta* **546**, 383.
Kessel, M., MacColl, R., Berns, D., and Edwards, M. R. (1973). *Canad. J. Microbiol.* **19**, 831.
Knoll, A. H., and Barghoorn, E. S. (1977). *Science* **198**, 396.
Koller, K. P., and Wehrmeyer, W. (1974). *Arch. Microbiol.* **100**, 253.
Koller, K. P., Wehrmeyer, W., and Schneider, H. (1977). *Arch. Microbiol.* **112**, 61.
Koller, K. P., Wehrmeyer, W., and Mörschel, E. (1978). *Eur. J. Biochem.* **91**, 57.
Lau, R. H., MacKenzie, M. M., and Doolittle, W. F. (1977). *J. Bacteriol.* **132**, 771.
Lefort, M. (1965). *C. R. Acad. Sci. (Paris)* **261**, 233.
Lefort-Tran, M., Cohen-Bazire, G., and Pouphile, M. (1973). *J. Ultrastruct Res.* **44**, 199.

Lemasson, C., Tandeau de Marsac, M., and Cohen-Bazire, G. (1973). *Proc. Natl. Acad. Sci. U.S.A.* **70**, 3130.

Ley, A. C., and Butler, W. L. (1977a). *In* "Photosynthetic Organelles." Issue of *Plant Cell Physiol.* **3**, 33.

Ley, A. C., and Butler, W. L. (1977b). *Biochim. Biophys. Acta* **462**, 290.

Ley, A. C., and Butler, W. L. (1979). Program 7th Ann. Meeting Photobiol. Soc., 180.

Ley, A. C., Butler, W. L., Bryant, D. A., and Glazer, A. N. (1977). *Plant Physiol.* **59**, 974.

Lichtlé, C. (1973). *C. R. Acad. Sci. (Paris)* **277**, 1865.

Lichtlé, C. (1978). Etude expérimentale d'une Cryptophycée analyse particulière de l'appariel photosynthetique comparaison avec quelques Rhodophycées. Ph.D. Thesis, Université Pierre et Marie Curie, Paris.

Lichtlé, C., and Giraud, G. (1969). *J. Microsc.* **8**, 867.

Lichtlé, C., and Giraud, G. (1970). *J. Phycol.* **6**, 281.

Lichtlé, C., and Thomas, J. C. (1976). *Phycologia* **15**, 393.

Miller, L. S., and Holt, S. C. (1977). *Arch. Microbiol.* **115**, 185.

Mimuro, M., and Fujita, Y. (1977). *Biochim. Biophys. Acta* **459**, 376.

Mörschel, E., Koller, K. P., Wehrmeyer, W., and Schneider, H. (1977). *Cytobiologie* **16**, 118.

Muir, M. D., and Grant, P. R. (1976). *In* "Early History of the Earth" (B. F. Windley, ed.), p. 595. Wiley, New York.

Myers, J., and Kratz, W. A. (1955). *J. Gen. Physiol.* **39**, 11.

Neushul, M. (1970). *Am. J. Bot.* **57**, 1231.

Neushul, M. (1971). *J. Ultrastruct. Res.* **37**, 532.

Newman, P. J., and Sherman, L. A. (1978). *Biochim Biophys. Acta* **503**, 343.

O'Cara, P., and O'Eocha, C. (1976). *In* "Chemistry and Biochemistry of Plant Pigments" (T. W. Goodwin, ed.), Vol. I, p. 328. Academic Press, New York.

Ohki, K., and Fujita, Y. (1978). *Plant Cell Physiol.* **19**, 7.

Peyriere, M. (1968). *C. R. Acad. Sci. (Paris)* **266**, 2253.

Reiman, S., and Thornber, J. P. (1979). *Biochim. Biophys. Acta* **547**, 188.

Ried, A., Hessenberg, B., Metzler, H., and Ziegler, R. (1977). *Biochim. Biophys. Acta* **459**, 175.

Rippka, R., Waterbury, J., and Cohen-Bazire, G. (1974). *Arch. Microbiol.* **100**, 419.

Rippka, R., Deruelles, J., Waterbury, J. B., Herdman, M., and Stanier, R. Y. (1979). *J. Gen. Microbiol.* **111**, 1.

Rüdiger, W. (1975). *Ber. Deutsch. Bot. Ges.* **88**, 125.

Searle, G. F. W., Barber, J., Porter, G., and Tredwell, C. J. (1978). *Biochim. Biophys. Acta* **501**, 246.

Sheath, R. G., Hellebust, J. A., and Sawa, T. (1977). *Phycologia* **16**, 265.

Staehelin, L., Giddings, T. H., Badami, P., and Krzymowski, W. W. (1978). *In* "Light Transducing Membranes" (D. Deamer, ed.), p. 335. Academic Press, New York.

Stanier, R. Y., and Cohen-Bazire, G. (1977). *Annu. Rev. Microbiol.* **31**, 225.

Stevens, C. L. R., and Myers, J. (1976). *J. Phycol.* **12**, 99.

Tandeau de Marsac, N. (1977). *J. Bacteriol.* **130**, 82.

Tandeau de Marsac, N., and Cohen-Bazire, G. (1977). *Proc. Natl. Acad. Sci. U.S.A.* **74**, 1635.

Teale, F. W. J., and Dale, R. E. (1970). *Biochem. J.* **116**, 161.

Tel-Or, E., and Malkin, S. (1977). *Biochim. Biophys. Acta* **459**, 157.

Thomas, J. C. (1972). *C. R. Acad. Sci. (Paris)* **274**, 2485.

Tomita, G., and Rabinowitch, E. (1962). *Biophys. J.* **2**, 483.

Troxler, R. F. (1972). *Biochemistry* **11**, 4235.

Troxler, R. F. (1977). *In* "Chemistry and Physiology of Bile Pigments" (P. D. Beck and N. I. Berlin, eds.), p. 431. Fogarty International Center Proc. No. 35. DHEW Publication No. (NIH) 77–1100.

Waaland, J. R., Waaland, S. D., and Bates, G. (1974). *J. Phycol.* **10**, 193.

Wang, R. J., Stevens, C. L. R., and Myers, J. (1977). *Photochem. Photobiol.* **25,** 103.

Wehrmeyer, W., and Schneider, H. (1975). *Biochem. Physiol. Pflanzen* **168,** 519.

Wildman, R. B., and Bowen, C. C. (1972). *In* "Proceedings of the Symposium on Taxonomy and Biology of Blue-green Algae" (T. V. Desikachary, ed.), p. 1. Bangalore Press, Bangalore.

Wildman, R. B., and Bowen, C. C. (1974). *J. Bacteriol.* **117,** 866.

Wollman, F.-A. (1979). *Plant Physiol.* **63,** 375.

Yamanaka, G., Glazer, A. N., and Williams, R. C. (1978). *J. Biol. Chem.* **253,** 8303.

Zilinskas, B. A., Zimmerman, B. K., and Gantt, E. (1978). *Photochem. Photobiol.* **27,** 587.

INTERNATIONAL REVIEW OF CYTOLOGY, VOL. 66

Structural Correlates of Gap Junction Permeation

Camillo Peracchia

Department of Physiology, University of Rochester, School of Medicine and Dentistry, Rochester, New York

I. Introduction

The definition of cells as independent units of tissue (Schleiden and Schwann, 1838) obviously implied the existence of a wall-like structure, the plasma membrane, surrounding the cells and separating them from the rest of the world. The main function of the plasma membrane was believed to be that of a protective barrier to prevent exchange of cellular material with neighboring cells or the surrounding extracellular medium. However, as early as in the last decade of the nineteenth century, it became apparent that the plasma membrane is also the site of a congested traffic of molecules, moving in and out of cells, selectively discriminated by the plasma membrane via fine transport mechanisms.

The earliest transport studies (Overton, 1895; Gryns, 1896; Hedin, 1897) emphasized the importance of the lipid solubility of a molecule (or rather its oil–water partition coefficient) in determining its capacity to move across the plasma membrane, a concept that prompted the formulation of hypotheses envisioning the plasma membrane as a structure composed mainly of "lipoid" (fatty

81

acids). However, as early as the beginning of this century, it became clear that
certain molecules such as ions, water, methanol, formamide, and ethylene glycol
(Bernstein, 1902; Jacobs, 1924; Collander, 1937, etc.) were transported across
the plasma membrane much faster than expected from their rather low oil–water
partition coefficient.

Hypotheses suggesting the existence of hydrophilic pores or channels in membranes were then formulated and embodied into a general concept picturing the
surface membrane as a mosaic of lipoidic and sievelike areas (Jacobs, 1924,
1935; Höber, 1936; Collander, 1937). Interestingly, already at that time fric-
tional properties and selectivity characteristics of the presumed channels clearly
indicated that they could not be large water-filled openings in the membrane, but
must be rather narrow tunnels with a diameter close to that of the permeant
molecules (Jacobs, 1935), and that they were probably "built up of some fibrous
protein" (Höber, 1936). Although it is now clear that in those days there was no
reason to believe that transport of hydrophilic molecules necessarily required the
existence of channels since other means not known at that time (e.g., carriers)
could have accounted for the observed phenomena, solid evidence for the existence of various hydrophilic channels in membranes is now available
(Armstrong, 1975; Loewenstein, 1975; Neher and Stevens, 1977).

A great deal is known about the physiological properties of membrane channels and about the physicochemical mechanisms that modulate their characteristics of permeability; however, in most cases, little is known about channel
composition and molecular architecture, as most often we have been unsuccessful in even localizing them in membranes. In this respect, the channels that
mediate direct intracellular exchange of small molecules are rather unique since it
has been possible to locate them unequivocally in specialized regions of the
plasma membrane (commonly known as gap junctions) and to study them with a
great variety of techniques.

Gap junction channels are often regarded by membranologists as an unusual
means of membrane permeability probably having little in common with other
membranous channels. Although it is true that gap junction channels establish
hydrophilic paths between two cells' interiors and are a rather "permissive"
type of channel, as they are bidirectionally and indiscriminately permeable to
charged and neutral molecules, it is clear that they have many structural and
functional similarities with other membranous channels.

Undoubtedly in the years to come, gap junction channels will be recognized as
one of the most useful and exciting model systems to be exploited in studying the
general properties of membranous channels. In fact, not only can they be easily
localized in intact cells and studied by ultrastructural methods such as thin
sections, lanthanum infiltration, freeze-fracture, and negative staining, they can
also be isolated in quite pure fractions and characterized chemically. Due to their
crystalline configuration in isolated fractions, they can be studied by laser,
X-ray, and electron diffraction techniques. Finally, their permeability properties

and the mechanisms of permeability modulation can be studied electrophysiologically, morphologically, and by tracer-transfer methods.

In the last two decades, gap junctions have been labeled with various names most often derived from their ultrastructural appearance (see Section IV,B). Recently, however, in view of our improved understanding of the physiological meaning of gap junctions, Simionescu et al. (1975) have suggested renaming them "maculae communicantes" or "communicating junctions." Although it is reasonable to use function-related terms, this name may not be the most appropriate as gap junctions are not the only junctions by which cells communicate with each other (others are, for instance, synapses and neuromuscular junctions).

A name such as "coupling junctions" seems more appropriate, as gap junctions indeed bidirectionally couple neighboring cells, creating communities with some syncytia-like properties. Moreover, this name would be consistent with a terminology widely used in reference to various functional aspects of direct cell-to-cell communication such as electrical "coupling," metabolic "coupling," "coupling" coefficient, and "uncoupling."

A number of reviews have been published in the last few years on gap junction structure and function. Among those concerned primarily with structural features are McNutt and Weinstein (1973), Staehelin (1974, 1978), Gilula (1976, 1978), Weinstein (1976), Flower (1977), Griepp and Revel (1977), Larsen (1977), McNutt (1977), Peracchia (1977b), Bennett and Goodenough (1978), and Lane (1978).

II. Direct Cell-to-Cell Communication

A. Historical Perspective

The concept that most neighboring cells are capable of freely exchanging small protoplasmic molecules with each other is the most recent and still the most mysterious one in our knowledge of cell communication. In fact, for more than a century after the cell theory was proposed, the only known mechanisms of communication between cells involved diffusible molecules transferred via blood and/or extracellular fluid from transmitting to receiving cells: some at great distances from each other (hormonal communication), others separated by an extracellular cleft a few nanometers wide (synaptic and neuromuscular communication).

Long ago, segmented axons of crayfish (Wiersma, 1947) and *Lumbricus* (Bullock, 1945) were known to transmit the electrical impulse in both directions across anatomically well-defined "septa" with a negligible delay—certainly a finding that is incompatible with the presence of chemically transmitting synapses. However, it was not until intracellular recording with microelectrodes became available that unequivocal data on electrical communication between cells were produced.

The first example was reported by Weidman (1952) in kid's Purkinje fibers. Here the low value of the specific resistance across the boundaries between adjacent cells clearly suggested the absence of ionic barriers of any importance. A few years later, Furshpan and Potter (1959) confirmed the existence of electrical coupling between median and motor giant fibers of crayfish. By the early sixties, electrical communication had been reported in another crayfish system, the lateral giant fiber (Watanabe and Grundfest, 1961), in lobster heart (Hagiwara et al., 1959), and in neurons of puffer fish (Bennett, 1960), Aplysia (Tauc, 1959), mormyrid electric fish (Bennett et al., 1963), leech (Hagiwara and Morita, 1962), goldfish (Furukawa and Furshpan, 1963), and chick (Martin and Pilar, 1963a,b).

Naturally all these examples involved excitable cells, as no one showed any interest in testing whether or not unexcitable cells also shared their ionic population with their neighbors. Electrical coupling seemed, in fact, unnecessary for cells incapable of generating and transmitting electrical impulses. Why should one have expected that liver cells, for instance, might have been electrically coupled? What for? Curiously, two decades later we are still somewhat puzzled by the presence of extensive communication between unexcitable cells. But knowing, as we do now, that cells not only communicate via small ions but also via a large variety of small molecules including metabolic intermediates, nucleotides, and second messengers, we certainly can appreciate (although still in a hypothetical way) the many possible reasons for the involvement of cell coupling in maintaining a harmonious cooperation among groups of cells.

In the early sixties, Loewenstein and Kanno (1963) were studying the electrical properties of the nuclear envelope. For this purpose, they chose salivary gland cells of insect larvae as experimental material because their large size (50–100 μm) (Fig. 16,A,i) allows their nucleus to be easily impaled with microelectrodes. Surprisingly, upon current injections into the nucleus of a cell, an electrical potential was recorded not only in the injected cell but also (and nearly as high) in the neighboring cells (Loewenstein and Kanno, 1964). In the same year, Kuffler and Potter made a similar observation in another type of unexcitable cell—the glial cells of leech. These observations opened a new and exciting chapter in cell biology. Thereafter, most cells could no longer be regarded as units independent from each other, but rather as members of communities intimately interrelated via small molecules.

B. Electrical Coupling

Knowledge of intercellular exchange of small ions comes from measurements of the electrical resistance across the boundaries between neighboring cells. In a typical experiment, a current microelectrode (I) is inserted into a cell (C_I) and two voltage microelectrodes are inserted, one in the same cell and the other in a neighboring cell (C_{II}) to record prejunctional (V_I) and postjunctional (V_{II}) voltages, respectively. If hyper- or depolarizing square current pulses are injected in C_I, steady state voltage changes will be obtained in both cells. From the current and voltage records, the resistance in C_I (input resistance, R_{xi}) and that in C_{II} (transfer resistance, R_{xt}) can be calculated:

$$R_{xi} = \frac{V_1}{I}$$

$$R_{xt} = \frac{V_2}{I}$$

These resistances contain components of cytoplasmic resistance, resistance across the surface membrane (R_{xm}), and resistance across the junctional membranes of the two cells (R_{xj}). R_{xm} and R_{xj} can be calculated (Watanabe and Grundfest 1961; Bennett, 1966) as follows:

$$R_{xm} = R_{xi} + R_{xt}$$

$$R_{xj} = \frac{R_{xi}^2 - R_{xt}^2}{R_{xt}}$$

In well coupled cells, the amplitude of the voltage changes is almost identical in both cells (Fig. 16,B), which indicates that the electrical resistance of the junctional membranes (R_{xj}) is very low and not much greater than that of the cytoplasm alone.

A low intercellular resistance clearly indicates a free exchange of small ions between coupled cells, but it does not allow one to discriminate between positively and negatively charged ions. Evidence that both types are in fact capable of easily crossing the cell boundaries has come from studies of intercellular diffusion of small ions such as $^{42}K^+$ (Weidman, 1966; Bennett et al., 1967), $^{86}Rb^+$ (Ledbetter and Lubin, 1979), $^{22}Na^+$, $^{36}Cl^-$, $^{125}I^-$, $^{35}SO_4^-$ (Bennett et al., 1967), Co^{2+} (Bennett, 1973a; Politoff et al., 1974), and tetraethyl ammonium ([1-^{14}C]TEA) (Weingart, 1974). In this respect, the intercellular channels differ drastically from other channels such as those of excitability and the postsynaptic channels, which are selectively permeable to certain cations only.

By combining electrical recordings and ultrastructure, many have attempted to calculate the specific electrical resistance of the junctional membrane. Early works, due to uncertainty in identifying the membrane specialization responsible for intercellular communication, have derived the specific resistance by using an estimate of the entire area of cell-to-cell apposition. Therefore, the reported values of specific resistance were rather high, ranging from 10 Ω/cm^2 in insect salivary glands (Wiener et al., 1964) to 1–3 Ω/cm^2 in mammalian heart (Woodbury and Crill, 1961; Weidman, 1966). An ingenious procedure for measuring the specific resistance was employed by Weidman (1966). Bundles of sheep ventricle were pulled through a hole to divide each one into two halves; one half was charged with ^{42}K while the other was washed with inactive saline. Several hours later, after a steady state was reached, the concentration of ^{42}K was measured at various distances from the labeled source and the resistance of the intercalated discs was estimated from the values of intracellular diffusion of ^{42}K.

In recent years, the improvements in our understanding of the structural basis of direct cell communication have prompted new estimates of the specific electrical resistance based only on gap junction surface area. In dog's heart, Spira (1971) has reported a value of 1.4 Ω/cm^2 for the specific gap junctional resistance. Kriebel (1968) has estimated a specific resistance of 0.2 Ω/cm^2 in tunicate heart cells and a similar value (0.3 Ωcm^2) has been reported recently for gap junctions of the septum between giant axons of earthworms (Brink and Barr, 1977; Brink and Dewey, 1978). Lower values (10^{-2} Ω/cm^2) have been reported by Ito et al. (1974) in giant newt embryo cells. Here the measurements were performed on single cell pairs that are joined by only three to four slender processes ~ 1 μm in diameter. Thus possible errors in estimating the surface area of junctional apposition are minimized.

In a number of studies, the electrical coupling resistance has been measured between cells growing in culture (Borek et al., 1969; Siegenbeck Van Henkelom et al., 1972; Jongsma and Van Rijn, 1972; Sheridan et al., 1978). As in other systems, early studies estimated the specific resistance on the basis of total area of cell contact (Borek et al., 1969; Siegenbeck Van Henkelom et al., 1972); thus, the values obtained were two or more orders of magnitude too high. In cultured heart cells from newborn rats, Jongsma and Van Rijn (1972) have indeed estimated the specific resistance on the basis of gap junctional area only. However, they have used values for the junctional area estimated by others on adult heart cells. Although, in some systems, the area does not change significantly after birth, in the heart, it may increase significantly during the first 2–3 days of life (Nakata and Page, 1978).

Reliable data have been obtained by Sheridan et al. (1978) on cultured Novikoff hepatoma cells. These cells are particularly suitable for studying the electrical properties of gap junctions because they assume shapes of simple geometry when grown in suspension, allowing one to correlate junctional and nonjunctional resistances with other parameters of the cell such as volume and surface. In this system, the specific junctional resistance has been estimated to be 0.3–1.2 × 10^{-2} Ω/cm^2; this is in agreement with a reliable estimate reported in another system (Ito et al., 1974). A specific resistance of the order of 10^{-2} Ω/cm^2 is certainly quite low for a membrance, as the specific resistance of nonjunctional membranes is of the order of 1000 Ω/cm^2 (Trautwein et al., 1956) to 2600 Ω/cm^2 (Van der Kloot and Dane, 1964).

Based on this value of specific junctional resistance and assuming that the channels are hexagonally packed at 10.0 nm spacing, Loewenstein (1975) has estimated the resistance of a single channel to be of the order of 10^{10} Ω. This value fits well with the resistance expected for a channel 1.0–2.0 nm in diameter and 20.0 nm in length—dimensions that are close to those suggested by ultra-structural, crystallographic, and tracer-transfer studies.

The resistance of a single channel is given (Loewenstein, 1975) by

$$\frac{l\tau}{\pi a^2} + \pi df(2a/d)$$

where the first term is the resistance of the isolated channel and the second term is the resistance due to electrostatic interaction between channels; a is the channel radius, l is the channel length, τ is the resistivity of the channel core, d is the channel spacing, and $f(2a/d)$ is the ratio of the resistance of the channel system to that of a single isolated channel.

The resistance of the gap junction channels is significantly lower than that of the channels of excitability, which is, for the sodium channels, 2.4×10^{11} Ω in squid (Conti *et al.*, 1975) and 1.5×10^{11} Ω in myelinated nerves (Conti *et al.*, 1976), and, for the potassium channel, 8×10^{10} Ω in squid (Conti *et al.*, 1975) and 2.5×10^{11} Ω in myelinated nerves (Begenisich and Stevens, 1975). However, it is similar to that of the acetylcholine-controlled postsynaptic channels (Katz and Miledi, 1972). Gap junction channels are $\sim$20 nm long, whereas postsynaptic channels are $\sim$7.5 nm long. Therefore, to account for similar values of resistance, the diameter of the former must be 50–75% larger than that of the latter (Sheridan *et al.*, 1978); this is consistent with the gap junction channel's permeability to molecules larger than small inorganic ions.

Selectivity and permeability properties of gap junction channels are certainly affected by both fixed charges and architectural characteristics of the channel walls. The question of how the mobility of a charged molecule is affected by the characteristics of the channels has recently been asked by Shneior Lifson (see Bennett and Goodenough, 1978). Assuming that the channels contain one negative charge per 100 Å^2 of inner surface and that the concentration of cytoplasmic cations is 0.1–0.2 M, Lifson has estimated the concentration of these ions in the channels to be of the order of 1–2 M and that of anions to be several orders of magnitude lower. This would imply that when current passes between excitable cells, a larger flow of cations and little flow of anions would take place, creating a depletion of cations on one side and of anions on the other side, followed by counter flow when the current stops.

C. Probes of Cell-to-Cell Coupling

The puzzling discovery of electrical coupling between nonexcitable cells (Loewenstein and Kanno, 1964) has prompted pertinent questions on the very meaning of cell coupling in addition to rapid transmission of electrical impulses. Although the capacity for exchanging small ions could be a sufficiently important aspect of cell communication to account for the presence of extensive electrical coupling in nonexcitable cells (since it provides a mean for equilibrating the ionic population of cell communities and for sharing the work for ionic transport among neighboring cells), the exciting possibility that other protoplasmic molecules (e.g., metabolic intermediates, second messengers, high energy compounds, and factors regulating cell growth and differentiation) could be transferred from cell to cell has stimulated a large number of studies aimed at probing the size limit of channel permeability by means of charged and uncharged tracers larger than small ions.

Fluorescein has been the first molecule to be tested (Loewenstein and Kanno, 1964). It is a small anion (MW 332) with excellent qualities as a tracer: it diffuses rapidly in the cytoplasm and can be detected in concentrations as low as 10^{-8} M by its fluorescence in ultraviolet light. Found at first to diffuse across the junctions of salivary gland cells of *Drosophila* (Loewenstein and Kanno, 1964), fluorescein has been later reported to cross gap junctions of a large variety of vertebrate and invertebrate cells (Pappas and Bennett, 1966; Furshpan and Potter, 1968; Rose, 1971; Johnson and Sheridan, 1971; Azarnia *et al.*, 1974; Pollack, 1976 etc.).

Fluorescein transfer also has been tested in various tumor cells. Furshpan and Potter (1968) have reported the rapid transfer of fluorescein between BHK cells transformed by polyoma virus and between Ehrlich's sarcoma 180 cells. A similar finding has been reported between reaggregated Novikoff hepatoma cells cultured *in vitro* (Johnson and Sheridan, 1971). In this latter system, however, not all coupled cells have been found to pass the dye. Cells that passed fluorescein were generally better coupled electrically than those that did not, although some well-coupled cells did not pass the tracer at early times after reaggregation. These results have been interpreted to suggest that the capacity to transfer fluorescein is indeed correlated, in most cases, with the degree of electrical coupling and that well-coupled cells that do not pass fluorescein could have been coupled by immature gap junctions (formation plaques) possibly permeable only to small ions. Alternatively, certain junctions could have been only partially permeable due to changes in the composition of the intracellular medium; in fact, the possibility that gap junctions undergo graded changes in their permeability in relation to small changes of intracellular ionized calcium has been suggested (Délèze and Loewenstein, 1976) (see Section V,A).

Fluorescein transfer has also been reported between normal cells and hybrids obtained by fusion of normal human cells and mouse malignant cells (Azarnia *et al.*, 1974). Interestingly also in this case, certain hybrids were electrically well-coupled but did not exchange fluorescein with their neighbors, indicating a condition of partial uncoupling, which might be related somehow to the mechanism of malignant transformation (Azarnia and Loewenstein, 1977).

Impermeability to fluorescein has been reported in rectifying junctions of crayfish (Keeter *et al.*, 1974). These junctions, which electrically couple median and lateral giant fibers to motor fibers, are the only known example of rectifying gap junctions (Furshpan and Potter, 1959). One of the authors (Bennett, 1978) feels that their impermeability to fluorescein is not surprising, as the tracer diffusion has been tested only in resting conditions. Since these junctions possess a voltage-sensitive gating mechanism, they have a relatively higher resistance at rest. Thus, their permeability to fluorescein should be retested during tetanic stimulation (Bennett, 1978). Other gap junctions with voltage-dependent conductance (e.g., those of *Ambystoma*) indeed pass another dye (lucifer yellow CH, MW 476) in their high conductance state (Spray *et al.*, 1978).

In the past, impermeability to fluorescein has also been reported in a variety of embryonal cells such as those of *Xenopus* (Slack and Palmer, 1969), *Asterias* (Tupper and Sanders, 1972), and *Fundulus* (Bennett *et al.*, 1972) at a blastula stage of development. However, in *Xenopus,* fluorescein passed at later stages (gastrula and neurula) (Sheridan, 1971). These data have been taken to indicate a limited degree of permeability in embryonic junctions and have stimulated hypotheses envisioning the possible role of permeability modulation in the mechanism of cell determination during development.

Recent studies, however, have warned about possible mistakes in the interpretation of the early data, as fluorescein and lucifer yellow CH have been successfully transferred between *Fundulus* blastomeres (Bennett *et al.*, 1977; Bennett, 1977). Early failures have been attributed to the rapid loss of fluorescein from the injected cell, as well as to the small junctional area available between embryonic cells. Thus, it seems that in order to demonstrate the intercellular movement of this dye in early embryos, well-coupled cells and long waiting times are necessary. In this respect, the successful transfer of lucifer yellow CH between *Fundulus* blastomeres is particularly significant because this dye (a 4-aminonaphthalimide substituted with two sulfonate groups), in contrast to fluorescein, does not pass easily through nonjunctional membranes (Stewart, 1978) and thus allows reliable measurements of its passage between coupled cells.

The high permeability of fluorescein across nonjunctional membranes makes this dye not ideal for intracellular tracer studies. Hence, in the late sixties, a variety of other tracers were tested. Procion yellow M4RS (MW 550) has been found to be more suitable than fluorescein in many respects. First used as a tracer of neuronal architecture (Stretton and Kravitz, 1968), Procion yellow was later successfully employed as an intercellular probe because it does not permeate significantly nonjunctional membranes, because it binds slowly to protoplasmic molecules and thus is preserved in place throughout the preparative procedure for histology, and because it can be revealed in sections by fluorescence microscopy. Junctions of crayfish septa (Payton *et al.*, 1969b), *Chironomus* gland cells (Rose, 1971), horizontal cells of dogfish retina (Kaneko, 1971), sheep and calf Purkinje fibers (Imanaga, 1974), and acinar cells of rat salivary glands (Hammer and Sheridan, 1978) are readily permeable to this dye.

In addition to fluorescein, procion yellow, and lucifer yellow CH, a number of other dyes and radioactively labeled molecules have been found to diffuse between coupled cells. Among the dyes tested are azure B (MW 305), orange G (MW 452), salantine turquoise (MW 700), trypan blue (MW 960), Evans blue (MW 961) (Kanno and Loewenstein, 1966), neutral red (MW 289) (Bennett *et al.*, 1967), Chicago blue 6B (MW 993) (Potter *et al.*, 1966; Sheridan, 1970), microperoxidase (MW 1800), and horseradish peroxidase (MW 40,000) (Bennett *et al.*, 1973). Among the radioactive tracers are [^{14}C]sucrose (MW 342) (Bennett and Dunham, 1970) and [1-^{14}C]tetraethylammonium (TEA, MW 130) (Weingart, 1974). Most of these molecules have been found to diffuse readily from cell to cell, the only exceptions being Chicago blue 6B, which passed between cells in culture (Potter *et al.*, 1966; Sheridan, 1970) but did not pass between heart cells (Imanaga, 1974), and both horseradish and microperoxidase, which failed to cross gap junctions unless the cells were fixed (Bennett, 1974).

In one case, also serum albumin (MW 69,000) labeled with fluorescein was believed to diffuse from cell to cell (Kanno and Loewenstein, 1966), although the possibility that small degradation fragments rather than the intact molecules diffused across cell boundaries could not be discarded. Indeed, lack of intercellular exchange of intact albumin has recently been reported in other systems (Bennett, 1978).

Recently, a systematic approach to the study of the size limit of permeant molecules has been carried out (Simpson *et al.*, 1977). To this purpose, a number of synthetic and natural peptides have been conjugated with fluorescent dyes to make negatively charged tracers ranging in MW from 251 to 4158. Fluorescein isothiocyanate (FITC), dansyl chloride (DANS), or lissamine rhodamine B (LRB) have been used as fluorescent dyes. Criteria for selecting the conjugates were: water solubility, nontoxicity, low cytoplasmic binding, and high fluorescent yield. The tracers were pressure injected into salivary gland cells of *Chironomus* and the spread of fluorescence was observed and photographed by dark field microscopy. Tracers up to 380 daltons crossed the first junctions within a few seconds while those between 593 and 1158 diffused more slowly (sometimes taking up to one minute). Larger tracers did not pass, the largest molecule to diffuse being LRB(Leu)$_3$(Glu)$_2$OH (MW 1158). From an estimate of the diameter of this tracer, the channel diameter was believed to be 10–14 Å.

In conclusion, from electrophysiological and tracer studies, one can assume that gap junctions are permeable to neutral and negatively charged molecules of a molecular weight not greater than 1200. The permeability of the junctions to small cations has also been proven, but their upper size limit has not been estimated as precisely as that of neutral and negatively charged molecules since cations make poor tracers due to their binding to cytoplasmic constituents (Bennett and Goodenough, 1978).

D. Metabolic Coupling

The tracer studies clearly indicate that charged and noncharged molecules up to approximately 1200 daltons can diffuse from cell to cell. Consequently one would expect that a variety of molecules of that molecular weight or smaller, normally present in the protoplasmic medium, should indeed be exchanged between neighboring cells. In the last decade, a number of studies have provided strong evidence that a great variety of metabolic intermediates such as nucleotides, amino acids, oligosaccharides, and second messengers are indeed directly transferred from cell to cell.

The first evidence for intercellular exchange of metabolic intermediates was provided by elegant experiments on the metabolic cooperation between cells cultured *in vitro*. Subak-Sharpe *et al.* (1966, 1969), while working on tissue cultures of hamster fibroblasts (BHK cells), noticed that certain mutant cells, which lack the enzyme hypoxanthine–guanine phosphoribosyl transferase (HGPRT; EC 2.4.2.8) and consequently are unable to incorporate hypoxanthine into their nucleic acids, become labeled if grown in mixed cultures with wild-type cells in a medium containing [^{3}H]hypoxanthine. Autoradiographic analysis showed that the mutant cells are labeled only if in contact directly, or through other mutant cells, with the wild type, and that incorporation of hypoxanthine stops if the cells are separated (Cox *et al.*, 1970; Pitts, 1971). This indicates that metabolic cooperation does not occur through the extracellular medium but requires direct cell-to-cell interaction. The intercellular exchange was so extensive that the two cell types rapidly became equally labeled (Bürk *et al.*, 1968). Similar experiments with cell lines unable to incorporate adenine produced the same results (Bürk *et al.*, 1968; Cox *et al.*, 1972).

These early experiments could not define unequivocally the type of labeled molecule exchanged between neighboring cells; possible candidates were the nucleotides, RNA, and the enzyme (HGPRT) itself (Subak-Sharpe *et al.*, 1966, 1969). The possible exchange of HGPRT was soon discarded by lack of evidence for an increase in HGPRT activity in mutant cells in contact with wild-type cells (Cox *et al.*, 1970; Pitts, 1971). However, the possibility that RNA (rather than nucleotides) was exchanged remained (Kolodny, 1971), although it seemed unlikely because metabolic cooperation was also found to occur between mutant BHK lacking thymidine kinase (TK; EC 2.7.1.21) and wild-type cells (Cox *et al.*, 1970).

More convincing evidence for the intercellular exchange of nucleotides, and other small metabolites, rather than macromolecules has been provided by recent studies using a new experimental approach (Simms, 1973; Pitts, 1976, 1977; Pitts and Simms, 1977; Pitts and Finhow, 1977). An obvious drawback in using mutant cells to test the metabolic cooperation between cells in culture is the limitation in the number of metabolites that can be tested, as only a limited variety of mutant lines are available. To overcome this limitation, Simms (1973) devised a new experimental procedure for prelabeling cells in culture by exposing them to a tritiated precursor. The precursor is taken up by the cells and then converted to an intermediate metabolite, which is retained in the cell. The cells are then washed to remove the precursor and cocultured with unlabeled recipient cells (Fig. 1a). The labeled cells contain the labeled intermediate and any labeled macromolecule in which the intermediate has been incorporated. However, if the

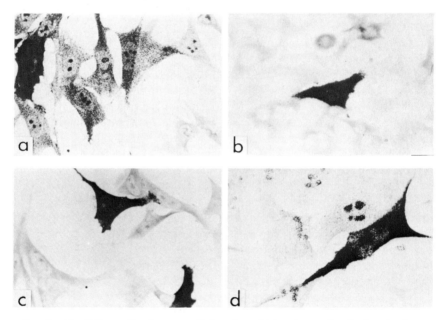

FIG. 1.   Metabolic coupling between cells in culture. Light micrographs of cultured cells processed for autoradiography. (a) Transfer of [³H]uridine nucleotides between BHK cells. (b) No transfer of [³H]uridine nucleotides, incorporated into RNA, between BHK cells. (c) No transfer of [³H]uridine nucleotides from L to BHK cells. (d) Transfer of [³H]uridine nucleotides between *Xenopus* and BHK cells. In (a), (c), and (d), the donors (black cells) have been labeled with [³H]uridine for 3 hours and washed with unlabeled medium. Unlabeled cells have been cocultured with the labeled cells for 3 hours before fixation and processing for autoradiography. In (b), the donors (black cell) have been labeled with [³H]uridine for 3 hours and then chased in unlabeled medium for 24 hours before coculturing with unlabeled cells; the donors contain labeled RNA but no labeled nucleotides. In (a) and (d), nucleotides were transferred via gap junctions to recipient cells in contact (light cells) but not to cells not in contact. (Courtesy of J. D. Pitts; Pitts, 1977.)

cells are cultured for 24 hours in unlabeled medium, before being cocultured with recipient cells, the labeled intermediate pool is chased into macromolecules.

When chased cells were cocultured with recipient cells, there was no transfer of labeled material (Fig. 1b), clearly indicating that RNA and protein did not cross the cell boundaries (Pitts, 1976; Pitts and Simms, 1977). These results were confirmed by experiments in which actinomycin D was employed to inhibit RNA synthesis of recipient cells mixed with labeled, nonchased donors (Pitts and Finbow, 1977). Due to the inhibition of RNA synthesis, the nucleotides transferred were not converted to RNA and thus very little activity was detected in recipient cells after acid fixation and autoradiography. Interestingly, nucleotides were not transferred between normal cells incapable of forming junctions (Fig. 1c) but passed readily between cells of different vertebrate classes (Fig. 1d) (Pitts and Simms, 1977; Pitts, 1977).

Using the prelabeling method, a large variety of metabolic intermediates were tested. Hypoxanthine, uridine (Fig. 1), thymidine, glucose, fucose, glucosamine, 2-deoxyglucose, choline, and various amino acids were found to pass readily from cell to cell (Pitts and Simms, 1977; Pitts and Finbow, 1977) at the high rate (for nucleotides, at least) of about 10^6 molecules/sec/cell pair. Evidence for transfer of proline and the vitamin-derived cofactor tetrahydrofolate was also obtained (Pitts, 1977). The active form of tetrahydrofolate, the tetraglutamate derivative, did not seem to be transferred, whereas the forms with fewer glutamate residues did, suggesting a cutoff size for the permeant molecules of approximately 960 daltons, which is quite close to the cut off size (~ 1200 daltons) determined with artificial probes in insect salivary gland cells (Simpson et al., 1977).

Among protoplasmic molecules of a size smaller than about 1000 daltons, there are compounds of great importance for the regulation of cell function. One of them is cAMP (cyclic adenosine monophosphate), a second messenger that translates signals produced at the cell surface by the interaction between hormones and the cell membrane and that modulates a large variety of cell activities. Free exchange of cAMP would be obviously advantageous for a cell community, as it would allow synchronization and potentiation of hormonal signals. The first convincing evidence for cAMP exchange has been produced by Tsien and Weingart (1974) in mammalian heart muscle fibers. Ventricle fibers were pulled through a hole separating two compartments (Weingart, 1974). In one compartment, containing a Ca^{2+}-free solution to prevent cell uncoupling (see Section V, A), the muscle fibers were cut. [^{3}H]cAMP was added to this compartment, and after sufficient time for uptake by the damaged fibers, it was washed out with a Ca^{2+}-containing solution, which allowed the cut fibers to seal. [^{3}H]cAMP diffusion into the muscle fibers of the unperfused compartment was determined by measuring the radioactivity of the fibers which were cut into 0.5 nm segments after freezing.

These data have recently been confirmed by the elegant work of Lawrence *et al.* (1978) on the transmission of hormonal stimulation between coupled cells of different origin. Mouse myocardial cells, electrically coupled with rat ovarian granulosa cells, have been found to respond to cell-specific hormones by a cAMP-dependent mechanism. Exposure of the cocultures to a hormone specific for one cell type caused the heterologous cells to respond only when cell-to-cell contact was present. Exposure of the cocultures to a hormone (FSH), which acts on specific receptors of the granulosa cells and results in cAMP synthesis, was followed by an increase in spontaneous beat frequency in the adjacent heart muscle cells. The change in frequency was believed to follow an increase in cAMP concentration because a similar effect is produced by treatment of heart muscle cells with noradrenalin, which acts via cAMP. The logical interpretation of the phenomenon is that cAMP, synthesized in the granulosa cells, has diffused to the muscle cells via cell coupling.

Little is known about metabolic cooperation between cancer cells. In Novikoff hepatoma, two amino acids labeled with a dye, dansyl-L-glutamate (MW 380) and dansyl-DL-aspartate (MW 366) have been found to diffuse to neighboring cells after iontophoretic injection into one cell (Johnson and Sheridan, 1971).

Few studies have been carried out in invertebrates. In leech ganglia, amino acids and RNA precursors have been found to diffuse between neuroglial cells and neurons (Globus *et al.*, 1973) and a variety of tritiated metabolites such as fucose, glucosamine, glycine, leucine, orotic acid, and uridine have been reported to cross the junctions between Retzius cells (Rieske *et al.*, 1975). In the latter study, the authors have been able to demonstrate that the substrates, but not their macromolecular products, are transferred, because puromycin injected into the transjunctional cell blocked labeling in that cell but not in the donor cell. In crayfish, Herman *et al.* (1975) have produced evidence for the exchange of two radioactive precursors, [^{3}H]glycine and [^{3}H]glucosamine, between lateral giant axons. Interestingly, only the lateral giant axon beyond the septum was labeled, whereas the motor fiber, coupled to the lateral giant fiber by a voltage-dependent, rectifying low resistance junction was not labeled. This observation confirmed similar results obtained with fluorescein (Keeter *et al.*, 1974) and demonstrated once again the peculiarity of these junctions.

An interesting aspect of metabolic cooperation between cells is the apparent capacity of certain cell lines resistant to ouabain, a blocker of active Na^+ transport, to transfer their ouabain resistance to normal ouabain-sensitive cells (Corsaro and Migeon, 1977). The ouabain resistance is linked to a genetic modification in the surface ATPase that results in a decreased affinity of this enzyme to ouabain. If ouabain-resistant and -sensitive cells are grown in mixed cultures, upon exposure to ouabain, sensitive cells in contact with resistant cells are not affected. Although it is possible that the acquisition of ouabain resistance by the normal cells is due to transfer of some unknown compounds from resistant to sensitive cells, the most likely explanation of the phenomenon is that Na^+ is transferred from normal to ouabain-resistant cells, which succeed in maintaining a low $[Na^+]_i$ in the ouabain-sensitive neighbors.

III. Gap Junctions Are Coupling Junctions

Nowadays an electronmicroscopist looking at the plasma membranes of neighboring cells would be convinced that the cells communicate directly with each other if he could see gap junctions. Similarly an electrophysiologist who has recorded a low coupling resistance between adjacent cells would feel confident that the cells possess gap junctions.

How has this structure–function correlation been established? Certainly neither by the unequivocal demonstration of anatomically definable channels spanning across neighboring plasma membranes nor by the observations of electron-opaque tracers crossing the cell boundaries at the gap junctions. Indeed only indirect evidence has been produced thus far, but the data have been so convincing that the correlation between cell-to-cell coupling and gap junctions is now generally accepted. Gap junctions, however, may not be the only structures that allow direct cell-to-cell communication, since in some special cases other forms of cell coupling have been suspected.

As early as 1953, Robertson noticed that the plasma membranes of electrically coupled crayfish median and motor axons were closely apposed, in some regions even "forming a single membrane of double thickness."

Soon after, regions of close membrane apposition were described between electrically coupled giant axons of earthworm (Hama, 1959) and lateral giant axons of crayfish (Hama, 1961).

The presence of a close membrane apposition seemed an obvious prerequisite of electrical coupling; however, tight membrane apposition was also observed in myelin (Robertson, 1957), which is a structure known to provide electrical insulation due to its high electrical resistivity. Thus, it is not surprising that in an early study of the junctions between cells of vertebrate heart, regions of close membrane apposition that were apparently similar to myelin lamellae were mistakenly interpreted as areas of high electrical resistivity (Sjöstrand et al., 1958). In most instances however, the consistent finding of regions of tight membrane apposition between cells that were known to be electrically coupled suggested that they could be the site of intercellular communication (Karrer, 1960a,b; Dewey and Barr, 1962; Robertson, 1963; Robertson et al., 1963; Bennett et al., 1963). These regions were named in various ways (see Section IV,B) but now they would all be referred to as gap junctions.

In salivary gland cells of insect larvae, a different type of junction, the septate junction, was proposed as the low resistance coupling structure (Wiener et al., 1964). This seemed a reasonable hypothesis because septate junctions were the only junctions seen between those cells, and their large extension (they occupy approximately a third of the surface of cell-to-cell apposition) seemed to fit well with the very low coupling resistance recorded between these cells (Loewenstein and Kanno, 1964; Rose, 1971). The possible involvement of septate junctions in cell-to-cell coupling was further suggested by more recent studies (Bullivant and Loewenstein, 1968; Gilula et al., 1970). However, due to the fact that these junctions have rarely been identified in vertebrates and that typical gap junctions have often been recognized in the same invertebrate cells where septate junctions are present (Hagopian, 1970; Rose, 1971; Flower, 1971; Hudspeth and Revel, 1971; Gilula and Satir, 1971; Hand and Gobel, 1972), an involvement of septate junctions in intercellular communication seems now unlikely, although it cannot be entirely discarded yet.

More convincing evidence for the role of gap junctions in cell coupling came from studies of Barr et al. (1965) on the reversible effects of hypertonic solutions on the electrical coupling and on the integrity of gap junctions in vertebrate heart. In hypertonic solutions, the membranes of gap junctions become separated, being mechanically pulled apart as a result of cell shrinkage. Simultaneously, the intercellular resistance increases to a complete cell uncoupling, suggesting that the integrity of gap junctions is necessary for the preservation of a normal cell-to-cell communication via small ions.

This interpretation was soon corroborated by a study on the effects of hypertonic solutions low in calcium or chlorides (Dreifuss et al., 1966; Kawamura and Konishi, 1967). In these solutions, the membranes of adjacent heart muscle cells became separated at desmosomal junctions, whereas gap junctions remained intact and cell coupling was maintained. Since desmosomal junctions (fasciae adherentes and spot desmosomes) and gap junctions are the only junctions linking these cells, the persistance of electrical coupling after disruption of the former leaves the latter as the only candidate for cell coupling.

Loss of cell coupling in parallel with disruption of gap junctions has also been demonstrated in smooth muscle (Barr et al., 1968) and in crayfish axons (Asada and Bennett, 1971; Pappas et al., 1971). In the latter, the junctions of the septum between lateral giant axons were pulled apart as a result of various treatments, e.g., exposure to solutions in which chlorides were substituted with larger, less permeable, anions (acetate, propionate, isothionate), rupture of one axon, and exposure to EDTA followed by return to normal saline solutions. More recently, finer modifications in the structure of gap junctions have been detected in parallel with changes in electrical coupling resistance (Peracchia and Dulhunty, 1974, 1976). They do not involve disruption of the junctional membrane and indicate that fine conformational rearrangements in the channel structure resulting in channel obliteration are likely to take place. These findings strongly support the correlation gap junction–cell coupling because these are the only type of junction seen at the septum between lateral giant axons (Hama, 1961; Payton et al., 1969b; Pappas et al., 1971; Peracchia, 1973a,b; Zampighi, 1978).

An elegant demonstration of the involvement of gap junctions in cell coupling came from the observation of induced synchrony in spontaneously beating mouse (Goshima, 1969) and chick (DeHaan and Hirakow, 1972) myocardial cells in culture. Individual cells beat at different frequency but become synchronized as soon as they come in contact with each other either directly or via heterologous cells. Synchronous beating is provided by the spread of current via newly formed gap junctions (Goshima, 1969, 1970; Hyde et al., 1969; DeHaan and Hirakow, 1972; Lawrence et al., 1978), because heart muscle cells connected via mouse L cells, which are incapable of forming gap junctions (Goshima, 1970), do not acquire synchrony (Goshima, 1969). Mouse L cells, a mutant line unable to incorporate hypoxanthine in their nucleic acids (Sanford et al., 1948), are also incapable of establishing metabolic cooperation with wild-type cells (Fig. 1c) (Pitts, 1971).

Taking advantage of cells of this type, Gilula et al. (1972) produced convincing evidence for the involvement of gap junctions in both metabolic and electrical coupling. Three types of cells were used: DON (a fibroblastic Chinese hamster line), DA (a clonal subline of DON), and A9 (a genetic variant of mouse L-cell line). DON cells contain HGPRT (see Section II,D), whereas DA and A9 do not and thus are incapable of incorporating hypoxanthine in their nucleic acids; only DON and DA are able to make gap junctions. In mixed cultures treated with [³H]hypoxanthine and studied electrophysiologically and morphologically, DA cells, but not A9 cells, became labeled (incorporated hypoxanthine) and established electrical coupling with neighboring DON cells. In the electron microscope, gap junctions were seen at DON-DA but not at DON-A9 or A9-A9 contacts.

Azarnia et al. (1974) further supported these results by fusing a malignant subline of mouse L cells (cl-1D) with normal human cells capable of electrical coupling and of making gap junctions. The resulting hybrid was able to establish electrical coupling and to form gap junctions with neighboring cells, but as soon as it lost the human chromosomes, clones appeared that were not coupled and did not make gap junctions. These data once more demonstrated the necessity of gap junctions for cell coupling, but also indicated that the incapacity of a cell to establish electrical coupling may result from a genetic defect.

A close temporal correlation between the formation of gap junctions, cell coupling and onset of cell fusion was reported in studies on myogenesis in vitro (Rash and Fambrough, 1973; Rash and Staehelin, 1974). Transient electrical coupling occurs shortly before fusion and, at the same time, gap junctions are seen; on the contrary cells in contact, but not coupled, do not show gap junctions. Similar conclusions were reached by Johnson et al. (1974) in a study on the parallel time sequence between the establishment of electrical coupling and gap junction formation between reaggregated Novikoff hepatoma cells.

Although the correlation between gap junctions and cell coupling is quite convincing, evidence that these are the only junctions that mediate direct cell-to-cell communication is still lacking. In invertebrates, different types of junctions resembling gap junctions have been described; in one case, two distinctly different types were present even in the same cells (Quick and Johnson, 1977). In vertebrates, gap junctions are more uniform; however in one study, two types were seen in close association with each other (Staehelin, 1972). Obviously, in these cases, one cannot decide which is the coupling junction. Could both types be coupling junctions? If so, why would a cell need two distinctly different junctions for communication? Even more puzzling are the cases in which electrically well coupled cells seem to lack gap junctions and any other known junction (Daniel et al., 1976). Thus far, absence of gap junctions in coupled cells has been reported only in certain smooth muscles such as intestinal longitudinal muscle and circular muscle (dense layer), myometrium, and fallopian tube. Interestingly, in pregnant uterus, gap junctions develop in great number at the beginning of labor, when synchronous contractions and the need for extensive electrical coupling arise (Garfield et al., 1977); although, even before the appearance of gap junctions, these cells show good electrical coupling (Kuriyama and Suzuki, 1976). However, one may wonder about the adequacy of our morphological methods for detecting gap junctions. In fact, although even small gap junctions can be detected by freeze–fracture, gap junctions composed of only two or three channels do exist and can hardly be recognized in the electron microscope (see Section IV,B). Moreover, in cases where gap junctions have not been detected by some investigators, some gap junctions, although very few, have been seen by others (Gabella, 1972a,b).

IV. Gap Junction Structure

Gap junctions are specialized regions of the cell surface in which two symmetrical patches of plasma membrane come in close contact with each other and establish well insulated cytoplasmic tunnels across the extracellular space. In the last few years, a number of studies have shown that the structural principles on which gap junction membranes are built do not differ significantly from those of other membrane regions; thus it seems reasonable to briefly review some general concepts of membrane structure before dealing with the structure of gap junctions.

A. MEMBRANE ARCHITECTURE

To account for a large variety of data accumulated throughout half a century of work on the membrane structure, Singer and Nicolson (1972) designed a membrane model which has been widely accepted, at least in its most important features. In this model, known as the "lipid globular protein mosaic model," the lipids are arranged in a fluid bilayer frequently interrupted by globular proteins (integral proteins) that penetrate and often span its thickness, interact hydrophobically and electrostatically with the lipids, and are able to migrate laterally in the plane of the membrane. Other proteins (peripheral proteins) are at the surface of the membrane and interact only electrostatically with the lipid polar groups.

The bilayer arrangement of the lipid (with the hydrocarbon chains pointing toward the center of the membrane and the charged groups at the surface) was clearly suggested in the past by a variety of experimental data obtained from measurements of the membrane surface area occupied by lipid (Gorter and Grendel, 1925), membrane surface tension determinations (Danielli and Davson, 1935), and studies using X-ray diffraction (Schmitt et al., 1935), polarizing microscopy (Schmidt, 1936), and electron microscopy (Robertson, 1960). The bilayer structure has been definitely proven in recent years by nuclear magnetic (Horwitz, 1972) and electron spin resonance (Hubbel and McConnell, 1971), differential scanning calorimetry (Chapman, 1973), and high resolution X-ray diffraction (Wilkins et al., 1971; Caspar and Kirschner, 1971) studies.

In a living cell, the lipids are in a semiliquid state (meso-phase) which can be changed to a crystalline state (gel-phase) by lowering the temperature below a critical value (phase transition temperature) determined by length and degree of saturation of the hydrocarbon chains (Chapman, 1973; McConnell, 1974). Mixtures of lipids in a membrane broaden the phase transition temperature by causing a "phase separation" into aggregates of crystalline and fluid lipid regions, and cholesterol, when present in high concentration, abolishes the temperature effect by restricing the mobility of those hydrocarbon chain regions close to the polar groups, yet maintaining the rest of the chains flexible enough to prevent their crystallization.

Evidence for asymmetry between inner and outer leaflets of the bilayer is also accumulating and indicates that glycolipids, phosphatidyl choline, and sphingomyelin are restricted to the outer leaflet (Steck and Dawson, 1974), whereas phosphatidyl serine and phosphatidyl ethanolamine are mostly at the inner half (Bretscher, 1972; Verkleij et al., 1973; Marinetti et al., 1974). Thus, at physiological pH, the cytoplasmic surface of the bilayer is likely to be negatively charged (Bretscher and Raff, 1975).

Somewhat slower and more controversial has been the development of our present understanding of the physicochemical characteristics of membrane proteins. Early membrane models restricting the proteins to the surfaces of the bilayer and suggesting their mainly electrostatic interaction with the lipid polar groups (Robertson, 1966, for a review) have been criticized in view of evidence for the existence of hydrophobic lipid–protein interactions (Singer, 1971) and the presence of α-helical structure in a number of membrane proteins (Wallack and Winzler, 1974); therefore it is now believed that most membrane proteins have a globular structure and penetrate the hydrophobic core of the bilayer.

A well-known example of globular membrane protein is the electrophoretic band 3 of red blood cells (Steck, 1974), a molecule (MW 100,000) believed to be involved in the transport of anions and glucose. However, not all proteins with α-helical structure are globular; glycophorin (Marchesi et al., 1972), for example, has a single α-helical chain of 23 amino acids within the membrane thickness (Segrest et al., 1972). Both band 3 and glycophorin have been shown to span the membrane thickness and protrude from both surfaces (Bretscher, 1971; Steck et al., 1971, 1976).

A variety of studies have clearly indicated that some membrane proteins are able to move laterally in the plane of the membrane (Frye and Edidin, 1970); however, movement is restricted (DePetris and Raff, 1973) and probably is regulated by the interaction between membrane protein and microfilaments, microtubules, and other cytoplasmic proteins (Revel et al., 1976). In addition there is evidence that certain proteins can rotate around an axis perpendicular to the plane of the membrane (Brown, 1972; Cone, 1972; Poo and Cone, 1974; Cherry et al., 1976).

B. GAP JUNCTION ARCHITECTURE

In 1957, Robertson produced the first high resolution electron micrographs of the profile of biological membranes; he described it as a trilaminar structure that was ~ 7.5 nm thick and was composed of an electron-transparent layer ~ 3.5 nm thick sandwiched between two electron-opaque layers ~ 2.0 nm thick. This image, which was believed to correspond to a double layer of lipids coated with proteins, was found in all membranes examined and became the cardinal element of a concept of membrane structure called "the unit membrane theory" (Robertson, 1960).

In various tissues, membranes with a typical trilaminar profile were seen surrounding neighboring cells separated by an extracellular space 10–15 nm wide. Soon, however, several investigators noticed small regions of the cell surface in which adjacent plasma membranes were unusually close to each other and provided the earliest descriptions of gap junctions (Sjöstrand et al., 1958; Karrer, 1960a,b; Dewey and Barr, 1962). In thin sections of vertebrate cells fixed with osmium tetroxide or potassium permanganate, gap junctions displayed a pentalaminar profile due to the tight apposition of two plasma membranes with obliteration of the extracellular space and apparent fusion of the outer electron-opaque layers. Regions with pentalaminar profile were reported in a variety of tissues such as heart (Sjöstrand et al., 1958; Karrer, 1960b; Dewey and Barr, 1962; Spiro and Sonnenblick, 1964), a number of epithelia (Karrer, 1960a;

Farquhar and Palade, 1963), and fish neurons (Robertson *et al.*, 1963; Robertson, 1963; Bennett *et al.*, 1963; Pappas and Bennett, 1966) and received various names such as longitudinal connecting surfaces (Sjöstrand *et al.*, 1958), quintuple layered cell interconnections (Karrer, 1960), nexuses (Dewey and Barr, 1962), tight junctions (Farquhar and Palade, 1963), and synaptic discs (Robertson, 1963).

In those years, gap junctions were frequently confused with another type of junction, the zonula occludens (tight junction) (Staehelin, 1974), since both displayed similar images of tight membrane apposition after osmium or permanganate fixation. Thus the term "tight junction" created a certain amount of confusion, as it was applied indiscriminately to either one.

A clear distinction between the two was obtained (Revel and Karnovsky, 1967) as a result of the introduction of "in block" staining with uranyl salts (Farquhar and Palade, 1965). The high electron opacity developed by membranes stained with uranyl allowed Revel and Karnovsky (1967) clearly to detect a 2.0–3.0 nm thick extracellular space (Fig. 2) between the membranes of gap junctions (Revel, 1968) [no space was seen at zonulae occludentes (tight junctions)] and to define the gap junction as a septilaminar structure 15.0–19.0 nm thick. The thin extracellular space at gap junctions was confirmed (Revel and Karnovsky, 1967) by its permeability to the electron-opaque element lanthanum (Doggenweiler and Frenk, 1965) used in colloidal form (lanthanum hydroxide) as an extracellular stain (Figs. 5B, 18, 32a). Indeed, the presence of an extracellular gap, which was previously undetected in vertebrates, has always been recognized as a typical feature of certain invertebrate gap junctions (Robertson, 1955, 1961; Hama, 1959, 1961; de Lorenzo, 1959; Coggeshall, 1965; Pappas and Bennett, 1966). This is due to the fact that the gap here is usually slightly larger (Figs. 10, 12) than in vertebrates and thus can be clearly seen in conventional preparations also.

At first the observation of an extracellular space at gap junctions seemed inconsistent with the assumption that the junctions provide intercellular paths insulated from the extracellular fluid; however, it is now clear that in spite of the

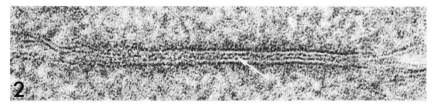

FIG. 2. Vertebrate gap junction (rat stomach). The cross-sectional profile of gap junctions displays two plasma membranes in close proximity with each other. An electron-transparent extracellular space (gap) of 2–3 nm (arrow) is interposed between the membranes giving the junction a septilaminar appearance (three electron-transparent and four electron-opaque layers). ×241,000.

apparent gap, the junctional membranes are indeed glued to each other in discrete spots where intramembrane particles, protruding from opposite membrane surfaces, come in contact with each other to bridge the gap (Fig. 10). Due to their low electron opacity, the protruding particles are only rarely seen in section (Figs. 10, 12) (Peracchia, 1973a), which explains the usual observation of an uninterrupted gap (Figs. 2, 5A).

A visualization of the particles protruding into the extracellular space was first obtained in mouse heart and liver gap junctions treated with lanthanum (Revel and Karnovsky, 1967). Colloidal lanthanum hydroxide, too large to penetrate into the cell, infiltrates the extracellular gap and surrounds and negatively stains the extracellular ends of the particles. The resulting face view image of the junctions is that of a polygonal array of 7.0–7.5 nm electron-transparent spots (the protruding particles) surrounded by an electron-opaque network (the gap filled with stain) (Figs. 5C, 14a,c, 18). A variety of vertebrate and invertebrate junctions displayed the same appearance after lanthanum, thus confirming the generality of gap junction architecture (Brightman and Reese, 1969; Payton et al., 1969b; McNutt and Weinstein, 1970; Rose, 1971; Hudspeth and Revel, 1971; Hand and Gobel, 1972; Friend and Gilula, 1972; Peracchia, 1973a; etc.).

Curiously, similar images had been obtained a few years earlier by Robertson (1963) in positively stained gap junctions of goldfish brain. These images may have resulted from the formation of some colloidal manganese dioxide during fixation and the consequential negative opacization of the junctions, like after lanthanum (McNutt and Weinstein, 1973). This seems to be a reasonable interpretation because similar images are produced in the same junctions after lanthanum impregnation (Brightman and Reese, 1969). Even though the images seen in goldfish could not be interpreted at first, Robertson's findings were remarkable indeed, since they provided for the first time evidence for a polygonal structure in the junctional membranes now believed to represent the framework of the intercellular channels. The structure was described by Robertson as a hexagonal array of facets ~9.0 nm in diameter, with a dense dot ~2.5 nm in the center. Similar dots (1.5 nm) (Fig. 14a,b) were also seen after lanthanum staining (Revel and Karnovsky, 1967) and were later interpreted as representing the intercellular channels partially filled with stain (Payton et al., 1969b; McNutt and Weinstein, 1970; Peracchia, 1973a). The reason why intracellular structures like the channels can be reached by extracellular stains is still unclear although the possibility that a portion of the channels become accessible to the extracellular medium under certain functional circumstances is supported by some data (see Section V,C).

Images similar to those of lanthanum-stained gap junctions have been seen in fragments of plasma membranes isolated from rat liver and negatively stained with sodium phosphotungstate (Na-PT) (Benedetti and Emmelot, 1965). These images (Figs. 3, 17 inset) at first believed (Benedetti and Emmelot, 1965) to

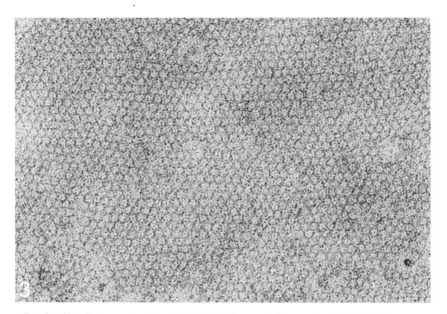

FIG. 3. Vertebrate gap junction (rat liver). A fragment of a gap junction isolated by sucrose gradient centrifugation and detergent treatment is shown here in face view, after negative staining with NaPT. The electron-opaque stain occupies the gap, surrounding the extracellular protrusions of the intramembrane particles and giving them the appearance of electron-transparent spots. Some of the spots contain a central electron-opaque dot, caused by the accumulation of some stain at the center of the two adjoined particles (the presumed location of the intercellular channels). The distribution of NaPT stain is similar to that of lanthanum (Fig. 5B and C). ×286,000.

represent a change in lipid phase from lamellar to hexagonal (Luzzati and Husson, 1962) brought about by exposure of plasma membranes to 37°C before staining, were later recognized to correspond to isolated gap junctions (Benedetti and Emmelot, 1968) because of their similarity to those of lanthanum-stained junctions (Revel and Karnovsky, 1967). Also, in isolated junctions, the polygonal array of electron-transparent spots was attributed to structures protruding from the external surface of the membrane because the cytoplasmic surface of the junctions appeared smooth in shadow casting experiments (Benedetti and Emmelot, 1968).

In addition to lanthanum and Na-PT, the extracellular gap can be infiltrated with horseradish peroxidase (MW 40,000) (Brightman and Reese, 1969; Goodenough and Revel, 1971) and K-pyroantimonate (Friend and Gilula, 1972). Curiously, in mouse liver, peroxidase seems to penetrate the gap of isolated junctions only, whereas catalase (MW 240,000) never does (Goodenough and Revel, 1971).

Data from thin sections and negative staining have indicated the presence of a polygonal array of particles in gap junctions but have not answered whether the particles are an integral part of the membrane structure or simply fixtures attached to the external membrane surface. This question and many others have been answered by freeze–fracture studies. In this technique (Steere, 1957; Moor *et al.*, 1961; reviewed by: Fisher and Branton, 1974; McNutt, 1977; Verkleij and Ververgaert, 1978), a small piece of tissue (usually fixed with aldehydes) is rapidly frozen after cryoprotective treatment and fractured in a vacuum. The fractured surface is replicated by depositing over it a platinum–carbon mixture at a 45° angle and carbon at a 90° angle. The tissue is then dissolved in sodium hypochlorite and the recovered replica is washed and examined in the electron microscope. In freeze–fracture, membranes oriented parallel or slightly obliquely to the fracturing plane are split down the middle, i.e., along a plane running between the two monolayers of lipids; thus, the internal aspects of either the protoplasmic (P face) or the exoplasmic (E face) membrane leaflets are exposed (Branton, 1966; Deamer and Branton, 1967).

In gap junctions, the fracture plane often steps from one to the other membrane of the same junction (Figs. 4, 5D, 5E, 7, 9, 11, 20, 22, 23, 25-28). In vertebrate gap junctions, the P face displays arrays of ~8 nm particles at an average center-to-center distance of ~10 nm (Figs. 4, 5D, 6, 22, 25) and the E face displays a

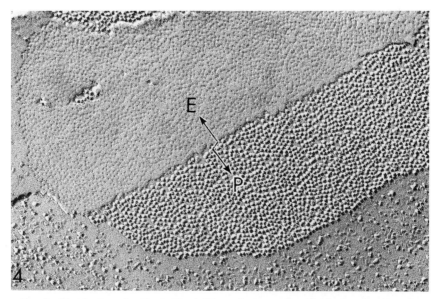

Fig. 4. Vertebrate gap junction (rat liver). Freeze-fracture replica. The fracture plane steps down (double-beaded arrow) from face E (E) to face P (P). Face P displays a disordered array of ~8.0 nm particles at an average center-to-center distance of ~10 nm; face E shows the complementary pits (see Fig. 5D and E). ×119,400. (From Peracchia, 1977a.)

similar array of pits (Figs. 4, 5D, 22, 25). Between particles or pits, the fractured surface is smooth.

These images arise from arrays of particles embedded in each membrane and protruding from it to make contact with similar particles of the opposite membrane; the bumps represent partially exposed particles, the pits represent the impression left by the particles cleaved away with the protoplasmic leaflet, and the smooth surface represents the central plane of a lipid bilayer (Fig. 5E). Early studies, however, interpreted these images in various other ways: in agreement with early hypotheses of membrane fracture behavior (Moor and Mühlethaler, 1963), some investigators proposed that particles and pits represent images of the true external surfaces of the junctional membranes (Kreutziger, 1968; Bullivant, 1969; McNutt and Weinstein, 1969); others believed these images to reflect a honeycomb-like membrane architecture (Sommer and Steere, 1969); in one case, the E face was not believed to contain pits but rather low relief protrusions (cobblestones) with a central depression (McNutt and Weinstein, 1970); in another study, the pits were believed to correspond to interparticle spaces (Goodenough and Revel, 1970); in another study, the images were attributed to a single layer of particles sandwiched between the two membranes (Spycher, 1970).

The controversy was finally settled by data obtained using the double replica technique (Chalcroft and Bullivant, 1970; Steere and Sommer, 1972). With this technique, both surfaces of a cleaved specimen are shadowed, allowing one to examine complementary images of the same cleaved structure. The complementary images of gap junctions obtained by Chalcroft and Bullivant (1970) clearly demonstrated that the membranes fracture internally and that particles arise from pits and vice versa. From these data, a model of gap junction architecture was designed showing particles that are embedded in each membrane and that become exposed on the P face after being cleaved away from the exoplasmic leaflet (Chalcroft and Bullivant, 1970). Although this feature of the model is still accepted, other details such as an asymmetrical fracture plane (closer to the cytoplasm) and a complete embedding of the particles within the membrane thickness do not agree with our present understanding of gap junction structure.

Before discussing finer details of gap junction structure and newer models of channel architecture designed according to recent data from ultrastructure, X-ray diffraction, and biochemistry (see Section IV,C), I shall deal briefly with general issues such as variability in gap junction morphology, relationship with cytoplasmic elements, and the nature of the cellular elements forming gap junctions. Typical gap junctions are usually envisioned as round membrane patches composed of particles that are homogeneous in size and shape and arranged in hexagonal or polygonal arrays. However, by scanning through the innumerable number of published micrographs of gap junctions, one cannot avoid being impressed by the degree of variability manifested by these structures and being

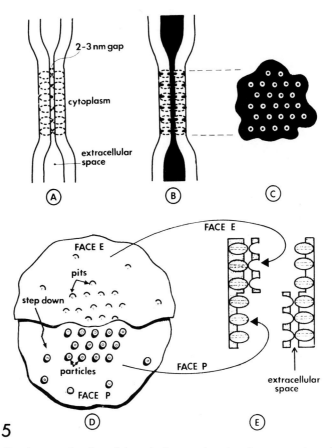

FIG. 5. Vertebrate gap junctions. Schematic diagram of gap junction structure in positively (A) or negatively (B, C) stained sections and in freeze–fracture (D, E). The junctions are shown in profile (A, B, E) and in face view (C, D). In (D) the fracture plane steps down from face E to face P following the path shown in (E). Each gap junction membrane contains a polygonal array of intramembrane particles matching precisely and binding extracellularly with a similar array in the adjoined membrane. The intramembrane particles are usually not seen in positively stained sections, a reason why the membranes appear separated by a continuous gap 2–3 nm thick (A). The gap appears electron opaque when colloidal lanthanum is mixed with the fixative, as the tracer penetrates between the membranes filling the extracellular space (B). By surrounding the particle protrusions, lanthanum produces face view images of electron-transparent rings (the particle protrusions) surrounded by an electron-opaque network (C). In freeze–fracture (D, E), the membranes split down the middle exposing a good portion of the intramembrane particles on the protoplasmic leaflet (face P) and their complementary pits on the exoplasmic leaflet (face E). (From Peracchia, 1977b.)

puzzled by the apparent contrast between structural variability and homogeneity in functional properties. Gap junction patches vary widely in size; they range from several micrometers in diameter, as in certain mammalian tissues such as liver, pancreas, and ovaries (Goodenough and Revel, 1970; Friend and Gilula, 1972; Albertini and Anderson, 1974) to a few nanometers in vertebrate photoreceptors (Raviola and Gilula, 1973), heart, (Mazet and Cartaud, 1976; Mazet, 1977; Kensler et al., 1977; Larsen, 1977), and ovaries (Gilula et al., 1978). In some cases, they are reduced to what looks like individual pairs of intramembrane particles (Mazet and Cartaud, 1976).

The packing arrangement of the particles also varies a great deal as the particles can be seen disorderly aggregated (Figs. 4, 6, 11, 20, 22, 25) in some preparations or packed into crystalline arrays (Figs. 7, 21, 23, 24, 26–28) in others. Curiously, both types of arrangements were seen in junctions from the same tissue (Figs. 4, 24; 6, 7; 20, 21; 22, 23; 25–28), suggesting that different particle arrays may correspond to different functional states at the time of fixation (Peracchia, 1973a). Recently, this hypothesis has been supported by studies on the effects of uncoupling treatments on gap junction structure (Peracchia, 1974, 1977a, 1978; Peracchia and Dulhunty, 1974, 1976), which indicate that only junctions with disorderly packed particle arrays have normal permeability properties (see Sections V,B and C). In crystalline junctions, the particles usually form hexagonal lattices; however, in one case, orthogonal and rhombic lattices (Fig. 28) have been seen also (Peracchia and Peracchia, 1978).

Most often, the particles of crystalline junctions form large patches fused together at different angles. There are cases, however, in which the particle arrays are subdivided into small crystalline aggregates a few particles wide and many particles long. The aggregates are separated from each other by a particle-free meshwork of aisles 10–15 nm wide (Pinto da Silva and Gilula, 1972; Albertini and Anderson, 1974; Kogon and Pappas, 1975; Elias and Friend, 1976; Herr, 1976; Baldwin, 1977, 1979). These junctions have been interpreted as stages of junction development (Albertini et al., 1975); however, junctions normally displaying disorderly packed particles (Fig. 6) can be induced to form crystalline arrays with aisles by certain uncoupling treatments such as exposure to Ca^{2+} ionophores (Fig. 7) (Peracchia, unpublished) or mechanical injury (Baldwin, 1977, 1979). Thus, it is reasonable to believe that this particle configuration may also correspond to uncoupled junctions. Baldwin (1977, 1979) has suggested that this crystalline configuration may represent an early stage of uncoupling, but it could also be related to a fast rate of junction crystallization, from which lipids, displaced by confluent particles, cannot flow to perijunctional regions being trapped within the junction (Peracchia, 1977a).

In certain junctions (Figs. 8, 9), the particle arrays are frequently interrupted by smooth discs 70–80 nm in diameter containing one or two particles roughly at

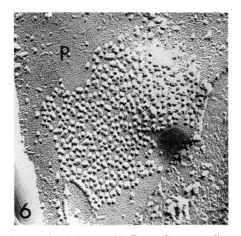

FIG. 6.   Vertebrate jap junction (rat stomach). Freeze–fracture replicas of gap junctions from control specimens display a disordered array of particles at an average center-to-center distance of 10 nm. P, face P. ×138,000.

their center (Landis *et al.,* 1974; Mazet and Cartaud, 1976; Mazet, 1977; Kensler *et al.,* 1977; Larsen, 1977). Interestingly, these junctions have been described thus far mainly in tissues which possess caveolae (spherical invaginations of the plasma membrane) (Figs. 8, 9). Since caveolae, when flattened, have a smooth disc-like appearance (Dulhunty and Franzini-Armstrong, 1975), the discs could reasonably be caveolar membranes trapped within the junctions. Even

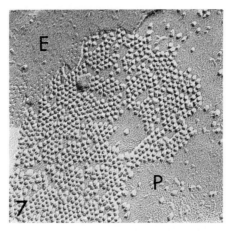

FIG. 7.   Vertebrate gap junction (rat stomach). Freeze–fracture replicas of junctions from specimens treated with a Ca^{2+} ionophore display particles and pits hexagonally aggregated into small patches surrounded by smooth aisles. P, face P; E, face E. ×138,000.

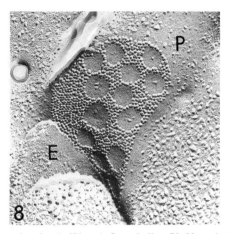

Fig. 8. Vertebrate gap junction (calf heart). Smooth discs 70-80 nm in diameter are a frequent feature of heart gap junctions. The disc surface is slightly concave on P face (P) and often contains one or two particles at its center. E, face E. ×100,900.

more unusual are gap junctions in which the particles are arranged in individual rows as seen in photoreceptor cells of vertebrate retina (Raviola and Gilula, 1973), mesangial and lacis cells of mammalian kidney (Pricam *et al.*, 1974), vertebrate heart (Mazet and Cartaud, 1976; Mazet, 1977; Kensler *et al.*, 1977; Larsen, 1977), ovaries (Gilula *et al.*, 1978) and developing tissues of insect

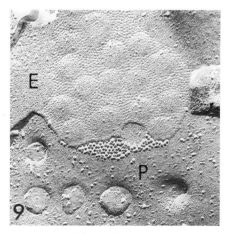

Fig. 9. Vertebrate gap junction (calf heart). Complementary image (face E) (E) of a junction similar to that shown in Fig. 8. The lower half of the micrograph shows the P face (P) of five open caveolae. Only the one on the right is exposed by the fracture in its entirety; the other four are filled with extracellular fluid. The similarity in size between caveolae and junctional discs suggests that the later could be caveolar membranes trapped within gap junctions. ×100,900.

(Lane and Swales, 1978a,b) and amphibian (Decker and Friend, 1974; Decker, 1976) nervous system, and mammalian eye lens (Benedetti et al., 1974).

Variable particle diameters have also been reported. In vertebrates, the particles have been seen to range from 6.0 to 10.0 nm in freeze-fractured and negatively stained preparations, whereas in invertebrates, they are 11.0 nm or larger (Staehelin, 1974; Gilula, 1976), which probably accounts for the wider gap of these junctions. In freeze-fracture, the range of size within each of the two groups could be due to differences in contamination and/or thickness of platinum–carbon deposit among different preparations. However, occasionally particles of different size can be seen in the same junctions (Goodenough and Gilula, 1974). In some cases, unevenness in particle diameter could be the result of plastic deformation during fracture (Bullivant, 1977); a mechanism also invoked to explain the higher degree of order in pitted than in particulated arrays (Chalcroft and Bullivant, 1970). Interestingly, in junctions which have assumed crystalline arrays, the particles are more uniform in size and shape and are slightly smaller than those of disordered arrays (Peracchia and Dulhunty, 1976; Peracchia, 1977a, 1978).

Coexistence in a junction of two types of particles, which are different in size and pattern of aggregation, has been reported in rat intestinal epithelium (Staehelin, 1972, 1974) and in crustacean neurons and intestine (Peracchia, 1971, 1972, 1973a; Graf, 1978). In crayfish neurons, these junctions (Fig. 18) have been interpreted as transitional stages of the uncoupling process (Peracchia, 1973a; Peracchia and Dulhunty, 1976), but this interpretation may not apply to rat intestinal junctions. In these junctions, in fact, conventional patches of disorderly packed 8.0–9.0 nm particles are seen adjacent to hexagonal arrays of 10.0–11.0 nm particles with unit cell dimensions of 19.0–20.0 nm which are much greater than those of uncoupled junctions (see Section V,B).

Particles distinctly different in size have also been reported in developing gap junctions (Johnson et al., 1974; Decker and Friend, 1974). Here, 10 nm particles, either scattered or in long rows, are seen in the vicinity of small clusters of conventional 8.5 nm gap junction particles on otherwise particle-free membrane patches. The 10 nm particles have been interpreted as precursors of the gap junction particles (Johnson et al., 1974).

In contrast to vertebrate and most invertebrate junctions, which fracture in such a way that all or most of the particles are retained with the protoplasmic leaflet, arthropods junctions (Flower, 1972; Gilula, 1972; Peracchia, 1972, 1973b) and some coelenterate, platyhelminth, and annellid junctions (Flower, 1977; Larsen, 1977; Wood, 1977) fracture in such a way that most of the particles remain attached to the exoplasmic leaflet and produce bumps on the E face and pits on the P face (Figs. 11, 13b, 17, 20, 21). In crayfish, whereas most of the particles appear on the E face, a few remain on the P face (Peracchia, 1973b) (Figs. 11a, 13a, 17). Interestingly, this alternate fracture behavior is occasionally also seen in vertebrates, e.g., the gap junctions between eye lens fibers (Peracchia, 1978) (Figs. 25, 26, 28) and those between hepatocytes after the membranes have been pulled apart by hypertonic sucrose solutions (Peracchia, 1977a). Recently, to complicate the issue, junctions with particles mostly on the P face have also been reported in arthropods (Graf, 1978) and annellids (Kensler et al., 1977). However, one should also be aware that in some cases unusual fracture properties may result from differences in preparative procedure; in other junctions (zonulae occludentes) in fact, an inversion of the fracture properties has been produced by treating unfixed specimens with cryoprotective agents (Staehelin, 1973).

A number of studies have reported the close association between gap junctions and cytoplasmic organelles. These include cisternae of either the smooth (Albertini and Anderson, 1975; Connell and Connell, 1977; Fry et al., 1977) or the rough (Nuñez, 1971; Garant, 1972) endoplasmic reticulum (ER), mitochondria (Nuñez, 1971), and in some cases free ribosomes (David-Ferreira and David-Ferreira, 1973; Merk et al., 1972, 1973; Garant, 1972; Larsen, 1977). In crayfish septa, gap junctions are bordered on both sides by layers of 70–100 nm vesicles (Figs. 10, 17), some of which seem to make contact with junctional particles (Peracchia, 1973a) (Fig. 10). The observation of such contacts has suggested the possible involvement of the vesicles in a form of intercellular communication via molecules confined in membrane bound compartments (Peracchia, 1973a), but it could be that the vesicles and possibly also the endoplasmic reticulum cisternae and mitochondria function as calcium buffering systems (Peracchia and Dulhunty, 1976). In fact, since an increase in intracellular calcium is known to block direct intercellular communications (Loewenstein, 1966), it would seem logical that calcium buffering structures be close to gap junctions to maintain a low calcium concentration in these regions at all times. In rectifying gap junctions of crayfish, vesicles are seen only on the presynaptic side (Keeter et al., 1974; Peracchia and Dulhunty, 1976; Hanna et al., 1978a). However, whether or not this asymmetry has a meaning in the phenomenon of rectification is not known.

It is generally known that gap junctions occur between cells of the same type; however, a few examples of gap junctions coupling cells of different type have also been reported. These have been called heterocellular gap junctions (Johnson et al., 1973) to distinguish them from the former, and most common, homocellular gap junctions. Heterocellular gap junctions, described in Limulus (Johnson et al., 1973), cockroaches (Noirot and Noirot-Timothée, 1976), lizards (Nadol et al., 1976), rabbits (Kogon and Pappas, 1975; Anderson and Albertini, 1976), rats (Gilula et al., 1978), and man (Fisher and Linberg, 1975) do not differ in structure from homocellular gap junctions. Interestingly, vertebrate cells can be induced to form gap junctions with cells of different origins in tissue culture, (Michalke and Loewenstein, 1971), but this interaction cannot be established between vertebrate cells and various cells from arthropods (Epstein and Gilula, 1977).

Surprisingly gap junctions have also been reported between membranes arising from the same cell. These junctions, known as "reflexive gap junctions" (Herr, 1976), have been seen in vertebrate smooth muscle (Iwayama, 1971), kidney (Pricam *et al.*, 1974), decidual cells of uterus and ovaries (Herr, 1976; Lawn *et al.*, 1971; Herr and Heidger, 1978), testis (Connell and Connell, 1977), and nerves (Schwann cells) (Sandri *et al.*, 1977; Schnapp and Mugniani, 1978; Rosenbluth, 1978). The meaning of these junctions is unclear, as it is difficult to conceive the need for a communication via gap junctions between two cellular regions that are indeed widely communicating with each other via cytoplasm. It has been suggested that they may have a significance in the secretory processes since they are often seen in actively secreting cells (Herr and Heidger, 1978). However, they may not have any special meaning at all, their formation possibly being the consequence of apposition between outer plasma membrane surfaces bearing gap junction precursors or potential for gap junction formation. These junctions are often seen between cell processes or between processes and cell soma; however, certain processes (microvilli, for instance) never interact via gap junctions even though they are close to each other. This could be explained by the absence of junction precursors in some luminal plasma membranes, as these membranes may be kept structurally different from the basolateral plasma membrane by a blockade to free intermixing of intramembrane components provided by the zonulae occludentes.

Cytoplasmic spheres made of double membranes with typical gap junction structure have been reported in a number of cells. These spheres, named "annular gap junctions" (Merk *et al.*, 1972) are, in some cases, invaginating gap junctions, but often they are truly intracellular structures as proven in serial sectioning (Merk *et al.*, 1972; Espey and Stutts, 1972) and by their inaccessibility to colloidal lanthanum (Garant, 1972; Merk *et al.*, 1973; Letourneau *et al.*, 1975; McNutt, 1977). Annular gap junctions are likely to represent an intermediate step in junction degradation following internalization of gap junctions by a process of membrane invagination and endocytosis (Merk *et al.*, 1973; Albertini and Anderson, 1974; Albertini *et al.*, 1975). A similar mechanism has been suggested for desmosomes (Overton, 1968) and zonulae occludentes (Staehelin, 1973). The observation of numerous annular gap junctions in granulosa cells of preovulatory follicles (Espey and Stutts, 1972; Merk *et al.*, 1972, 1973; Albertini and Anderson, 1974; Zamboni, 1974; Albertini *et al.*, 1975) and other findings (Larsen, 1975) have also suggested hypotheses envisioning gap junctions as the site of hormone receptor–adenylate cyclo complexes (Larsen, 1975; Azarnia and Larsen, 1977). Interestingly, isolated gap junctions can be induced to assume spherical configurations by trypsin digestion (Goodenough, 1976);

thus, the possibility that gap junction internalization follows the activation of proteolytic enzymes should also be kept in mind.

C. CHANNEL FRAMEWORK

The core of the gap junction problem is the intercellular channel. Functional studies have provided convincing evidence that the free exchange of small molecules between cells takes place via channels. Structure–function studies have pointed out unequivocally that gap junctions are the structures that house the intercellular channels. But what do we really know about the structure of the channels? Where are they located within the gap junction? What is their molecular framework? What is the composition of their building blocks? Do they change in size and structure to modulate the degree of intercellular communication? These and many other relevant questions are only partially answered at present, but undoubtedly the great deal of attention devoted to these problems in the past decade has given us a pretty good idea of the general principles of channel architecture, composition, and functional plasticity.

In the late sixties and early seventies, most gap junction models depicted the channels as cylindrical empty spaces that spanned the junctional membranes and the interposed extracellular gap and that were localized at the center of intramembrane particles. The particles, which were designed as six subunit structures, were shown to occupy the membrane thickness, to protrude from both membrane surfaces, and to precisely match and bind to similar particles of the adjoined membrane (Payton et al., 1969b; McNutt and Weinstein, 1970; Pappas et al., 1971). Indeed, many of these features were based on assumptions (Peracchia, 1973a,b); however, in recent years most of them have been proven true by data obtained from ultrastructure, X-ray crystallography, and biochemistry.

In crayfish gap junctions fixed with glutaraldehyde–H_2O_2 (Peracchia and Mittler, 1972), the intramembrane particles were seen in positively stained sections and were unequivocally shown to protrude from both membrane surfaces and to bind precisely with similar particles of the adjoined membranes (Peracchia, 1973a) (Figs. 10, 12). Occasionally the particles appeared spanning the membrane thickness (Fig. 12), but more convincing evidence for this came from freeze–fracture observations (Peracchia, 1973b). Crayfish junctions fracture in such a way that although most of the particles remain with the exoplasmic leaflet, as in other arthropods, a few fracture with the protoplasmic leaflet, thus allowing one to see the morphology of both the cytoplasmic (Fig. 11c) and the external (Fig. 11a) end of the particles. Estimates of the height of the particles confirmed the hypothesis that they indeed span the membrane thickness and protrude from both membrane surfaces (Peracchia, 1973b; Hanna et al., 1978a). Both the cytoplasmic and the extracellular end of the particles exposed by freeze–fracture

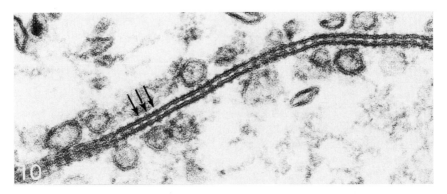

FIG. 10.   Invertebrate gap junction (crayfish neurons). In interneuronal gap junctions of crayfish, fixed with glutaraldehyde-H_2O_2, the intramembrane particles are seen as electron-opaque beads repeating every $\sim$ 20 nm and binding precisely to similar particles of the adjoined membrane (arrows). Fifty to eighty nm vesicles border both sides of the junctions. The vesicles often appear physically attached to the junctions and frequently display electron-opaque beads in their membrane. $\times$133,500. (From Peracchia, 1973a.)

clearly displayed a central depression $\sim$2.5 nm in size (Figs. 11a and c, 13). This image was matched by the complementary images of a central bump (Figs. 11b, 13) on the opposite fracture face and was interpreted to represent the internal and the external opening of the channels, respectively. It thus confirmed the existence and location of the channels spanning the membranes at the center of pairs of intramembrane particles (Peracchia, 1973b) (Figs. 13, 29, 31a).

Goodenough (1976) and Hertzberg and Gilula (1979) have described electron-opaque lines in edge views of NaPT-stained junctions of mammalian liver. These lines extend from the center of the gap into the membrane thickness and have been interpreted as side views of the channels filled with stain. This interpretation has been criticized (Larsen *et al.*, 1976; Robertson, in Bennett and Goodenough, 1978) mainly because these images could be tilt-overlap artifacts, although similar images have been seen also in lanthanum-stained sections (Peracchia, 1973a) (Fig. 32a) where the membranes were sectioned precisely across. It is not clear, however, whether or not the electron-opaque lines extend all the way through the junctional membrane to reach the cytoplasmic surface. In the crayfish, these lines extend for $\sim$5 nm from the center of the gap (Fig. 32a) and thus, only 2–3 nm into the thickness of each of the two membranes, which is consistent with the hypothesis that these junctions are in an uncoupled state (see Sections V,B and C).

In tangential sections through lanthanum stained junctions (Peracchia, 1973a) (Fig. 14a and c), as well as in isolated, NaPT-stained (Peracchia, 1973b) (Fig. 14b and d) and freeze–fractured (Peracchia, 1973b, 1974) (Fig. 15a and b) junctions, the intramembrane particles were shown to be composed of six main

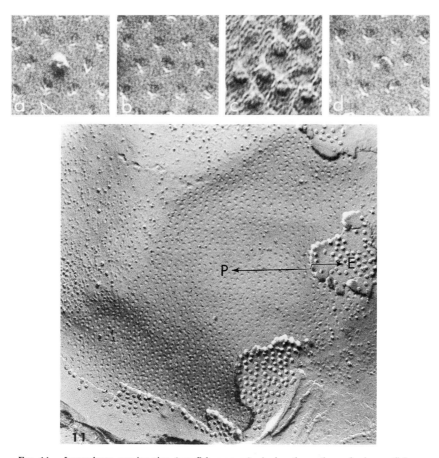

FIG. 11. Invertebrate gap junction (crayfish neurons). As in other arthropods, in crayfish gap junctions most of the particles fracture away with the exoplasmic leaflet (E). The particles, 12–15 nm in size, are packed disorderly at an average center-to-center distance of 20 nm. On the P face (P), some of the pits are occupied by particles of various size; the smaller ones are probably particle fragments (see Figs. 13c and 20). The double-headed arrow indicates the step-like path followed by the fracture plane (from E face down to P face). Inset (a) shows, at the center, the extracellular end of a particle that has remained on the P face (notice its central depression, believed to represent the external mouth of the channel). A similar volcano-like appearance is displayed by the cytoplasmic end of the particles shown in E face, in inset (c); their complementary P face image is that of a pit with a central bump, as shown most clearly at the center of inset (b). The (P face) pit at the center of inset (d) is partially occupied by a small particle, probably a fragment of an intramembrane particle fractured as shown in Fig. 13c. ×95,000. Insets, ×400,000.

subunits arranged in a circle around the presumed channel. Indirectly confirming the subunit composition of the particles is the frequent observation of P faces of crayfish junctions in which some of the pits are partially occupied by small

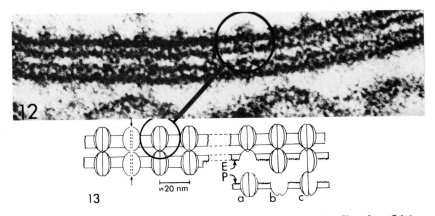

FIG. 12. Invertebrate gap junction (crayfish neurons). Cross-sectional profiles of crayfish junctions often display images of intramembrane particles clearly spanning the trilaminar structure of the
membrane. The edges of one such particle appear, in the encircled area, as electron-opaque lines
crossing the central layer of the membrane. ×500,000. (From Peracchia and Dulhunty, 1976.)

FIG. 13. Schematic diagram of the architecture and fracture properties of crayfish gap junctions.
Intramembrane particles composed of six main subunits are shown spanning the membrane thickness,
protruding from the membrane surfaces and binding to similar particles of the adjoined membrane.
The intercellular channels are believed to span the junctional thickness at the center of pairs of
particles (arrows). The right portion of the diagram shows the fracture behavior of these junctions;
although most of the particles remain with the exoplasmic leaflet (b), a few fracture with the
protoplasmic one (a), and some (c) may be fragmented in such a way that some of the subunits remain
on P face and others on E face. P, P face; E, E face.

particles (Peracchia, 1974) (Figs. 11d, 20). These images were interpreted to
indicate that some of the particles are fractured in the middle along a plane
perpendicular to the membrane surface such that some of the subunits remain
attached to the protoplasmic leaflet and the others to the exoplasmic one (Peracchia, 1974) (Fig. 13c). In vertebrates, a five- or six-subunit composition of the
particles has recently been shown by high resolution freeze–fractures of goldfish
and guinea pig gap junctions (Hama and Saito, 1977). Although most often the
particles have been shown to be composed of six subunits, in isolated junctions
of vertebrate eye lens incubated at low pH the particles are likely to be composed
of four subunits (Peracchia and Peracchia, 1978).

Interesting data on the architecture of gap junctions have come from X-ray
diffraction studies of isolated gap junctions. Diffraction patterns cannot be obtained from individual gap junctions; thus, it is necessary to isolate them in large
numbers and arrange them in an orderly fashion. Gap junctions isolated in pure
fractions from mouse liver (Goodenough, 1974) and partially oriented by high
speed centrifugation have been exposed on edge to the X-ray beam (Caspar *et*

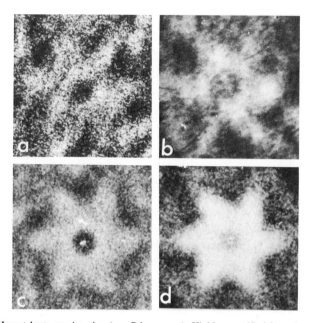

FIG. 14.   Invertebrate gap junction (crayfish neurons). Highly magnified face views of the junctional particles negatively stained with lanthanum (a, c) or NaPT (b, d), display images of six-pointed stars with an electron-opaque center (believed to represent the intercellular channel filled with stain). The star-like images suggest that the particles are composed of six main subunits. A six-step photographic rotation, (c) and (d), of the images seen in (a) and (b), respectively, confirms the number of subunits. ×2,500,000. (From Peracchia, 1973a,b.)

al., 1977; Makowski *et al.*, 1977). The resulting diffraction patterns are characterized by a series of broad maxima on the meridian and a series of short arcs on the equator. The meridional diffraction carries information on the electron density distribution along the profile of the membranes, while the equatorial ones reflect the orderly array of the intramembrane particles in the plane of the membrane and index the reciprocal of the lattice periodicity.

From meridional diffraction patterns Makowski *et al.* (1977) have succeeded in calculating the electron density profile of the junction. Aside from confirming the presence of a lipid bilayer (the matrix of the junctional membranes) with the phosphate groups separated by 4.2 nm across each bilayer and 4.4 nm across the gap, this study has indicated the presence of an electron density minimum in the hydrocarbon region of the bilayers similar to that of water ($0.332 \ e/Å^3$), thus much higher than that expected from hydrocarbon chains alone ($0.27 \ e/Å^3$) but consistent with the presence of protein in the hydrophobic region. This information has confirmed previous evidence for the presence of macromolecules, at the intramembrane particles, penetrating into the hydrophobic membrane region.

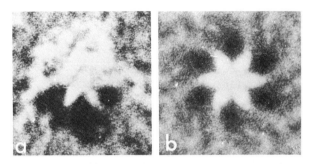

FIG. 15. Invertebrate gap junction (crayfish neuron). In freeze–fracture replicas, some of the intramembrane particles show clear images of subunits. One such image (E face) is seen in (a); due to the obliquity of the shadowing direction only four of the six subunits are visible. A six-step photographic rotation (b) confirms the six-subunit composition of the particle. ×1,600,000.

Protein material was also found in the gap, supporting electron microscopic evidence for extracellular particle protrusions (Peracchia, 1973a).

From equatorial diffraction patterns, Makowski *et al.* (1977) have derived data to support the presence of intramembrane particles arranged in a hexagonal lattice with a unit cell dimension of 8.67 nm (8.20 nm after trypsin treatment). The particles are ~6.0 nm in diameter and show a deep electron density minimum at the center, which is presumably the location of the hydrophilic core. This confirms once more the previous hypotheses on channel location (Payton *et al.*, 1969b; McNutt and Weinstein, 1970; Peracchia, 1973a,b).

In a recent X-ray diffraction study on gap junctions isolated with a combination of detergents (*n*-lauryl sarcosine, sodium deoxycholate, and Brij 58 at pH 9) to achieve a high degree of particle crystallinity, Goodenough *et al.* (1978) have succeeded in improving the resolution of the equatorial reflections out to 1.0 nm and in resolving maxima centered at 0.47 nm on the meridian. The meridional diffraction at 0.47 nm has been interpreted to indicate the presence of a significant amoung of β structure in the junctional protein at the hydrophobic membrane region.

D. GAP JUNCTION COMPOSITION

There is little doubt that the protein material found in the hydrophobic region of the membrane makes the framework of the channels. Hence, it is crucial for achieving a full understanding of channel structure and behavior to define the chemical structure of gap junction protein. Gap junction membranes represent, in most tissues, only a small fraction of plasma membrane, yet they are quite suitable for chemical studies due to a peculiar characteristic: they are extremely resistant to detergents. Because of this feature, as early as a decade ago, Benedetti and Emmelot (1968) succeeded in isolating purified fractions of gap junctions from rat liver. Sodium deoxycholate was used at first (Benedetti and Emmelot, 1968), but later other detergents such as n-lauryl sarcosine (Goodenough and Stoekenius, 1972; Evans and Gurd, 1972), Lubrol WX (Zampighi and Robertson, 1977), and Brij 58 (Goodenough et al., 1978) have been employed either alone or in various combinations. One, n-lauryl sarcosine, has been widely used as it allows one to isolate highly purified fractions of gap junctions (Goodenough, 1974, 1976).

Most commonly, tissues rich in gap junctions such as mammalian liver, lens, or heart are homogenized in bicarbonate buffer, washed repeatedly to remove soluble proteins, and centrifuged in a sucrose gradient to isolate plasma membrane fractions. Enriched gap junction fractions are obtained by treating the plasma membrane fractions with detergents, and highly purified gap junction material is subsequently isolated by a second sucrose density gradient centrifugation. The polypeptide composition of gap junction proteins is studied by polyacrylamide gel electrophoresis of junctions solubilized with sodium dodecyl sulfate (SDS-PAGE); the lipids are analyzed by thin layer chromatography.

In most procedures, various contaminants such as mucopolysaccharides and collagen migrate with the junctions and have to be removed by means of hyaluronidase and collagenase treatment. Commercial enzymes often contain other hydrolytic enzymes which, after prolonged treatment, partially digest the junctional protein. Proteolysis either by collagenase contaminants or endogenous proteases has been considered the major reason for the disagreement among various investigators on the number and molecular weight of the polypeptide components. Polypeptides of MW ranging from 10,000 to 47,000 have been reported, but it is presently agreed that components smaller than 25,000 are likely to be products of proteolytic digestion.

Polypeptides of the following molecular weights have been reported. In mouse liver: (a) 20,000 (Goodenough and Stoekenius, 1972); (b) 34,000, 18,000, and two 10,000 (Goodenough, 1974); (c) 35,500 and 25,500 (Duguid and Revel, 1975); (d) 34,000 (Ehrhart and Chauveau, 1977); (e) 40,000 and 38,000 (Culvenor and Evans, 1977); (f) 26,000 and minor 21,000 (Henderson et al., 1979). In rat liver: (a) 20,000 and 10,000 (Gilula, 1974); (b) 47,000 and 27,000 (Herztberg and Gilula, 1979). In calf lens: (a) 34,000 and (minor) 26,000 (Dunia et al., 1974; Benedetti et al., 1976; Bloemendal et al., 1977); (b) 26,000–27,000 (Alcalá et al., 1975; Broekhuyse et al., 1976). In chicken lens: (a) 26,000–27,000 (Maisel et al., 1976). Excluding the components smaller than 25,000 that have been seen only after prolonged enzymatic treatment, one may recognize a certain consistency in the reports regarding a component of 26,000–27,000. Less consistent has been the observation of a larger component, which, if present, could be of a molecular weight somewhere between 35,000 and 47,000. In any event, it is remarkable to notice that the gap junction proteins may be composed of only one or two polypeptides.

Assuming from the various data that the gap junction particles are $\sim$6.0 nm in diameter, contain a central empty space $\sim$1.5 nm in diameter, and are composed of six subunits, one can estimate the cross-sectional area of the protein material to be $\sim$26.0 nm^2. Since the cross-sectional area of a polypeptide with α-helical conformation is close to 1 nm^2, approximately 25 polypeptide chains with α-helical structure could fit in each particle. If this were the case, each of the 6 subunits could be composed of 4 polypeptide chains oriented perpendicular to the membrane surface.

The composition of the lipid bilayer that occupies the spaces between particles has been the object of a limited number of studies. Early results indicating a phosphatidylcholine/phosphatidylethanolamine ratio higher in gap junctions than in plasma membranes (Goodenough and Stoekenius, 1972) have not been confirmed (Hertzberg and Gilula, 1979). However, there is general agreement for the presence of a substantial amount of cholesterol (Evans and Gurd, 1972; Caspar et al., 1977; Hertzberg and Gilula, 1979; Henderson et al., 1979). Interestingly, gap junctions seem to lack entirely sphingomyelin (Hertzberg and Gilula, 1979).

V. Modulation of Gap Junction Permeability

A. Cell Uncoupling

The extent of direct cell-to-cell communication can be decreased down to a complete uncoupling of neighboring cells by a variety of treatments that modify the intracellular homeostasis. Interestingly, more than a century ago, much before the concept of direct cell communication had developed, Engelmann (1877) noticed that cardiac cells, in direct contact with each other during life, became independent as they died. This phenomenon, known as "healing over" was believed to depend on ionic barriers being newly formed at the boundaries between injured and noninjured cells. The phenomenon of healing over, recognized in a variety of vertebrate hearts (Rothschuh, 1951; Weidman, 1952; Délèze, 1965; DeMello et al., 1969; etc.), was not observed in skeletal muscles. This was a surprising finding for early physiologists, but it is now clearly explained by the absence of gap junctions and electrical coupling in skeletal muscle.

As soon as the widespread existence of cell-to-cell communication was discovered (Loewenstein and Kanno, 1964; Kuffler and Potter, 1964), a large number of studies were carried out on the phenomenon of cell uncoupling. Most of the work has been done on embryonal salivary gland cells of an insect (*Chironomus*) (Loewenstein, 1975), on crayfish giant axons (Bennett, 1978), and on mammalian heart (DeMello, 1977), but successful uncoupling experiments have been performed on a variety of vertebrate and invertebrate cell systems.

In *Chironomus* salivary gland cells, uncoupling has been obtained by lowering $[Ca^{2+}]_0$ and $[Mg^{2+}]_0$ with ethylenediamine tetraacetate (EDTA) or just $[Ca^{2+}]_0$ with ethyleneglycol-bis (β-aminoethyl ether) N,N'-tetraacetate (EGTA) (Loewenstein, 1966, 1967; Nakas *et al.*, 1966); by injecting Ca^{2+} iontophoretically into the cells (Loewenstein *et al.*, 1967; Rose and Loewenstein, 1975, 1976; Délèze and Loewenstein, 1976; Loewenstein *et al.*, 1978) (Fig. 16) or by letting Ca^{2+}, or other divalent cations (Mg^{2+}, Sr^{2+}, Ba^{2+}, Mn^{2+}), enter the cells through holes in the plasma membrane (Loewenstein *et al.*, 1967; Oliveira-Castro and Loewenstein, 1971) or via Ca^{2+} ionophores (Rose and Loewenstein, 1974, 1975, 1976); by inhibiting the metabolism with 2,4-dinitrophenol (DNP), N-ethylmaleimide (NEM), prolonged cooling, cyanide, or oligomycin (Politoff *et al.*, 1969); by substituting Li^+ for external Na^+ (Rose and Loewenstein, 1971); by depolarizing the surface membrane (Socolar and Politoff, 1971); by treating the cells with trypsin or saline solutions made hypertonic with sucrose or buffered to high pH (Loewenstein, 1966; Loewenstein *et al.*, 1967; Rose and Rick, 1978).

In crayfish, the gap junctions between lateral giant axons have been successfully uncoupled by replacing extracellular Cl^- with impermeant anions such as acetate, propionate, and isothionate, or by mechanically injuring one of the axons (Asada *et al.*, 1967; Asada and Bennett, 1971). Like *Chironomus*, crayfish giant axons can be uncoupled by exposure to Ca^{2+}-, Mg^{2+}-free solutions or these solutions followed by return to saline of normal $[Ca^{2+}]$ (Asada and Bennett, 1971; Peracchia and Dulhunty, 1974, 1976) (Fig. 19), as well as by treatment with DNP (Politoff and Pappas, 1972; Peracchia and Dulhunty, 1974, 1976), low temperature (Payton *et al.*, 1969 a), and aldehyde fixation (Politoff and Pappas, 1972; Bennett, 1973a,b), or by intracellular injection of Co^{2+} (Politoff *et al.*, 1974).

In mammalian heart, uncoupling has been obtained by substituting external Cl^- with propionate (Dreifuss *et al.*, 1966); by treating the cells with hypertonic solutions (Barr *et al.*, 1965); by lowering the intracellular pH (Weingart and Reber, 1979); by intracellular injection of Ca^{2+} or Sr^{2+} (DeMello, 1975); by raising $[Na^+]_i$, either via intracellular injection or by means of ouabain (a blocker of the Na^+,K^+ pump) (DeMello, 1975, 1976; Weingart, 1977); by exposing the cells to DNP (DeMello, 1979). As previously mentioned, uncoupling of heart cells also follows mechanical injury ("healing over") (Engelmann, 1877; Rothschuh, 1951; Weidman, 1952; Délèze, 1965, 1970; DeMello *et al.*, 1969; DeMello, 1972; Nishiye, 1977; etc.).

In rats and mice, the acinar cells of pancreas and the lacrimal gland cells uncouple momentarily when secretion is stimulated with acetylcholine (Iwatsuki and Petersen, 1978a,b,c) or the intracellular pH is lowered with CO_2 (Iwatsuki and Petersen, 1979), and skin cells are readily uncoupled by mechanical injury (Loewenstein and Penn, 1967). In tissue cultures, human lymphocytes electrically coupled as a result of stimulation with plant lectins (PHA), can be un-

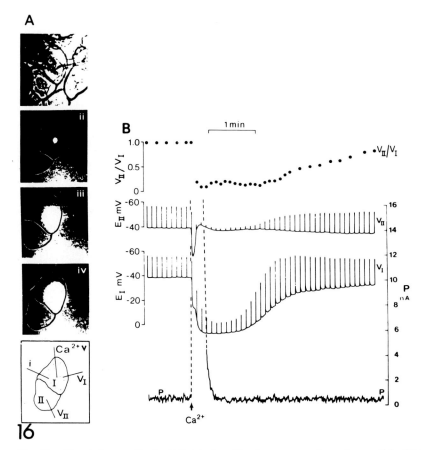

FIG. 16.   Electrical uncoupling by Ca^{2+} injection. Simultaneous recordings of aequorin luminescence, membrane potentials (E_I, E_{II}), and steady state potential changes (V_I, V_{II}) produced by
pulsing current injections ($i = 4 \times 10^{-8}$ A) in salivary gland cells of *Chironomus*. At the arrow (B),
cell I is injected intracellularly with Ca^{2+}. The aequorin luminescence (A:ii, iii, iv) is viewed (on a
television camera coupled to an image intensifier) through a darkfield microscope and is measured
(B) by recording current changes (*P*) with a photomultiplier. A current (*i*) and two voltage (V_I, V_{II})
microelectrodes, as well as Ca^{2+} injection micropipette are inserted into the neighboring cells as
shown in (A:v). The cell outlines (A:ii, iii, iv) are traced by superposition of a brightfield video
picture (A:i) on the darkfield micrographs. Injections of small amounts of Ca^{2+} (A:ii, iii), causing
only limited areas of luminescence, do not uncouple the cells, whereas injections, large enough to
extend the area of luminescence to the junctional region (A:iv), cause transient uncoupling (B), as
indicated by the drastic change in coupling ratio (V_{II}/V_I). A simultaneous depolarization of the
injected cell is noticeable. (From Rose and Loewenstein, 1975.)

coupled by intracellular injection of Ca^{2+} (Oliveira-Castro and Barcinski, 1974; Oliveira-Castro and Dos Reis, 1977), and 3T3 cells are uncoupled (although after a long delay) by treatment with a Ca^{2+} ionophore (Gilula and Epstein, 1976). Curiously, the same treatment quickly uncoupled cultured insect cell (TN) but not mouse myocardial cells (Gilula and Epstein, 1976).

In amphibians, embryonal cells of *Xenopus laevis* (Turin and Warner, 1977) and *Amblyostoma mexicanum* (axolate) (Hanna *et al.*, 1978b) have been uncoupled by exposure to 100% CO_2, the former also by intracellular injection of Ca^{2+} (Loewenstein *et al.*, 1978), and horizontal cells of salamander retina, by low [Na^+] solutions (Miller, 1978). In fish, cells of *Fundulus* (killifish) blastomeres can be uncoupled by glutaraldehyde fixation (Bennett *et al.*, 1972), exposure to 100% CO_2 (Bennett *et al.*, 1978) and transjunctional polarization (in either direction) produced by voltage clamp (Harris *et al.*, 1978; Spray *et al.*, 1979). Finally, in worms, uncoupling of nerve cells of Retzius (the leech, *Hirundo medicinalis*) has been obtained by exposure to low Ca^{2+} (Penn and Loewenstein, 1966).

Scanning superficially through the large variety of treatments capable of uncoupling cells, one may assume that the uncoupling mechanism can be triggered by many factors. On the contrary, there are reasons to believe that only a few factors are involved. Délèze (1965) noticed that cut heart fibers do not heal in the absence of extracellular Ca^{2+}, but do so quickly when Ca^{2+} is supplied; this observation suggested for the first time the possible role of Ca^{2+} in uncoupling. Aside from the obvious increase in [Ca^{2+}]$_i$ resulting from cell rupture, intracellular Ca^{2+} injections, or treatment with ionophores, an increase in [Ca^{2+}]$_i$ is expected after most uncoupling treatments. In fact, chelation of external divalent cations, inhibition of the metabolism, and treatment with ionophores are believed to increase [Ca^{2+}]$_i$ as a result of Ca^{2+} release from mitochondria, an increase in Ca^{2+} influx, a decrease in Ca^{2+} efflux, or various combinations of these mechanisms (Politoff *et al.*, 1969; Rose and Loewenstein, 1971; DeMello, 1976; Weingart, 1977). Similarly, an increase in [Ca^{2+}]$_i$ is expected to follow the treatment of lacrimal and pancreatic cells with acetylcholine in view of the well known relationship between [Ca^{2+}]$_i$ and acetylcholine-evoked secretion (Iwatsuki and Petersen, 1978a,b,c,). Indeed in most of these cases, an increase in [Ca^{2+}]$_i$ has been proven to take place simultaneously with cell uncoupling by elegant experiments in which the changes in [Ca^{2+}]$_i$ have been monitored with an intracellular Ca^{2+} indicator (aequorin) simultaneously with the electrical recording of changes in junctional resistance (Rose and Loewenstein, 1975, 1976) (Fig. 16).

In a number of cases, a correlation between [Ca^{2+}]$_i$ and uncoupling has not been clearly established. Treatments with hypertonic sucrose and substitution of Cl^- with nonpermeable anions may cause uncoupling by mechanically pulling the cells apart; however, whether cell separation is the cause of uncoupling, or an effect of it, has not been established. Indeed, changes in $[Ca^{2+}]_i$ have not been detected with aequorin after substitution of propionate for Cl^- or Li^+ for Na^+ (Rose and Loewenstein, 1976). Equally unclear is the mechanism of action of trypsin, glutaraldehyde, and transjunctional polarization.

A challenge to the calcium hypothesis has been posed recently by a study on the uncoupling effects of exposure to 100% CO_2 (Turin and Warner, 1977). In this study, the cells of early *Xenopus* embryos were uncoupled reversibly by lowering the intracellular pH to values ranging from 6.4 to 6.2. The possibility that an increase in $[Ca^{2+}]_i$ had taken place simultaneously with the decrease in pH was believed unlikely because no signs of contraction of the cortical cytoplasm were detected.

Recently, the effects of pH on coupling have been investigated in depth by Rose and Rick (1978) on *Chironomus* salivary gland cells in which pH_i, $[Ca^{2+}]_i$, and electrical coupling were monitored simultaneously. In this study, the intracellular pH was decreased by exposing the cells to 100% CO_2, by injecting HCl intracellularly, and by exposing cells treated with an H^+ ionophore (Nigericine) to saline at pH 6.5. In all cases, uncoupling readily occurred, and in most experiments, an increase in $[Ca^{2+}]_i$ was also detected. After exposure to 100% CO_2, $[Ca^{2+}]_i$ increased in only about 40% of the experiments. However, this was attributed more to a decrease in aequorin response at low pH than to inconsistent changes in $[Ca^{2+}]_i$. After intracellular injection of Ca-EDTA solutions at pH 4 with a $[Ca^{2+}]$ of $<10^{-6}$ M, no uncoupling was detected, whereas pH_i decreased to 6.8. Coupling recovered quickly after removal of CO_2: in some cases, starting at a pH_i as low as 6.3. In other experiments, uncoupling was paralleled by an increase in $[Ca^{2+}]_i$ with little or no change in pH_i; these included exposure of the cells to CN, DNP, and Ca-ionophores. Interestingly, an increase in $[Ca^{2+}]_i$ probably brought about by increased Ca^{2+} influx (Baker and Honerjager, 1978) was also detected in uncoupling produced by alkalinization of the cytoplasm to pH > 7.8. Undoubtedly these data strongly support the hypothesis that Ca^{2+}_i is capable of uncoupling cells independently from H^+_i; however, the question as to whether or not uncoupling could also be caused independently by an increase in $[H^+]_i$ still remains. Still unanswered also is the possible independent role of $[Mg^{2+}]_i$ in uncoupling. Since Mg^{2+} binds to aequorin with an affinity similar to that of Ca^{2+}, without producing luminescence (Shimomura *et al.*, 1962), changes in $[Mg^{2+}]_i$ sufficient to cause uncoupling would not be detected by the aequorin technique.

Is uncoupling an all-or-none or a graded phenomenon? Délèze and Loewenstein (1976), asked this question in experiments in which changes in electrical coupling and fluorescein transfer were monitored in *Chironomus* gland cells during intracellular injections of variable quantities of Ca^{2+}. Interestingly, injections of small quantities of Ca^{2+} insufficient to depolarize the cells did not affect the electrical coupling but blocked fluorescein transfer (partial uncoupling).

These data, which were confirmed with other tracer molecules (Rose et al., 1977), suggested the possibility that uncoupling is a graded phenomenon in which the channel size and/or the ratio between open and closed channels could be finely regulated by $[Ca^{2+}]$ between 10^{-7} and 5.10^{-5} M.

In view of the recent data of quantum jumps in cell-to-cell conductance during opening and closing of junctional channels in early Xenopus embryo cells (Loewenstein et al., 1978), the hypothesis that partial uncoupling might be due to changes in the fraction of open channel seems more plausible. Evidence for quantal steps of increasing conductance during junction development (Loewenstein et al., 1978) also supports once more and most directly the hypothesis that electrical coupling is indeed related to the presence of intercellular channels.

Many cell functions are known to be finely modulated by the competitive action of stimulatory and inhibitory factors, yet in the uncoupling mechanism, only stimulatory factors have been considered thus far. Although it is possible that uncoupling is entirely dependent on changes in the concentration of certain cations such as Ca^{2+}, Mg^{2+}, and H^+ it seems reasonable to consider the possibility that the action of these uncoupling agents is antagonized by inhibitory factors. Interestingly, there is evidence (Weingart, 1977) that coupling can be improved, as well as decreased, by certain treatments (early phase of ouabain effects) and that uncoupling can be inhibited by certain hormones (glucocorticoids) or stimulated by others (mineral corticoids) (Suzuki and Higashino, 1977). A decrease of $[Ca^{2+}]_i$ below normal values could account for the early decrease in coupling resistance after ouabain. However, in view of the fact that low doses of ouabain can stimulate the Na pump before inhibiting it (reviewed by Weingart, 1977), an increase in $[K^+]_i/[Na^+]_i$ might have caused the phenomenon. Changes in $[K^+]_i$ could also be involved in the effects of corticosteroids on theoretical grounds (Suzuki and Higashino, 1977). Indeed various characteristics of the potassium ion could make it a weak calcium competitor for sites on the junctional proteins (see Section V,C).

The most obvious meaning of uncoupling is that of a safety device by which coupled members of a cell community can be isolated from a damaged cell, but it is very likely that a temporary interruption or a partial decrease in the extent of intercellular communication takes place physiologically in healthy cells in relation to developmental, metabolic, or other functional needs. Moreover, uncoupling may be involved in various pathological conditions such as malignant transformation, metastatization, cardiac arrhythmias, and abnormal regenerative processes.

B. CHANGES IN GAP JUNCTION STRUCTURE WITH FUNCTIONAL UNCOUPLING

Knowledge that intercellular communication can be decreased down to a complete cell uncoupling has stimulated a number of studies aimed at defining the structural basis of channel permeability modulation. For almost a decade since electron microscopists became interested in this problem, cell uncoupling was believed to follow a complete disruption of the gap junctions. This concept originated with the work of Barr *et al.* (1965, 1968) on the uncoupling effects of hypertonic sucrose or NaCl solutions on heart and smooth muscle cells. In both cases, the membranes of gap junctions became completely separated after being pulled apart by the shrinking cells. Soon after, similar disruptions in gap junctions were described in crayfish lateral giant axons uncoupled by mechanical injury, replacement of Cl^- with less permeable anions and exposure to Ca^{2+}-, Mg^{2+}-free solutions followed by return to normal saline (Pappas *et al.*, 1967; Asada *et al.*, 1967; Asada and Bennett, 1971; Pappas *et al.*, 1971). However, at least three exceptions to this rule were reported: the junctional resistance increased significantly in crayfish septate axons exposed to low temperature (5°C) Payton *et al.*, 1969a), Ca^{2+}, Mg^{2+}-free saline (Pappas *et al.*, 1971), or DNP (Politoff and Pappas, 1972) without obvious separation of the gap junction membranes. In these cases, the possibility that finer changes in the structure of the junctions may have caused the decreased permeability of the channels seemed likely (Peracchia, 1973a). But could the changes be detected by electron microscopy?

Early in this decade, two puzzling observations were made on the structure of crayfish axo-axonal gap junctions. The center-to-center spacing between neighboring particles was ~20 nm in thin-sectioned and freeze–fractured junctions, but it decreased to ~15 nm in isolated, negatively stained junctions (Peracchia, 1973a,b; Peracchia and Dulhunty, 1974, 1976) (Fig. 17). In lanthanum-stained preparations, two particle arrays were reported (Peracchia, 1971, 1972, 1973a): in one, the particles were packed at a center-to-center distance of 18–20 nm, in the other, they were tightly packed at distances as small as 12.5 nm (Fig. 18); interestingly, the two arrays were continuous with each other and often appeared intermixed. These observations suggested that the junctions possess peculiar dynamic properties possibly related to different functional states and, as a hypothesis, it was proposed that the changes in particle array from a loose to a tight pattern may reflect changes in junction permeability from low to high coupling resistance (Peracchia, 1973a). This seemed reasonable because isolated junctions, which always display tightly packed particle arrays (Figs. 3 and 17, inset), are indeed believed to be in an uncoupled state as a result of cell rupture during tissue homogenization.

To test this hypothesis, the electrical resistance and structure were studied in the same crayfish gap junctions after uncoupling with EDTA (Fig. 19) or DNP (Peracchia and Dulhunty, 1974, 1976). The junctions of the septum between lateral giant axons, on which the electrical recordings were made were studied in thin sections, while other axo-axonal gap junctions were freeze–fractured. The hypothesis was confirmed, as the majority of the junctions fixed when the junc-

FIG. 17. Invertebrate gap junction (crayfish neurons). The particle periodicity of gap junctions fixed in intact (well coupled) cells is ~20 nm (see freeze-fractured junction, face P), whereas that of isolated, Na-PT-stained junctions (inset) is ~15 nm. Isolated junctions are believed to be in an uncoupled state as a result of cell rupture during tissue homogenization. ×250,000. (From Peracchia and Dulhunty, 1976.)

tional resistance was high displayed tightly and hexagonally packed particle arrays (Fig. 21), while in controls (Fig. 20) as well as in samples followed to recovery of normal junctional permeability, most of the junctions had loosely packed particles. Other structural changes included a decrease in gap thickness (from 20 to 18 nm) and possibly in particle diameter (from 15.2 ± 0.97 to 12.5 ± 1.25 nm) in tightly packed junctions. These changes were believed to reflect conformational rearrangements in the particle protein resulting in obliteration of the intercellular channels (Peracchia, 1974; Peracchia and Dulhunty, 1974, 1976) (Fig. 30).

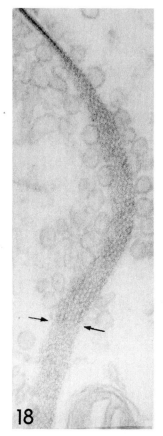

FIG. 18.   Invertebrate gap junction (crayfish neurons). Face view images of obliquely sectioned
gap junctions, negatively stained with colloidal lanthanum, often show two types of particle packing
in direct continuity with each other. In the micrograph, loosely packed particles at a center-to-center
distance of ~20 nm become tightly packed, below the arrows, with a minimum center-to-center
distance of 12.5 nm. In the upper left corner, the junction is cross-sectioned; notice the electron-
opaque central layer representing the extracellular gap filled with lanthanum. ×90,000. (From
Peracchia, 1973a.)

   Similar changes in particle packing and particle size were reported in gap
junctions of rat stomach (Fig. 23) and liver exposed to uncoupling treatments
such as DNP or hypoxia (Peracchia, 1977a), confirming the generality of the
phenomenon. Here disorderly packed arrays with particles at a center-to-center
distance of 10–11 nm (Fig. 22), typical of control junctions, changed, after the
treatments, into regularly hexagonal ones in which the particles were spaced at
~8.5 nm (Fig. 23), and the particle diameter had decreased by ~1 nm. Crystalline particle packings were also obtained in gap junctions of rabbit ciliary

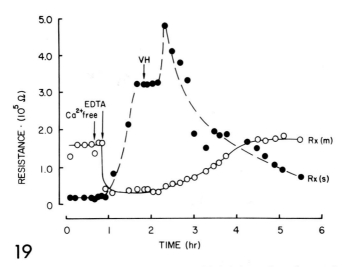

FIG. 19. Time course of junctional resistance $R_{x(s)}$ (filled circles) and membrane resistance $R_{x(m)}$ (open circles) at a septum between lateral giant axons of crayfish (abdominal nerve cord) during treatment with a Ca^{2+} chelator (EDTA) followed by Van Harreveld's saline (VH). EDTA treatment is preceded by a brief wash in a Ca^{2+}-, Mg^{2+}-free solution. In EDTA, $R_{x(s)}$ increases. Upon return to a normal $[Ca^{2+}]_0$ $[Mg^{2+}]_0$ (VH), $R_{x(s)}$ increases further and then returns to control values. Meanwhile $R_{x(m)}$ decreases sharply in EDTA and then returns to normal values after exposure to VH. (From Peracchia and Dulhunty, 1976.)

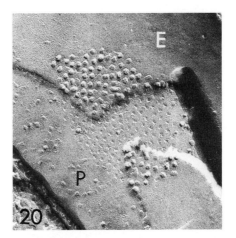

FIG. 20. Invertebrate gap junction (crayfish neurons). Junctions of untreated ganglia have particles and pits disorderly packed at an average center-to-center distance of ~20 nm. Notice that one of the pits is partially occupied by a small particle, probably a fragment of an intramembrane particle (see also Figs. 11d and 13c). P, face P; E, face E. ×124,000.

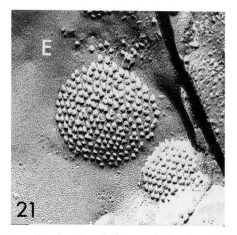

FIG. 21.   Invertebrate gap junctions (crayfish neurons). Junctions of ganglia, subjected to an uncoupling treatment (EDTA followed by Van Harreveld's solution) and fixed when the resistance of the septal junction $R_{x(s)}$, is high (Fig. 19), display tightly and hexagonally packed particle arrays with an average center-to-center distance between neighboring particles as small as 15 nm. Most of these junctions are curved (as suggested here by the dome-like appearance). E, face E. ×124,000. (From Peracchia and Dulhunty, 1976.)

epithelia made hypoxic with N_2 (Raviola *et al.*, 1978). In this study the specimens were prepared for freeze–fracture by rapid freezing (Heuser *et al.*, 1976) without glutaraldehyde fixation and cryoprotective treatment. Indeed glutaraldehyde has been shown to uncouple cells (Bennett *et al.*, 1972; Politoff and Pappas, 1972; Bennett, 1973a,b), but this is likely to be more the result of a

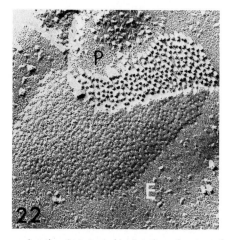

FIG. 22.   Vertebrate gap junction (rat stomach). Junction of untreated epithelial cells display particles and pits disorderly packed at an average center-to-center distance of 10 nm. P, face P; E, face E. ×124,000. (From Peracchia, 1977a.)

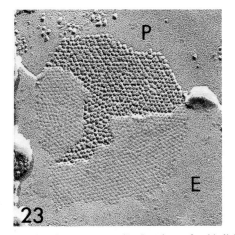

Fig. 23.   Vertebrate gap junction (rat stomach). Junctions of epithelial cells subjected to an uncoupling treatment (exposure to $5 \times 10^{-3}\ M$ DNP in well oxygenated Tyrode's solution at 37°C for 1 hour) show regularly hexagonal particle packings in which the average center-to-center distance between particles is 8.5 nm. P, face P; E, face E. ×124,000. (From Peracchia, 1977a.)

damage to the junctional protein brought about by the crosslinking and denaturing action of the fixative, than the selective activation of a physiological mechanism of cell uncoupling, as glutaraldehyde does not raise $[Ca^{2+}]_i$, affects the junctional permeability irreversibly and is known to alter the permeability properties of other channels as well (Shrager *et al.*, 1969; Horn *et al.*, 1978).

Tight hexagonal packings were also seen in cardiac cells mechanically injured (healing over) (Baldwin, 1977, 1979) and in individual junctional membranes pulled apart by vascular perfusion with hypertonic sucrose (Peracchia, 1977a) (Fig. 24), which is reasonable because the membranes of junctions pulled apart are known to acquire high electrical resistivity (Barr *et al.*, 1965; Asada and Bennett, 1971). Increased crystallinity was noticed earlier by Goodenough and Gilula (1972) in junctions still intact after exposure to hypertonic sucrose. However, in this study, the crystallinity was not correlated with changes in junction permeability and this observation was not mentioned in a subsequent and more complete report of this work (Goodenough and Gilula, 1974).

Although the structural changes appear well correlated with changes in permeability, one may wonder whether they are part of the uncoupling process or simply a parallel phenomenon. Indeed, recent findings on the effects of a calcium ionophore on junctions of intact cells (Peracchia, unpublished) (Fig. 7) and of the effects of divalent cations (Figs. 26, 27) and H^+ (Fig. 28) on isolated junctions (Peracchia, 1977a, 1978; Peracchia and Peracchia, 1978) strongly support a close relationship between the two phenomena.

Gap junctions between calf lens fibers maintain disorderly packed particle arrays typical of control (coupled) junctions if they are isolated in the presence

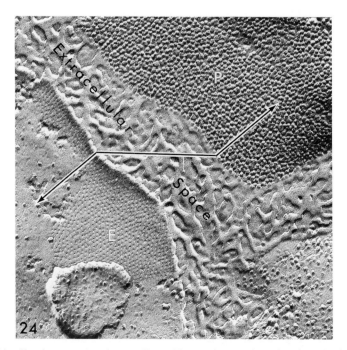

FIG. 24. Vertebrate gap junction (rat liver). The membranes of gap junctions that have been pulled apart by vascular perfusion with hypertonic sucrose solutions show particles and pits tightly packed into hexagonal arrays. Notice that the pits are more regularly packed than the particles; this is believed to result from plastic deformation of the particulated leaflet during freeze–fracture. The double-headed arrow shows the path followed by the fracture plane, from face E (E) down to face P (P), across an extracellular space greatly widened by the newly separated junctional membrane of two shrinking cells. ×117,500.

of EDTA (Fig. 25), but change into crystalline particle packings if they are incubated in solutions in which Ca^{2+} (Fig. 26), Mg^{2+} (Fig. 27), and H^+ (Fig. 28) are increased independently. The switch from a disorderly to a crystalline particle packing occurs at a $[Ca^{2+}]$ of 5.0×10^{-7} M (Peracchia, 1977a, 1978), a $[Mg^{2+}]$ $> 1.0 \times 10^{-3}$ M (Peracchia, unpublished) and a $[H^+]$ of 3.0×10^{-7} M (Peracchia and Peracchia, 1978). Since these values are very close to those believed to affect electrical coupling (Loewenstein, 1975; Weingart, 1977; Turin and Warner, 1977), it seems reasonable to assume that the structural changes are part of the uncoupling mechanism (Figs. 29, 30).

C. UNCOUPLING MECHANISM: AN HYPOTHESIS

By examining the various data accumulated on the uncoupling phenomenon, several important facts emerge. First of all, the uncoupling process can be ini-

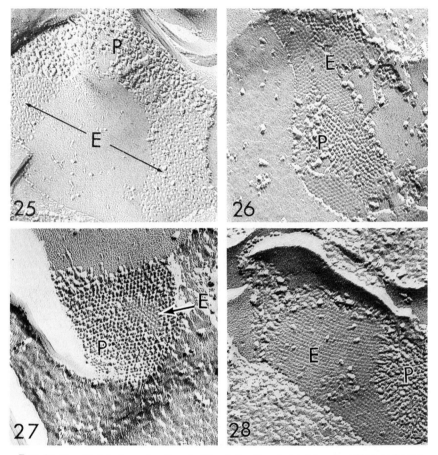

FIGS. 25–28.   Vertebrate gap junctions (calf lens). Junctions isolated from lens fibers in the presence of EDTA (Fig. 25) maintain disordered particle packings typical of control (coupled) junctions
fixed in intact cells. Exposure of similar junctions, isolated in the presence of EDTA, to solutions of 5
$\times$ 10^{-7} M Ca^{2+} (Fig. 26), 5 $\times$ 10^{-3} M Mg^{2+} (Fig. 27), or 3 $\times$ 10^{-7} M H^{+} (Fig. 28) causes the
particles to pack into crystalline arrays. Notice that the particles in one case (Fig. 28) form orthogonal
rather than hexagonal arrays. P, face P; E, face E. $\times$134,000.

tiated by more than one factor, since all the divalent cations tested and possibly
also a monovalent one are effective. This indicates that the ions may not act in a
very specific way, but most likely just by neutralizing negative charges. However, the minimum effective concentration is not the same for all divalent cations. Oliveira-Castro and Loewenstein (1971) found that Ca^{2+} acts at 4–7 $\times$
10^{-5} M, Mg^{2+} at 1.2–6 $\times$ 10^{-4} M, Sr^{2+} at 3–5 $\times$ 10^{-3} M, and Ba^{2+} at 0.7–1 $\times$
10^{-2} M, indicating the following sequence of effectiveness: Ca^{2+} > Mg^{2+} > Sr^{2+}

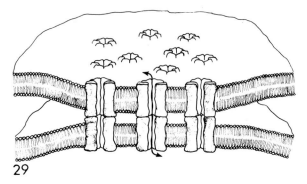

FIG. 29. Tentative model of gap junction architecture (coupled junction). Intramembrane parti-cles, spanning the membrane thickness and binding to similar particles of the adjoined membrane, house cylindrical channels, ~ 1.5 nm in diameter, (double-headed arrow), well insulated from the extracellular medium to provide free exchange of small molecules between neighboring cells. There are reasons to believe that in most gap junctions the intramembrane particles are composed of six main subunits, each one made of several polypeptide chains. The particles are aggregated disorderly at an average center-to-center distance of 10 nm (vertebrates) or larger (invertebrates). A lipid bilayer fills the interparticle spaces $\times 1,000,000$.

$> Ba^{2+}$. Consistent with this is also the observation that Ca^{2+} triggers crystallin-ity in isolated gap junctions at a much lower concentration than Mg^{2+} (see Section V,B).

The various degrees of effectiveness of divalent cations may suggest dif-ferences in their affinity to negatively charged sites on the junctional proteins.

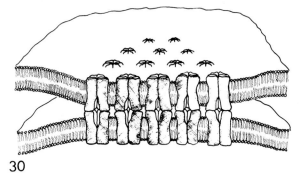

FIG. 30. Tentative model of gap junction architecture (uncoupled junction). Uncoupling agents such as divalent cations (and possibly H^+) are believed to act directly on the junctional molecules, producing a narrowing of the intercellular channels at their cytoplasmic end (Fig. 32) and a simul-taneous clumping of the intramembrane particles into crystalline (most often hexagonal) arrays in which the average center-to-center distance between particles is 8.5 nm (vertebrates). There are reasons to believe that a decrease in particle diameter and gap thickness may also take place. Channel narrowing and particle clumping are believed to follow neutralization of negative charges on the proteins (Fig. 31). $\times 1,000,000$.

Although one should keep in mind that these differences could depend on complex physicochemical circumstances, there are no reasons to believe that this phenomenon could not be explained on the basis of simple steric mechanisms (Peracchia *et al.*, 1979). Since Ca^{2+} is the most effective cation, one may assume that the size of the protein site is similar to the diameter of Ca^{2+} (~ 2 Å, anhydrous). Since Sr^{2+} and Ba^{2+} are larger and Mg^{2+} is smaller than Ca^{2+} (see Table I), their interaction with the proteins would require a slight deformation of the sites in either direction. Since Sr^{2+} is closer in size to Ca^{2+}, one would expect it to be more effective than Ba^{2+}, as indeed it is. From these considerations, one may envision a functional site of 2 ± 0.7 Å endowed with two net negative charges (Fig. 31).

Hydrogen ions, if effective, could function in pairs. Being very small indeed, two H^+ should have easy access to the site, which could be consistent with the postulated uncoupling effect of H^+ at concentrations lower than 1×10^{-6} M (Turin and Warner, 1977) and its effect on isolated junctions at a concentration of 3×10^{-7} M (Peracchia and Peracchia, 1978). Larger monovalent cations normally present in cells such as K^+ and Na^+ could act as weak inhibitors of uncoupling (Peracchia *et al.*, 1979). In fact, being too large to fit in the sites in pairs (see Table I) they would occupy them individually, and by neutralizing only one charge, would fail to trigger uncoupling (Fig. 31a).

Neutralization of negative charges by divalent cations could induce conformational changes causing the protein subunits to associate more tightly around the central channel, narrowing the channel bore (Peracchia *et al.*, 1979) (Figs. 30, 31b). Consistent with this suggestion could be the findings of decreased particle size in uncoupled junction (see Section V,B).

Makowsky *et al.* (1971) have proposed that the narrowing of the channels takes place at their extracellular mouth; they based their model on the X-ray

TABLE I

CRYSTAL DIAMETER OF IONS IN ANGSTRTOMS[a]

Li+	1.2	Be2+	0.62	Ti2+	1.7
Na+	1.9	Mg2+ [b]	1.3	Pt2+	1.92
Cu+	1.92	Cu2+	1.44	Cd2+	1.98
Ag+	2.52	Ni2+	1.46	Ca2+ [b]	1.98
K+	2.66	Co2+ [b]	1.48	Hg2+	2.2
Au+	2.74	Zn2+	1.48	Su2+	2.2
Rb+	2.96	Fe2+	1.52	Sr2+ [b]	2.26
Tl+	3.18	Mn2+ [b]	1.56	Pb2+	2.54
Cs+	3.38	Cr2+	1.6	Ba2+ [b]	2.7

[a] From Pauling (1960).

[b] Ions tested as uncouplers.

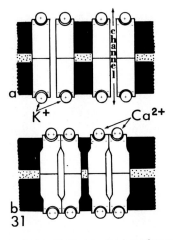

FIG. 31.   Schematic diagram of a hypothetical mechanism of uncoupling. Gap junction proteins are believed to contain sites endowed with two net negative charges. Uncoupling is believed to follow neutralization of such charges by divalent cations (b). Since Ca^{2+} is the most effective uncoupler ($Ca^{2+} > Mg^{2+} > Sr^{2+} > Ba^{2+}$), the protein site could be similar in size to the crystal diameter of Ca^{2+} (~ 2 Å). Hydrogen ions, if effective, could act in pairs, whereas other monovalent cations, too large to act in pairs (see Table I), would fit in the sites individually (a) and, blocking only one charge, would fail to uncouple. In coupled junctions (a), charge repulsions would keep the particles separated. Neutralization of the negative charges (b) by divalent cations (or H^+) would cause both a narrowing of the channels and a clumping of the particles into crystalline arrays. The lipid bilayers are drawn in black. The extracellular space is dotted.

diffraction finding of an increase in protein density in the gap and a decrease in gap thickness after trypsin digestion. However, there are no reasons to believe that these changes are necessarily associated with a narrowing of the channel extracellularly or anywhere else. Moreover, whether or not trypsin digestion of isolated gap junctions promotes changes in the proteins comparable to those of uncoupling is debatable.

After lanthanum infiltration, junctions with tight particle arrays, typical of the uncoupled state, display some lanthanum accumulation at the center of two closely joined particles (Fig. 32a) (Peracchia, 1973a); lanthanum probably reached the intercellular channels from the gap. Here, however, the tracer does not span the entire membrane thickness but stops abruptly at a depth of 4–6 nm from the center of the gap; this indicates that the channels are narrowed somewhere near their cytoplasmic mouth (Figs. 30, 31b, 32b). Together with changes in channel size, a weakening of the seal between channels and extracellular space may occur as suggested by the accessibility of the channels to colloidal lanthanum.

The loose particle packings of control (presumably coupled) junctions may

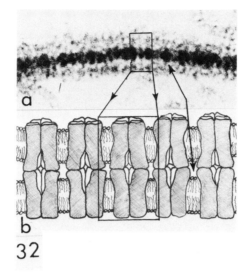

FIG. 32. Invertebrate gap junction (crayfish neurons). The extracellular gap of junctions with tight particle packings displays (a) (when cross sectioned after lanthanum) rows of electron-opaque beads. The beads correspond to the spaces between neighboring pairs of particles (double-headed arrow). Often moderately electron-opaque lines are visible between two beads (at center of rectangle). These lines are located at the center of pairs of particles and are believed to represent the intercellular channels partially filled with lanthanum. Since the lines do not span the entire membrane thickness, but stop abruptly at 4–6 nm from the center of the gap, the channels are believed to be narrowed somewhere closer to their cytoplasmic end (b). (a):×390,000. (b):×1,400,000. [(a), From Peracchia, 1973a.]

result from repulsions between negative charges of neighboring particles (Fig. 31a), while the clumping of the particles after uncoupling treatments (Fig. 30) could reflect the natural tendency of the particles to aggregate (and interact hydrophobically) after neutralization of the repulsive charges by divalent cations (Peracchia *et al.*, 1979) (Fig. 31b). A similar mechanism has been suggested to explain the clumping of the intramembrane particles of red blood cells by low pH (Elgsaeter and Branton, 1974).

Formation of more stable chemical bonds between the gap junction proteins in the clumped state seems unlikely in view of the variability of the unit cell dimensions of hexagonal arrays in different junctions and of the short range disorder always seen in crystalline junctions (see Section IV,C). Thus, it seems reasonable to believe that the geometry of the crystalline arrays simply reflects the configuration of individual particles: hexagonal arrays probably result from the tight packing of hexameres (Fig. 30), and orthogonal or rhombic arrays result (Peracchia and Peracchia, 1978) from the packing of tetrameres.

ACKNOWLEDGMENTS

I am indebted to my wife Lillian for her excellent technical assistance and for her comments and suggestions in preparing the manuscript and to Dr. Giovanni Bernardini for his most helpful criticism. This work was supported by a grant from the National Institutes of Health (GM-20113).

REFERENCES

Albertini, D. F., and Anderson, E. (1974). *J. Cell Biol.* **63,** 234.
Albertini, D. F., and Anderson, E. (1975). *Anat. Rec.* **181,** 171.
Albertini, D. F., Fawcett, D. W., and Olds, P. J. (1975). *Tissue Cell* **7,** 389.
Alcalá, J., Lieska, N., and Maisel, H. (1975). *Exp. Eye Res.* **21,** 581.
Anderson, E., and Albertini, D. F. (1976). *J. Cell Biol.* **71,** 680.
Armstrong, C. M. (1975). *Q. Rev. Biophys.* **7,** 179.
Asada, Y., and Bennett, M. V. L. (1971). *J. Cell Biol.* **49,** 159.
Asada, Y., Pappas, G. D., and Bennett, M. V. L. (1967). *Fed. Proc. Fed. Am. Soc. Exp. Biol.* **26,** 330.
Azarnia, R., and Larsen, W. J. (1977). *In* "Intercellular Communication" (W. C. DeMello, ed.), pp. 145-172. Plenum, New York.
Azarnia, R., and Loewenstein, W. R. (1977). *J. Membr. Biol.* **34,** 1.
Azarnia, R., Larsen, W. J., and Loewenstein, W. R. (1974). *Proc. Natl. Acad. Sci. U.S.A.* **71,** 880.
Baker, P. F., and Honerjäger, P. (1978). *Nature (London)* **273,** 160.
Baldwin, K. M. (1977). *J. Mol. Cell. Cardiol.* **9,** 959.
Baldwin, K. M. (1979). *J. Cell Biol.* **82,** 66.
Barr, L., Dewey, M. M., and Berger, W. (1965). *J. Gen. Physiol.* **48,** 797.
Barr, L., Berger, W., and Dewey, M. M. (1968). *J. Gen. Physiol.* **51,** 347.
Begenisich, T., and Stevens, C. F. (1975). *Biophys. J.* **15,** 843.
Benedetti, E. L., and Emmelot, P. (1965). *J. Cell Biol.* **26,** 299.
Benedetti, E. L., and Emmelot, P. (1968). *J. Cell Biol.* **38,** 15.
Benedetti, E. L., Dunia, I., and Bloemendal, H. (1974). *Proc. Natl. Acad. Sci. U.S.A.* **71,** 5074.
Benedetti, E. L., Dunia, I., Bentzel, C. J., Vermorken, A. J. M., Kibbelaar, M., and Bloemendal, H. (1976). *Biochim. Biophys. Acta* **457,** 353.
Bennett, M. V. L. (1960). *Fed. Proc. Fed. Am Soc. Exp. Biol.* **19,** 282.
Bennett, M. V. L. (1966). *Ann. N. Y. Acad. Sci.* **137,** 509.
Bennett, M. V. L. (1973a). *Fed. Proc. Fed. Am. Soc. Exp. Biol.* **32,** 65.
Bennett, M. V. L. (1973b). *In* "Intracellular Staining Techniques in Neurobiology" (S. D. Kater and C. Nicholson, eds.), pp. 115-133. Elsevier, Amsterdam.
Bennett, M. V. L. (1974). *In* "Synaptic Transmission and Neuronal Interaction" (M. V. L. Bennett, ed.), pp. 153-178. Raven, New York.
Bennett, M. V. L. (1977). *In* "Handbook of Physiology" (J. M. Brookhart and V. B. Montcastle, eds.), Vol. I, pp. 357-416. American Physiological Society, Bethesda, Maryland.
Bennett, M. V. L. (1978). *In* "Intercellular Junctions and Synapses" (J. Feldman, N. B. Gilula and J. D. Pitts, eds.), pp. 25-36. Chapman Hall, London.
Bennett, M. V. L., and Dunham, P. B. (1970). *Biophys. J.* **10,** 114a.
Bennett, M. V. L., and Goodenough, D. A. (1978). *Neurosci. Res. Prog. Bull.* **16,** 373.
Bennett, M. V. L., Aljure, E., Nakajima, Y., and Pappas, G. D. (1963). *Science* **141,** 262.
Bennett, M. V. L., Dunham, P. B., and Pappas, G. D. (1967). *J. Gen. Physiol.* **50,** 1094a.
Bennett, M. V. L., Spira, M. E., and Pappas, G. D. (1972). *Dev. Biol.* **29,** 419.

Bennett, M. V. L., Feder, N., Reese, T. S., and Stewart, W. (1973). *J. Gen. Physiol.* **61**, 254.

Bennett, M. V. L., Spira, M., and Spray, D. C. (1977). *J. Cell Biol.* **75**, 64a.

Bennett, M. V. L., Brown, J. E., Harris, A. L., and Spray, D. C. (1978). *Biol. Bull.* **155**, 428a.

Bernstein, J. (1902). *Pflügers Arch.* **92**, 521.

Bloemendal, H., Vermorken, A. J. M., Kibbelaar, M., Dunia, I., and Benedetti, E. L. (1977). *Exp. Eye Res.* **24**, 413.

Borek, C., Higashino, S., and Loewenstein, W. R. (1969). *J. Membr. Biol.* **1**, 274.

Branton, D. (1966). *Proc. Natl. Acad. Sci. U.S.A.* **55**, 1048.

Bretscher, M. S. (1971). *J. Mol. Biol.* **59**, 351.

Bretscher, M. S. (1972). *J. Mol. Biol.* **71**, 523.

Bretscher, M. S., and Raff, M. C. (1975). *Nature (London)* **258**, 43.

Brightman, M. W., and Reese, T. S. (1969). *J. Cell Biol.* **40**, 648.

Brink, P., and Barr, L. (1977). *J. Gen. Physiol.* **69**, 517.

Brink, P., and Dewey, M. (1978). *J. Gen. Physiol.* **72**, 67.

Broekhuyse, R. M., Kuhlman, E. D., and Stols, A. L. H. (1976). *Exp. Eye Res.* **23**, 365.

Brown, P. K. (1972). *Nature (London) New Biol.* **236**, 35.

Bullivant, S. (1969). *Micron* **1**, 46.

Bullivant, S. (1977). *J. Micros.* **111**, 101.

Bullivant, S., and Loewenstein, W. R. (1968). *J. Cell Biol.* **37**, 621.

Bullock, T. H. (1945). *J. Neurophysiol.* **8**, 55.

Bürk, R. R., Pitts, J. D., and Subak-Sharpe, H. (1968). *Exp. Cell Res.* **53**, 297.

Caspar, D. L. D., and Kirschner, D. A. (1971). *Nature (London) New Biol.* **231**, 46.

Caspar, D. L. D., and Goodenough, D. A., Makowski, L., and Phillips, W. C. (1977). *J. Cell Biol.* **74**, 605.

Chalcroft, J. P., and Bullivant, S. (1970). *J. Cell Biol.* **47**, 49.

Chapman, D. (1973). *In* "Biological Membranes" (D. Chapman and D. F. H. Wallach, eds.), Vol. II, pp. 91–144. Academic Press, New York.

Cherry, R. J., Bürkli, A., Busslinger, M., Schneider, G., and Parish, G. R. (1976). *Nature (London)* **263**, 389.

Coggeshall, R. E. (1965). *J. Comp. Neurol.* **125**, 393.

Collander, R. (1937). *Trans. Faraday Soc.* **33**, 1023.

Cone, R. A. (1972). *Nature (London) New Biol.* **236**, 39.

Connell, C. J., and Connell, G. M. (1977). *In* "The Testis" (A. D. Johnsen and W. R. Gomes, eds.), Vol. IV, pp. 333–369. Academic Press, New York.

Conti, F., DeFelice, L. J., and Wanke, E. (1975). *J. Physiol. (London)* **248**, 45.

Conti, R., Hille, B., Neumeke, B., Nonner, W., and Stämpfli, R. (1976). *J. Physiol.* **262**, 729.

Corsaro, C. M., and Migeon, B. R. (1977). *Nature (London)* **268**, 737.

Cox, R. P., Krauss, M. R., Balis, M. E., and Dancis, J. (1970). *Proc. Natl. Acad. Sci. U.S.A.* **67**, 1573.

Cox, R. P., Krauss, M. J., Balis, M. E., and Dancis, J. (1972). *Exp. Cell Res.* **74**, 251.

Culvenor, J. C., and Evans, W. H. (1977). *Biochem. J.* **168**, 475.

Daniel, E. E., Daniel, V. P., Duchon, G., Garfield, R. E., Nichols, M., Malho ra, S. K., and Oki, M. (1976). *J. Membr. Biol.* **28**, 207.

Danielli, J. F., and Davson, H. A. (1935). *J. Cell. Comp. Physiol.* **5**, 495.

David-Ferreira, J. F., and David-Ferreira, K. L. (1973). *J. Cell Biol.* **58**, 226.

Deamer, D. W., and Branton, D. (1967). *Science* **158**, 655.

Decker, R. S. (1976). *J. Cell Biol.* **69**, 669.

Decker, R. S., and Friend, D. S. (1974). *J. Cell Biol.* **62**, 32.

DeHaan, R. L., and Hirakow, R. (1972). *Exp. Cell Res.* **70**, 214.

DeHaan, R. L., and Sachs, H. G. (1972). *Curr. Top. Dev. Biol.* **7**, 193.

Délèze, J. (1965). *In* "Electrophysiology of the Heart" (B. Taccardi and G. Marchetti, eds.), pp. 147–148. Pergamon, Oxford.

Délèze, J. (1970). *J. Physiol.* **208,** 547.

Délèze, J., and Loewenstein, W. R. (1976). *J. Membr. Biol.* **28,** 71.

DeLorenzo, A. Y. (1959). *Biol. Bull.* **117,** 390.

DeMello, W. C. (1972). *Experientia* **28,** 832.

DeMello, W. C. (1975). *J. Physiol.* **250,** 231.

DeMello, W. C. (1976). *J. Physiol.* **263,** 171.

DeMello, W. C. (1977). *In* "Intercellular Communication" (W. C. DeMello, ed.), pp. 87–125. Plenum, New York.

DeMello, W. C. (1979). *Pflügers Arch.* **380,** 267.

DeMello, W. C., Motta, G., and Chapeau, M. (1969). *Circ. Res.* **24,** 475.

DePetris, S., and Raff, M. C. (1973). *Nature (London) New Biol.* **241,** 257.

Dewey, M. M., and Barr, L. (1962). *Science* **137,** 670.

Doggenweiler, C. F., and Frenk, S. (1965). *Proc. Natl. Acad. Sci. U.S.A.* **53,** 425.

Dreifuss, J. J., Girardier, L., and Forssman, W. G. (1966). *Pflügers Arch.* **292,** 13.

Duguid, J. R., and Revel, J. P. (1975). *Cold Spring Harbor Symp. Quant. Biol.* **40,** 45.

Dulhunty, A. F., and Franzini-Armstrong, C. (1975). *J. Physiol.* **250,** 513.

Dunia, I., Sen Ghosh, C., Benedetti, E. L., Zweers, A., and Bloemendal, H. (1974). *FEBS Lett.* **45,** 139.

Ehrhart, J. C., and Chauveau, J. (1977). *FEBS Lett.* **78,** 295.

Elgsaeter, A. and Branton, D. (1974). *J. Cell Biol.* **63,** 1018.

Elias, P. M., and Friend, D. S. (1976). *J. Cell Biol.* **68,** 173.

Engelmann, T. W. (1877). *Pflügers Arch.* **15,** 116.

Epstein, M. L., and Gilula, N. B. (1977). *J. Cell Biol.* **75,** 769.

Espey, L. L., and Stutts, R. H. (1972). *Biol. Reprod.* **6,** 168.

Evans, W. H., and Gurd, J. W. (1972). *Biochem. J.* **128,** 691.

Farquhar, M. G., and Palade, G. E. (1963). *J. Cell Biol.* **17,** 375.

Farquhar, M. G., and Palade, G. E. (1965). *J. Cell Biol.* **26,** 263.

Fisher, K., and Branton, D. (1974). *In* "Methods in Enzymology" (S. Fleischer and L. Packer, eds.), Vol. XXXII, pp. 35–44. Academic Press, New York.

Fisher, S. K., and Linberg, K. A. (1975). *J. Ultrastruct. Res.* **51,** 59.

Flower, N. E. (1971). *J. Ultrastruct. Res.* **37,** 259.

Flower, N. E. (1972). *J. Cell Sci.* **10,** 683.

Flower, N. E. (1977). *J. Cell Sci.* **25,** 163.

Friend, D. S., and Gilula, N. B. (1972). *J. Cell Biol.* **53,** 758.

Fry, G., Devine, C., and Burnstock, G. (1977). *J. Cell Biol.* **72,** 26.

Frye, L. D., and Edidin, M. (1970). *J. Cell Sci.* **7,** 319.

Furshpan, E. J., and Potter, D. D. (1959). *J. Physiol.* **145,** 289.

Furshpan, E. J., and Potter, D. D. (1968). *Curr. Top. Dev. Biol.* **3,** 95.

Furukawa, T., and Furshpan, E. J. (1963). *J. Neurophysiol.* **26,** 140.

Gabella, G. (1972a). *J. Neurocytol.* **1,** 341.

Gabella, G. (1972b). *Z. Zellforsch. Mikrosk. Anat.* **125,** 191.

Garant, P. R. (1972). *J. Ultrastruct. Res.* **40,** 333.

Garfield, R. E., Sims, S., and Daniel, E. E. (1977). *Science* **198,** 958.

Gilula, N. B. (1972). *J. Ultrastruct. Res.* **38,** 215a.

Gilula, N. B. (1974). *J. Cell Biol.* **63,** 111a.

Gilula, N. B. (1976). *In* "Cell Biology" (P. L. Altman and D. Dittmer Katz, eds.), pp. 138–141. Federation of American Societies for Experimental Biology, Bethesda, Maryland.

Gilula, N. B. (1978). *In* "Intercellular Junctions and Synapses" (J. Feldman, N. B. Gilula and J. D. Pitts, eds.), pp. 1–22. Chapman & Hall, London.

Gilula, N. B., and Epstein, M. L. (1976). *Soc. Exp. Biol. Symp.* **30**, 257.

Gilula, N. B., and Satir, P. (1971). *J. Cell Biol.* **51**, 869.

Gilula, N. B., Branton, D., and Satir, P. (1970). *Proc. Natl. Acad. Sci. U.S.A.* **67**, 213.

Gilula, N. B., Reeves, O. R., and Steinbach, A. (1972). *Nature (London)* **235**, 262.

Gilula, N. B., Epstein, M. L., and Beers, W. H. (1978). *J. Cell Biol.* **78**, 58.

Globus, A., Lux, H. D., and Schubert, P. (1973). *Exp. Neurol.* **40**, 104.

Goodenough, D. A. (1974). *J. Cell Biol.* **61**, 557.

Goodenough, D. A. (1976). *J. Cell Biol.* **68**, 220.

Goodenough, D. A., and Gilula, N. B. (1972). *In* "Membranes and Viruses in Immunopathology" (S. B. Day and R. A. Good, eds.), pp. 155–168. Academic Press, New York.

Goodenough, D. A., and Gilula, N. B. (1974). *J. Cell Biol.* **61**, 575.

Goodenough, D. A., and Revel, J. P. (1970). *J. Cell Biol.* **45**, 272.

Goodenough, D. A., and Revel, J. P. (1971). *J. Cell Biol.* **50**, 81.

Goodenough, D. A., and Stoeckenius, W. (1972). *J. Cell Biol.* **54**, 646.

Goodenough, D. A., Caspar, D. L. D., Phillips, W. C., and Makowski, L. (1978). *J. Cell Biol.* **79**, 223a.

Gorter, E., and Grendel, R. (1925). *J. Exp. Med.* **41**, 439.

Goshima, K. (1969). *Exp. Cell Res.* **58**, 420.

Goshima, K. (1970). *Exp. Cell Res.* **63**, 124.

Graf, F. (1978). *C. R. Acad. Sci. (Paris)* **287**, 41.

Griepp, E. B., and Revel, J. P. (1977). *In* "Intercellular Communication" (W. C. DeMello, ed.), pp. 1–32. Plenum, New York.

Gryns, G. (1896). *Pflugers Arch.* **63**, 86.

Hagiwara, S., and Morita, H. (1962). *J. Neurophysiol.* **25**, 721.

Hagiwara, S., Watanabe, A., and Saito, N. (1959). *J. Neurophysiol.* **22**, 554.

Hagopian, M. (1970). *J. Ultrastruct. Res.* **33**, 233.

Hama, K. (1959). *J. Biophys. Biochem. Cytol.* **6**, 61.

Hama, K. (1961). *Anat. Rec.* **141**, 275.

Hama, K., and Saito, K. (1977). *J. Neurocytol.* **6**, 1.

Hammer, M. G., and Sheridan, J. D. (1978). *J. Physiol.* **275**, 495.

Hand, A. R., and Gobel, S. (1972). *J. Cell. Biol.* **52**, 397.

Hanna, R. B., Keeter, J. S., and Pappas, G. D. (1978a). *J. Cell Biol.* **79**, 764.

Hanna, R. B., Spray, D. C., Model, P. G., Harris, A. L., and Bennett, M. V. L. (1978b). *Biol. Bull.* **155**, 442.

Harris, A. L., Spray, D. C., Bennett, M. V. L., and Hanna, R. B. (1978). *Soc. Neurosci.* **4**, 235a.

Hedin, S. G. (1897). *Pflügers Arch.* **68**, 229.

Henderson, D., Eibl, H., and Weber, K. (1979). *J. Mol. Biol.* *132*, 193.

Herman, A., Rieske, E., Kreutzberg, G. W., and Lux, H. D. (1975). *Brain Res.* **95**, 125.

Herr, J. C. (1976). *J. Cell Biol.* **69**, 495.

Herr, J. C., and Heidger, P. M. (1978). *Am. J. Anat.* **152**, 29.

Hertzberg, E. L., and Gilula, N. B. (1979). *J. Biol. Chem.* **254**, 2138.

Heuser, J. E., Reese, T. S., and Landis, D. M. D. (1975). *Cold Spring Harbor Symp. Quant. Biol.* **40**, 17.

Höber, R. (1936). *J. Cell Comp. Physiol.* **7**, 367.

Horn, R., Brodwick, M. S., and Eaton, D. C. (1978). *Biophys. J.* **21**, 42a.

Horwitz, A. F. (1972). *In* "Membrane Molecular Biology" (C. F. Fox and A. D. Keith, eds.), pp. 164–191. Sinauer, Stamford, Connecticut.

Hubbell, W. L., and McConnell, H. M. (1971). *J. Am. Chem. Soc.* **93**, 314.
Hudspeth, A. J., and Revel, J. P. (1971). *J. Cell Biol.* **50**, 92.
Hyde, A., Blondel, B., Matter, A., Cheneval, J. P., Filloux, B., and Girardier, L. (1969). *Prog. Brain Res.* **31**, 283.
Imanaga, I. (1974). *J. Membr. Biol.* **16**, 381.
Ito, S., Sato, E., and Loewenstein, W. R. (1974). *J. Membr. Biol.* **19**, 339.
Iwatsuki, N., and Petersen, O. H. (1978a). *J. Physiol.* **274**, 81.
Iwatsuki, N., and Petersen, O. H. (1978b). *J. Physiol.* **275**, 507.
Iwatsuki, N., and Petersen, O. H. (1978c). *J. Cell Biol.* **79**, 533.
Iwatsuki, N., and Petersen, O. H. (1979). *J. Physiol.* **291**, 317.
Iwayama, T. (1971). *J. Cell Biol.* **49**, 521.
Jacobs, M. H. (1924). *Am. J. Physiol.* **68**, 134.
Jacobs, M. H. (1935). *Ergeb. Biol.* **12**, 1.
Johnson, R. G., and Sheridan, J. D. (1971). *Science* **174**, 717.
Johnson, R. G., Herman, W. S., and Preus, D. M. (1973). *J. Ultrastruct. Res.* **43**, 298.
Johnson, R. G., Hammer, M., Sheridan, J., and Revel, J. P. (1974). *Proc. Natl. Acad. Sci. U.S.A.* **71**, 4536.
Jongsma, H. J., and VanRijn, H. E. (1972). *J. Membr. Biol.* **9**, 341.
Kaneko, A. (1971). *J. Physiol.* **213**, 95.
Kanno, Y., and Loewenstein, W. R. (1966). *Nature (London)* **212**, 629.
Karrer, H. E. (1960a). *J. Biophys. Biochem. Cytol.* **7**, 181.
Karrer, H. E. (1960b). *J. Biophpys. Biochem. Cytol.* **8**, 135.
Katz, B., and Miledi, R. (1972). *J. Physiol.* **224**, 665.
Kawamura, K., and Konishi, T. (1967). *Jpn Circ. J.* **31**, 1533.
Keeter, J. S., Dechênes, M., Pappas, G. D., and Bennett, M. V. L. (1974). *Biol. Bull.* **147**, 485.
Kensler, R. W., Brink, P., and Dewey, M. M. (1977). *J. Cell Biol.* **73**, 768.
Kogon, M., and Pappas, G. D. (1975). *J. Cell Biol.* **66**, 671.
Kolodny, G. M. (1971). *Exp. Cell Res.* **65**, 313.
Kreutziger, G. O. (1968). *Proc. Electron Microsc. Soc. Am.* **26**, 234.
Kriebel, M. E. (1968). *J. Gen. Physiol.* **54**, 46.
Kuffler, S. W., and Potter, D. D. (1964). *J. Neurophysiol.* **27**, 290.
Kuriyama, H., and Suzuki, H. (1976). *J. Physiol.* **260**, 315.
Landis, D. M. D., Reese, T. S., and Raviola, E. (1974). *J. Comp. Neurol.* **155**, 67.
Lane, N. J. (1978). *In* "Electron Microscopy: State of the Art. Symposia" (J. M. Sturges, ed.), Vol. 3, pp. 673–691. Imperial Press, Cooksville, Ontario, Canada.
Lane, N. J., and Swales, L. S. (1978a). *Dev. Biol.* **62**, 389.
Lane, N. J., and Swales, L. S. (1978b). *Dev. Biol.* **62**, 415.
Larsen, W. J. (1975). *J. Cell Biol.* **67**, 801.
Larsen, W. J. (1977). *Tissue Cell* **9**, 373.
Larsen, W. J., Heidger, P. M., Jr., and Herr, J. C. (1976). *J. Cell Biol.* **71**, 333.
Lawn, A. M., Wilson, E. W., and Finn, C. A. (1971). *J. Reprod. Fertil.* **26**, 85.
Lawrence, T. S., Beers, W. H., and Gilula, N. B. (1978). *Nature (London)* **272**, 501.
Ledbetter, M. L., and Lubin, M. (1979). *J. Cell Biol.* **80**, 150.
Letourneau, R. J., Li, J. J., Rosen, S., and Villee, C. A. (1975). *Cancer Res.* **35**, 6.
Loewenstein, W. R. (1966). *Ann. N. Y. Acad. Sci.* **137**, 441.
Loewenstein, W. R. (1967). *J. Colloid Sci.* **25**, 34.
Loewenstein, W. R. (1975). *Cold Spring Harbor Symp. Quant. Biol.* **40**, 49.
Loewenstein, W. R., and Kanno, Y. (1963). *J. Gen. Physiol.* **46**, 1123.
Loewenstein, W. R., and Kanno, Y. (1964). *J. Cell Biol.* **22**, 565.
Loewenstein, W. R., and Penn, R. D. (1967). *J. Cell Biol.* **33**, 235.

Loewenstein, W. R., Nakas, M., and Socolar, S. J. (1967). *J. Gen. Physiol.* **50,** 1865.

Loewenstein, W. R., Kanno, Y., and Socolar, S. J. (1978). *Nature (London)* **274,** 133.

Luzzati, V., and Husson, F. (1962). *J. Cell Biol.* **12,** 207.

Maisel, H., Alcalá, J., and Lieska, N. (1976). *Doc. Opthamol. Proc. Prog. Lens Res.* **8,** 121.

Makowski, L., Caspar, D. L. D., Phillips, W. C., and Goodenough, D. A. (1977). *J. Cell Biol.* **74,** 629.

Marchesi, V. T., Tillack, T. W., Jackson, R. L., Segrest, J. P., and Scott, R. E. (1972). *Proc. Natl. Acad. Sci. U.S.A.* **69,** 1445.

Marinetti, G. V., Sheeley, S., Baumgarten, R., and Love, R. (1974). *Biochem. Biophys. Res. Commun.* **59,** 502.

Martin, A. R., and Pilar, G. (1963a). *J. Physiol.* **168,** 443.

Martin, A. R., and Pilar, G. (1963b). *J. Physiol.* **168,** 464.

Mazet, R. (1977). *Dev. Biol.* **60,** 139.

Mazet, R., and Cartaud, J. (1976). *J. Cell Sci.* **22,** 427.

McConnell, H. M. (1974). *Adv. Exp. Med. Biol.* **51,** 103.

McNutt, N. S. (1977). *In* "Dynamic Aspects of Cell Surface Organization" (G. Poste and G. L. Nicolsen, eds.), pp. 75–126. Elsevier, Amsterdam.

McNutt, N. S., and Weinstein, R. S. (1969). *Proc. Electron Microsc. Soc. Am.* **27,** 330.

McNutt, N. S., and Weinstein, R. S. (1970). *J. Cell Biol.* **47,** 666.

McNutt, N. S., and Weinstein, R. S. (1973). *Prog. Biophys. Mol. Biol.* **26,** 45.

Merk, F. B., Botticelli, C. R., and Albright, J. J. (1972). *Endocrinology* **90,** 992.

Merk, F. B., Albright, J. T., and Botticelli, C. R. (1973). *Anat. Rec.* **175,** 107.

Michalke, W., and Loewenstein, W. R. (1971). *Nature (London)* **232,** 121.

Miller, R. F. (1978). *Brain Res.* **139,** 178.

Moor, H., and Mühlethaler, K. (1963). *J. Cell Biol.* **17,** 609.

Moor, H., Mühlethaler, K., Waldner, H., and Frey-Wyssling, A. (1961). *J. Biophys. Biochem. Cytol.* **10,** 1.

Nadol, J. B., Mulroy, M. J., Goodenough, D. A., and Weiss, T. F. (1976). *Am. J. Anat.* **147,** 281.

Nakas, M., Higashino, S., and Loewenstein, W. R. (1966). *Science* **151,** 89.

Nakata, K., and Page, E. (1978). *J. Cell Biol.* **79,** 330a.

Neher, E., and Stevens, C. F. (1977). *Annu. Rev. Biophys. Bioeng.* **6,** 345.

Nishiye, H. (1977). *Jpn. J. Physiol.* **27,** 451.

Noirot, C., and Noirot-Timothée, C. (1976). *Tissue Cell* **8,** 345.

Nuñez, E. A. (1971). *Am. J. Anat.* **131,** 227.

Oliveira-Castro, G. M., and Barcinski, M. A. (1974). *Biochim. Biophys. Acta* **352,** 338.

Oliveira-Castro, G. M., and Dos Reis, G. A. (1977). *In* "Intercellular Communication" (W. C. DeMello, ed.), pp. 201–230. Plenum, New York.

Oliveira-Castro, G. M., and Loewenstein, W. R. (1971). *J. Membr. Biol.* **5,** 51.

Overton, E. (1895). *Vjschr. Naturf. Ges. Zurich.* **40,** 159.

Overton, J. (1968). *J. Exp. Zool.* **168,** 203.

Pappas, G. D., and Bennett, M. V. L. (1966). *Ann. N. Y. Acad. Sci.* **137,** 495.

Pappas, G. D., Asada, Y., and Bennett, M. V. L. (1971). *J. Cell Biol.* **49,** 173.

Pauling, L. C. (1960). "Nature of the Chemical Bond." Cornell Univ. Press, Ithaca, New York.

Payton, B. W., Bennett, M. V. L., and Pappas, G. D. (1969a). *Science* **165,** 594.

Payton, B. W., Bennett, M. V. L., and Pappas, G. D. (1969b). *Science* **166,** 1641.

Penn, R. D., and Loewenstein, W. R. (1966). *Science* **151,** 88.

Peracchia, C. (1971). *Proc. Am. Soc. Cell Biol.* **11,** 221.

Peracchia, C. (1972). *J. Cell Biol.* **55,** 202a.

Peracchia, C. (1973a). *J. Cell Biol.* **57,** 54.

Peracchia, C. (1973b). *J. Cell Biol.* **57,** 66.

Peracchia, C. (1974). *Proc. Int. Cong. Electron Microsc., 8th* Canberra, **II**, 226.
Peracchia, C. (1977a). *J. Cell Biol.* **72**, 628.
Peracchia, C. (1977b). *Trends Biochem. Sci.* **2**, 26.
Peracchia, C. (1978). *Nature (London)* **271**, 669.
Peracchia, C., and Dulhunty, A. F. (1974). *J. Cell Biol.* **63**, 263a.
Peracchia, C., and Dulhunty, A. F. (1976). *J. Cell Biol.* **70**, 419.
Peracchia, C., and Mittler, B. S. (1972). *J. Cell Biol.* **53**, 234.
Peracchia, C., and Peracchia, L. L. (1978). *J. Cell Biol.* **79**, 217a.
Peracchia, C., Bernardini, G., and Peracchia, L. L. (1979). *J. Cell Biol.* **83**, 86a.
Pinto da Silva, P., and Gilula, N. B. (1972). *Exp. Cell Res.* **71**, 393.
Pitts, J. D. (1971). *In* "Growth Control in Cell Cultures" (G. E. W. Wolstenholme and J. Knight, eds.), pp. 89–105. Churchill Livingston, London.
Pitts, J. D. (1976). *In* "Developmental Biology of Plants and Animals" (C. F. Graham and P. F. Wareing, eds.), pp. 96–110. Blackwell, Oxford.
Pitts, J. D. (1977). *In* "International Cell Biology" (B. R. Brinkley and K. R. Porter, eds.), pp. 43–49. Rockfeller Univ. Press, New York.
Pitts, J. D., and Finbow, M. E. (1977). *In* "Intercellular Communication" (W. C. DeMello, ed.), pp. 61–86. Plenum, New York.
Pitts, J. D., and Simms, J. W. (1977). *Exp. Cell Res.* **104**, 153.
Politoff, A., and Pappas, G. D. (1972). *Anat. Rec.* **172**, 384.
Politoff, A. L., Socolar, S. J., and Loewenstein, W. R. (1969), *J. Gen. Physiol.* **53**, 498.
Politoff, A., Pappas, G. D., and Bennett, M. V. L. (1974). *Brain Res.* **76**, 343.
Pollack, G. H. (1976). *J. Physiol.* **255**, 275.
Poo, M. M., and Cone, R. A. (1974). *Nature (London)* **247**, 438.
Potter, D. D., Furshpan, E. J., and Lennox, E. S. (1966). *Proc. Natl. Acad. Sci. U.S.A.* **55**, 328.
Pricam, C., Humbert, F., Perrelet, A., and Orci, L. (1974). *J. Cell Biol.* **63**, 349.
Quick, D. C., and Johnson, R. G. (1977). *J. Ultrastruct. Res.* **60**, 348.
Rash, J. E., and Fambrough, D. (1973). *Dev. Biol.* **30**, 166.
Rash, J. E., and Staehelin, L. A. (1974). *Dev. Biol.* **36**, 455.
Raviola, E., and Gilula, N. B. (1973). *Proc. Natl. Acad. Sci. U.S.A.* **70**, 1677.
Raviola, E., Goodenough, D. A., and Raviola, G. (1978). *J. Cell Biol.* **79**, 229a.
Revel, J. P. (1968). *Proc. Electron Microsc. Soc. Am.* **26**, 40.
Revel, J. P., and Karnovsky, M. J. (1967). *J. Cell Biol.* **33**, C7.
Revel, J. P., Henning, V., and Fox, C. F., eds. (1976). "Cell Shape and Surface Architecture: Progress in Clinical and Biological Research," Vol. 17. Liss, New York.
Rieske, E., Schubert, P., and Kreutzberg, G. W. (1975). *Brain Res.* **84**, 365.
Robertson, J. D. (1953). *Proc. Soc. Exp. Biol.* **82**, 219.
Robertson, J. D. (1955). *Exp. Cell Res.* **8**, 226.
Robertson, J. D. (1957). *J. Biophys. Biochem. Cytol.* **3**, 1043.
Robertson, J. D. (1960). *Prog. Biophys. Biophys. Chem.* **10**, 344.
Robertson, J. D. (1961). *Ann. N.Y. Acad. Sci.* **94**, 339.
Robertson, J. D. (1963). *J. Cell Biol.* **19**, 201.
Robertson, J. D. (1966). *In* "Principles of Biomolecular Organization" (G. E. W. Wolstenholme and M. O'Connor, eds.), pp. 357–408. Churchill, London.
Robertson, J. D., Bodenheimer, T. S., and Stage, D. E. (1963). *J. Cell Biol.* **19**, 159.
Rose, B. (1971). *J. Membr. Biol.* **5**, 1.
Rose, B. and Loewenstein, W. R. (1971). *J. Membr. Biol.* **5**, 20.
Rose, B., and Loewenstein, W. R. (1974). *Fed. Proc. Fed. Am. Soc. Exp. Biol.* **33**, 1340.
Rose, B., and Loewenstein, W. R. (1975). *Nature (London)* **254**, 250.
Rose, B., and Loewenstein, W. R. (1976). *J. Membr. Biol.* **28**, 87.

Rose, B., and Rick, R. (1978). *J. Membr. Biol.* **44**, 377.

Rose, B., Simpson, I., and Loewenstein, W. R. (1977). *Nature (London)* **267**, 625.

Rosenbluth, J. (1978). *J. Neurocytol.* **7**, 709.

Rothschuh, K. E. (1951). *Pflugers Arch.* **253**, 238.

Sandri, C., Van Buren, J. M., and Akert, K. (1977). *Prog. Brain Res.* **46**, 1.

Sanford, K. K., Earle, W. R., and Likely, G. D. (1948). *J. Natl. Cancer Inst.* **9**, 229.

Schleiden, M. J., and Schwann, T. (1838). "Beiträge Zur Phytogenesis" translated with Schwann's "Mikroscopische Untersuchungen" by H. Smith for the Sydenham Society, London (1847).

Schmidt, W. J. (1936). *Z. Zellforsch. Mikrosk, Anat.* **23**, 657.

Schmitt, F. O., Bear, R. S., and Clark, G. L. (1935). *Radiology* **25**, 131.

Schnapp, B., and Mugnaini, E. (1978). *In* "Physiology and Pathobiology of Axons" (S. G. Waxman, ed.), pp. 83–122. Raven, New York.

Segrest, J. P., Jackson, R. L., Marchesi, V. T., Guyer, R. B., and Terry, W. (1972). *Biochem. Biophys. Res. Commun.* **49**, 964.

Sheridan, J. D. (1970). *J. Cell Biol.* **41**, 91.

Sheridan, J. D. (1971). *Dev. Biol.* **26**, 627.

Sheridan, J. D., Hammer-Wilson, M., Preus, D., and Johnson, R. G. (1978). *J. Cell Biol.* **76**, 532.

Shimomura, O., Johnson, F. H., and Saiga, Y. (1962). *J. Cell Comp. Physiol.* **59**, 223.

Shrager, P. G., Strickholm, A., and Macey, R. I. (1969). *J. Cell Physiol.* **74**, 91.

Siegenbeck Van Henkelom, J., Van Der Gon, J. J. D., and Prop, F. J. A. (1972). *J. Membr. Biol.* **7**, 88.

Simionescu, M., Simionescu, N., and Palade, G. E. (1975). *J. Cell Biol.* **67**, 863.

Simms, J. W. (1973). Ph.D. Thesis, University of Glasgow, Scotland.

Simpson, I., Rose, B., and Loewenstein, W. R. (1977). *Science* **195**, 294.

Singer, S. J. (1971). *In* "The Structure and Function of Biological Membranes" (L. I. Rothfield, ed.), pp. 145–122. Academic Press, New York.

Singer, S. J., and Nicolson, G. L. (1972). *Science* **175**, 720.

Sjöstrand, F. S., Andersson-Cedergren, E., and Dewey, M. M. (1958). *J. Ultrastruct. Res.* **1**, 271.

Slack, C., and Palmer, J. F. (1969). *Exp. Cell Res.* **55**, 416.

Socolar, S. J., and Politoff, A. L. (1971). *Science* **172**, 492.

Sommer, J. R., and Steere, R. L. (1969). *J. Cell Biol.* **43**, 136a.

Spira, A. W. (1971). *J. Ultrastruct. Res.* **34**, 409.

Spiro, D., and Sonnenblick, E. H. (1964). *Circ. Res.* **15**, 14.

Spray, D. C., Harris, A. L., Bennett, M. V. L., and Model, P. G. (1978). *Soc. Neurosci.* **4**, 238a.

Spray, D. C., Harris, A. L., and Bennett, M. V. L. (1979). *Science* **204**, 432.

Spycher, M. A. (1970). *Z. Zellforsch. Mikr. Anat.* **111**, 64.

Staehelin, L. A. (1972). *Proc. Natl. Acad. Sci. U.S.A.* **69**, 1318.

Staehelin, L. A. (1973). *J. Cell Sci.* **13**, 763.

Staehelin, L. A. (1974). *Int. Rev. Cytol.* **39**, 191.

Staehelin, L. A. (1978). *Sci. Am.* **238**, 140.

Steck, T. L. (1974). *J. Cell Biol.* **62**, 1.

Steck, T. L., and Dawson, G. (1974). *J. Biol. Chem.* **249**, 2135.

Steck, T. L., Fairbanks, G., and Wallack, D. F. H. (1971). *Biochemistry* **10**, 2617.

Steck, T. L., Ramos, B., and Strapazon, E. (1976). *Biochemistry* **15**, 1154.

Steere, R. L. (1957). *J. Biophys. Biochem. Cytol.* **3**, 45.

Steere, R. L., and Sommer, J. R. (1972). *J. Microsc.* **15**, 205.

Stewart, W. W. (1978). *Cell* **14**, 741.

Stretton, A. O. W., and Kravitz, E. A. (1968). *Science* **162**, 132.

Subak-Sharpe, H., Bürk, R. R., and Pitts, J. D. (1966). *Heredity* **21**, 342.

Subak-Sharpe, H., Bürk, R. R., and Pitts, J. D. (1969). *J. Cell Sci.* **4**, 353.

Suzuki, K., and Higashino, S. (1977). *Exp. Cell Res.* **109**, 263.
Tauc, L. (1959). *C. R. Acad. Sci. (Paris)* **248**, 1857.
Trautwein, W., Kuffler, S. W., and Edwards, C. (1956). *J. Gen. Physiol.* **40**, 135.
Tsien, R. W., and Weingart, R. (1974). *J. Physiol.* **242**, 95P.
Tupper, J. T., and Sanders, J. W., Jr. (1972). *Dev. Biol.* **27**, 546.
Turin, L., and Warner, A. (1977). *Nature (London)* **270**, 56.
Van der Kloot, W. C., and Dane, B. (1964). *Science* **146**, 74.
Verkleij, A. J., and Ververgaert, P. H. J. Th. (1978). *Biochim. Biophys. Acta* **515**, 303.
Verkleij, A. J., Zwaal, R. F. A., Roelofsen, B., Comfurius, P., Kasteleijn, D., and van Deenen, L. L. M. (1973). *Biochim. Biophys. Acta* **323**, 178.
Wallack, D. F. H., and Winzler, R. J. (1974). "Evolving Strategies and Tactics in Membrane Research." Springer-Verlag, Berlin and New York.
Watanabe, A., and Grundfest, H. (1961). *J. Gen. Physiol.* **45**, 267.
Weidmann, S. (1952). *J. Physiol.* **118**, 348.
Weidmann, S. (1966). *J. Physiol.* **187**, 323.
Weingart, R. (1974). *J. Physiol.* **240**, 741.
Weingart, R. (1977). *J. Physiol.* **264**, 341.
Weingart, R., and Reber, W. (1979). *Experientia* **35**, 17.
Weinstein, R. S. (1976). *Adv. Cancer Res.* **23**, 23.
Wiener, J., Spiro, D., and Loewenstein, W. R. (1964). *J. Cell Biol.* **22**, 587.
Wiersma, C. A. G. (1947). *J. Neurophysiol.* **10**, 23.
Wilkins, M. H. F., Blaurock, A. E., and Engelman, D. M. (1971). *Nature (London) New Biol.* **230**, 72.
Wood, R. L. (1977). *J. Ultrastruct. Res.* **58**, 299.
Woodbury, J. W., and Crill, W. E. (1961). *In* "Nervous Inhibition" (E. Florey, ed.), pp. 124–135. Pergamon, Oxford.
Zamboni, L. (1974). *Biol. Reprod.* **10**, 125.
Zampighi, G. (1978). *Tissue Cell* **10**, 413.
Zampighi, G., and Robertson, J. D. (1977). *Biophys. J.* **17**, 31a.

INTERNATIONAL REVIEW OF CYTOLOGY, VOL. 66

The Kinetics and Metabolism of the Cells of Hibernating Animals during Hibernation

S. G. Kolaeva, L. I. Kramarova, E. N. Ilyasova, and F. E. Ilyasov

Laboratory of the Biophysics of Living Structures, Institute of Biological Physics, Academy of Sciences of the USSR, Biological Center of the Academy of Sciences of the USSR, Pushchino, Moscow Region, USSR

I. Introduction

The potential value of hibernating animals as objects for the study of energy metabolism, enzymatic kinetics, biophysical changes accompanying the physiological processes, and the seasonal rhythms of the physiological functions has been widely discussed in several excellent reviews over the past 10 years (Hoffman, 1968; Willis, 1967; Willis *et al.*, 1972). Apparently the difference in the degree and selectivity of the changes in many systems at the tissue and organism level in the hibernating animals during their life phases make these animals unique objects for biological studies.

Almost all the physiological changes occurring during hibernation demand a study at a cellular level. In this connection, it is interesting to consider the data on the kinetic, structural, and metabolic peculiarities of the cells in these animals during hibernation.

At present, limited but sufficiently informative published data permit us to form hypotheses about hibernation mechanism. A knowledge of these mechanisms will enable us to extend our conceptions of mitotic and intracellular homeostatis during hibernation, and the latter hypotheses can play a significant role in understanding the mechanism underlying the adaptation of cells of hibernating animals for survival in unfavorable conditions of the *milieu ambiant*.

II. Kinetics of Cells of Hibernating Animals during Hibernation

A. PROLIFERATIVE ACTIVITY OF CELLS OF DIFFERENT TISSUES IN HIBERNATING ANIMALS DURING HIBERNATION

The mitotic activity of cells in different tissues is markedly decreased during hibernation (Mayer and Bernick, 1958; Adelstein *et al.,* 1966; Suomalainen and Oja, 1967; Unker and Alekseeva, 1974). During hibernation, mitoses are either absent or encountered in atypical forms even in tissues with high proliferative activity, e.g., the epithelium of the small intestine (Table I). In hibernators, the erythropoietic response to massive hemorrhage is considerably decreased (Lyman *et al.,* 1957); seminal vesicles do not react to injection of testosterone (Lyman and Dempsey, 1951); and lymphopoiesis is reduced in the spleen and thymus (Brace, 1952). Suomalainen and Oja (1967) made a detailed study of the mitotic activity during all the periods of hibernation, including short periodic awakenings. All the investigations were carried out on several groups of hedgehogs (Table II). It appears that the number of mitoses sharply falls just when the animal enters hibernation. This decline accompanies a decrease in body temperature. Mitoses in the prophase are the first to disappear; this indicates a cessation of entrance of new cells into the mitotic cycle. Until the fifth to tenth day of hibernation, mitoses are almost entirely absent and only pyknotic metaphases are encountered. During short-term awakenings, the first mitoses appear

TABLE I

EXAMPLES OF LOW MITOTIC ACTIVITY IN SMALL INTESTINE EPITHELIUM DURING HIBERNATION IN VARIOUS SPECIES OF HIBERNATIVE MAMMALS

Species	Active period	Hibernation	Reference
Erinaceus	100	0	Suomalainen and Oja (1967)
Glis glis	50	17	Adelstein *et al.* (1966)
C. tridecemlineatus	130	10	Jaroslow *et al.* (1976)
C. suslicus	110	0	Kolaeva *et al.* (1978)

TABLE II

THE DIFFERENTIAL COUNT OF PHASES OF MITOSES[a]

	Prophase (%)	Metaphase (%)	Anaphase (%)	Telophase (%)
Group I	13	57	7	23
Awake (37°C)	8	60	6	10
	24	51	18	7
	36	37	18	9
	11	42	9	38
	10	51	10	29
	14	61	2	8
Group 2	3	64	0	23
Entering	1	62	0	24
hypothermia (10°C)	6	68	0	4
	4	74	4	16
	5	43	10	32
	3	67	0	26
Group 3	1	41	0	25
In deep	2	28	1	38
hypothermia (6°C)	0	24	0	20
	0	15	0	20
	2	22	2	9
	4	44	2	6
	4	8	0	16
Group 4	0	0	0	0
In deep hypothermia	0	0	0	0
(4–5 days of	0	0	0	0
hibernation)	0	0	0	1
	0	0	0	3
	0	0	0	7
	0	0	0	9
Group 5	0	0	0	0
In deep hypothermia	0	0	0	0
(8–10 days of	0	1	0	3
hibernation)	0	2	0	4
	0	0	0	1
	0	0	0	1
	0	0	0	0
Group 6	0	1	0	6
Arousing (15°C)	0	0	0	0
	0	0	0	0
	0	0	0	0
	0	0	0	0
	0	0	0	0

[a] From Suomalainen and Oja (1967).

when the body temperature is 26°C. However, the total number of mitoses does not attain the value typical of the active animals. It has been found that in ground squirrels (*Citellus tridecemlineautus*) aroused from hibernation the mitotic index reaches a level typical of the active state in only 18–24 hours. This suggests that during the entire period of hibernation the cells of the tissues studied are accumulated in the G_1 phase (Jaroslow *et al.*, 1976) of the cell cycle.

B. PECULIARITIES OF [³H]THYMIDINE INCORPORATION INTO THE DNA OF CELLS OF HIBERNATING ANIMALS DURING HIBERNATION

The kinetics of cell proliferation during hibernation has not yet been studied since [³H]thymidine is not incorporated in many species of the hibernating animals (the golden hamster, ground squirrels, and the dormouse) (Tables III, IV, and V). Therefore, all studies of kinetics of cell proliferation have been carried out on arousing animals. The level of the label incorporation on arousal reaches the value typical of the cells of active animals only after 24 hours; this observation is also in agreement with the conception of accumulation of cells in the early G_1 phase (Manasek *et al.*, 1965).

Accumulation of cells in the G_1 phase is not a result of action of low temperatures. The autoradiographic studies of Manasek *et al.* (1965), which utilized [³H]thymidine incorporation at 37°C into the cells of lymphoid tissue derived from hibernating and active golden hamsters, showed marked differences even 6 hours after initiation of incorporation (Fig. 1). Reduction of DNA synthesis during hibernation does not depend on a decrease in the pool of precursors. This was also shown by Manasek *et al* (1965).

There is also indirect data showing that, in animals such as dormouse *Glis glis* (Adelstein *et al.*, 1966) and *Citellus erythrogenus* (Kolaeva and Lutsenko, 1972), the cells can also accumulate during the period of G_2 phase. Thus, autoradiographic studies of [³H]thymidine incorporation into DNA of the cells of the lymphoid tissue of the dormouse (Adelstein *et al.*, 1966) have shown that the unlabeled mitoses first appear as the temperature reaches 37°C, while the labeled

TABLE III

SPECIFIC ACTIVITY OF DNA OF THE SMALL INTESTINE OF ACTIVE AND HIBERNATING ANIMALS 4 HOURS AFTER THE INJECTION OF [³H]THYMIDINE[a]

State	Number of animals	Specific activity (cpm/gm DNA)	Coefficient of variance (%)
Active	8	572	64
Hibernating	5	24	53

[a] From Fisher (1967).

TABLE IV

COMPARISON OF BIOCHEMICAL AND AUTORADIOGRAPHIC RESULTS ON THE INCORPORATION OF THYMIDINE BY CELL SUSPENSIONS FROM HIBERNATING AND ACTIVE HAMSTERS[a]

Tissue	Experiment number	Specific activity (cpm/μg DNA)			Labeled lymphocytes (%)[b]		
		Hibernating	Active	Active/hibernating	Hibernating	Active	Active/hibernating
Thymus	I	933	1967	2.1	3.7	9.7	2.4
	II	544	1240	2.3	4.0	6.3	1.6
Spleen	I	138	1663	12.1	1.2	9.1	7.6
	II	155	1683	10.8	0.7	9.0	12.9

[a] Data from Manasek et al. (1965).
[b] Cells were incubated at 37°C for 2 hours with [³H]thymidine in Hanks solution containing 50% horse serum.

TABLE V

Specific Activity of DNA of the Small Intestine of
Ground Squirrel (*Citellus suslicus*) and Rat[a,b]

Species	Specific activity (cpm/gm DNA)	
	Active	Hibernating
Ground squirrel	1.21	1.56
Rat	88.5	—

[a] From Kolaeva *et al.* (1977a).
[b] Specific activity was determined 2 hours after injection of [³H]thymi-
dine. Concentration of injected thymidine was 2.5 μCi/gm tissue weight
and specific activity was 13 μCi/mmole.

mitoses appear 2.5 hours later (Fig. 2). This enabled Adelstein *et al.* (1966) to propose the model diagrammed in Fig. 3.

Like the intermitotic block, blocking of the cycle in the G_2 phase is considered by the authors to be due not to the action of low temperatures, but to the hibernation itself.

C. Peculiarities of [³H]Thymidine Incorporation into the DNA of Cells of Hibernating Animals in the Active Period of Their Life

In 1964, Adelstein *et al.* (1964) studied the possibility of using [³H]thymidine as a precursor of DNA synthesis by cells of different species of rodents (both hibernating and nonhibernating) during the active period of their life.

The following species of rodents were used: mouse (*Mus musculus*), golden hamster (*Mesocricetus auratus*), dormouse (*Glis glis, Eliomys quercinus*), gray and red squirrels (*Sciurus carolinensis pennsylvanicus, Tamiasciurus hudsonicus loquax*), chipmunk (*Tamias striatus lysteri*), woodchuck (*Marmota monax pre-blorum*), and two species of ground squirrels (*Citellus tridecemlineatus, Citellus lateralis*). [³H]Thymidine incorporation into DNA was studied in cells of the small intestine, the tongue, and the spleen.

Considerable variability was demonstrated in the ability of several rodent species to utilize thymidine as a precursor for DNA during the active period of their life. It is noteworthy that the smallest incorporation of thymidine occurred in woodchucks and ground squirrels, which are representatives of the squirrel family (*Sciuromorpha*) (Table III), whereas in chipmunks, [³H]thymidine is successfully used as precursor of DNA biosynthesis. The data of Kolaeva *et al.* (1977) obtained on other representatives of the squirrel family [on hibernating

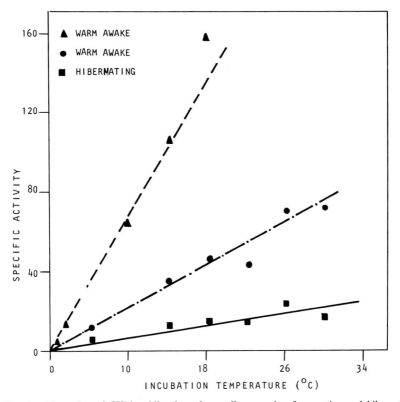

FIG. 1.   Absorption of [³H]thymidine by spleen cell suspension from active and hibernating hamsters. The cells were incubated with [³H]thymidine at 37°C. (From Manasedk *et al.*, 1965.)

and active ground squirrels (*Citellus suslicus*)], showed similar results (Table V). Analysis of the data enabled Adelstein *et al.* (1964) to conclude that there is no relationship whatsoever between the ability to utilize [³H]thymidine as a precursor and the ability to hibernate, since a great number of the representative hibernators incorporate [³H]thymidine (Fig. 4).

Individual species of the squirrel family, that have the ability to utilize thymidine as a precursor of DNA synthesis, may also have some common morphological features of different organs and systems (Adelstein *et al.*, 1964).

In an attempt to account for the data, Adelstein *et al.* assumed that nonutilization of [³H]thymidine as a precursor of DNA biosynthesis in certain representatives of the squirrel family during the active period of their life may be due to: (1) poor cellular permeability; (2) the increase of the precursor pool; and (3) decrease in the thymidine kinase activity. [Bianchi (1962) has shown that the decrease in

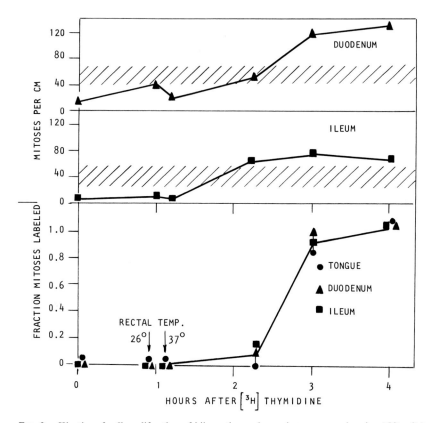

Fig. 2.   Kinetics of cell proliferation of hibernating and arousing common dormice (*Glis glis*). Animals were killed 1, 1.1, 2.3, 3, and 4 hours after [³H]thymidine injection, when the rectal temperature reached 26, 27, and 37°C for 1 hour, 37°C for 2 hours, and 37°C for 3 hours, respectively. In the upper graphs, the shaded area shows the range of values for active animals. (From Adelstein *et al.*, 1966.)

[³H]thymidine incorporation into human leukocytes was caused by a considerable thymidine kinase insufficiency.]

To understand the mechanisms underlying the differences in [³H]thymidine incorporation among different species of hibernating animals, Adelstein and Lyman in 1968 undertook a comparative biochemical study of the differences in thymidine metabolism between cells of golden hamster (*Mesocricetus auratus*) and ground squirrel (*Citellus lateralis*) during the active period of their life. These species were chosen on the basis of the differences in their ability to use [³H]thymidine (Adelstein *et al.*, 1964) (Table VI).

These authors made a detailed comparative study of stages in DNA synthesis based on the following scheme (Adelstein *et al.*, 1968).

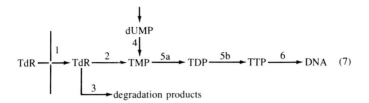

$$\text{TdR} \xrightarrow{\text{1}} \text{TdR} \xrightarrow{\text{2}} \text{TMP} \xrightarrow{\text{5a}} \text{TDP} \xrightarrow{\text{5b}} \text{TTP} \xrightarrow{\text{6}} \text{DNA} \quad (7)$$

They studied

1. Whether thymidine penetrates cells.
2. Whether there is phosphorylation up to thymidine monophosphate by thymidine kinase.
3. The activity of the catabolic processes.
4. The use of other ways of biosynthesis through deoxyuridine monophosphate.
5. The activity of thymidylate kinase.
6. The activity of DNA polymerase.

HIBERNATING

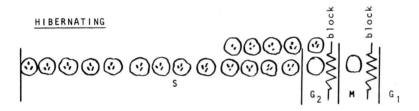

AROUSAL   3 hr after $\left[^{3}\text{H}\right]$Td R, 2 hr after temp. 37°

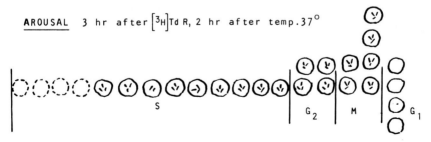

FIG. 3.   The scheme of the proposed blocks in mitotic cycle found in hibernation and as a result of arousal. (From Adelstein *et al.*, 1966.)

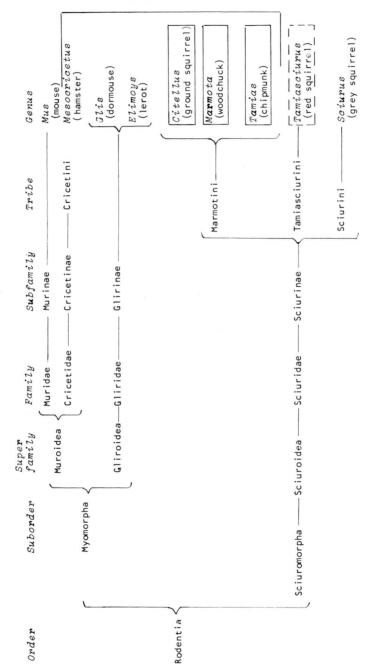

FIG. 4. Diagram of the phylogenetic relationships of rodents. Species severely limited in their ability to utilize thymidine are surrounded by solid lines, those with relatively limited ability are surrounded by dashed lines. (From Adelstein *et al.*, 1964.)

TABLE VI

In Vivo Incorporation of [^{3}H]Thymidine into DNA of
Gut and Sleen[a,b]

Species	Specific activity (cpm/gm DNA)	
	Intestine	Spleen
Mus	374	139
Mesocricetus	487	178
Glis	392	262
Eliomus	415	206
Sciurus[c]	96	23
Tamiasciurus	13	28
Tamias	44	7.5
Marmota[c]	1.8	0.3
C. tridecemlineatus	1.0	0.7
C. lateralis	1.6	0.2

[a] From Adelstein *et al.* (1964).
[b] Tissues were removed 4 hours after interperitoneal injection of [^{3}H]-thymidine (4 Ci/gm).
[c] Intraperitoneal injection of 2 Ci/gm.

These authors came to the conclusion that nonincorporation of [^{3}H]thymidine is not related to changes in the permeability of the plasma membrane of cells for thymidine, nor to the activity of thymidine kinase, nor to the size of the pool of DNA precursors. The activity of thymidylate kinase and DNA polymerase also did not differ in the two species studied. In both species, the biosynthetic pathway through deoxyuridine monophosphate can be utilized. (In ground squirrels, it is probably the main pathway.)

A difference was recorded only in the degree of activity of catabolic processes. Chromatography of acid-soluble and acid-insoluble fractions of renal and splenic cells of ground squirrels and hamsters showed that in hamsters the greater part of [^{3}H]thymidine penetrates the cells and is incorporated into DNA, whereas in ground squirrels, intracellular thymidine rapidly degradates to products that have no direct relation to DNA biosynthesis. Similar differences have also been detected *in vitro*.

The authors also concluded that there is a divergence in the catabolic pathways in different species of hibernating animals. They considered it necessary to study the level of [^{3}H]thymidine incorporation into the liver cells of golden hamsters and ground squirrels, since it is known that in rodents the greater part of [^{3}H]thymidine is rapidly degradated in the liver (Littlefield, 1965). *In vitro*

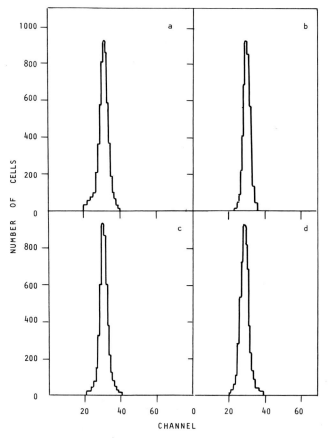

Fig. 5. DNA distribution obtained by flow microfluorometric analysis of lymphoid (a), corneal (b), adrenal (c), and thyroid (d) cells in *Citellus suslicus* during hibernation (stained by ethidium bromide). Cells tend to pile up in the G_1 phase of the cell cycle. (From Kolaeva *et al.*, 1977b.)

experiments (Adelstein and Lyman, 1968) showed that the rate of disappearance of DNA precursors from the cultivating medium was the same and that the volatile activity of the main product of catabolism of [^{14}C]thymidine, $^{14}CO_2$, was also identical in both species.

Taking into account that liver cells belong to tissues with low proliferative activity, the authors suggested that the level of catabolic activity of the enzymes involved in biosynthesis of DNA is related to the degree of proliferative activity of the tissue. The data of Potter (1962) on the marked increase of thymidine phosphorylation and on the decrease of its breakdown rate in the regenerating liver confirm this view.

D. Distribution of Cells According to the Cell Cycle Phases in
Hibernating Animals during Hibernation

Modern methods of flow microfluorometry permit the study of the distribution
of cells among the phases of the cell cycle in an organism with a limited ability to
utilize [^{3}H]thymidine as a precursor of DNA.

The study of cell distribution in different tissues of the hibernating ground
squirrel (*Citellus suslicus*) with the flow fluorometric system (Kolaeva *et al.*,
1977b) revealed the accumulation of cells of different organs in G_1 phase of the
cell cycle during the entire period of hibernation (Fig. 5).

III. Structural and Metabolic Properties of Cells of Hibernating Animals during Hibernation

It should be noted that the structure and metabolism of cells of different tissues
during hibernation have been studied infrequently and, on the whole, without
dealing with the specific features of different tissues. The available data indicate
a decrease in the synthesis of DNA, RNA, and protein and a decrease in the
number of elements of the protein synthetic system of the cell (Lutsenko *et al.*,
1975; Kolpakov *et al.*, 1974). The cells of various tissues from hibernating
animals during hibernation are probably heterogeneous in their structural and
metabolic properties. In this connection, it would be interesting to review the
data obtained by Lutsenko and co-workers (1973) in studies of the hibernating
ground squirrel (*Citellus erythrogenus*). It was found that the cells of different
zones of the adrenal cortex show distinct seasonal changes in the volume of
nucleoli (Table VII). There is a maximum in May, a decrease in June, and a
further decrease in the period of deep hibernation. The alterations in the volume
of nucleoli correlate with the available data on the production of corticosteroids
(Table VII).

Lutsenko *et al.* (1973) also examined the question of whether the whole
population of nucleoli reacts in a similar way to the seasonal change. If, irrespective of the size, the whole population displays a similar reaction to the change of
the season, only the mean volume of this structural unit will be altered, while the
total range of volumes remains constant. The change in the total range of volumes suggests a heterogeneity of cells involved in the reaction to the season. The
size of the nucleolus determines the functional activity of the synthetic cell
apparatus. Therefore, the decrease in the range of the nucleolus volume from
May to January (Table VII) may indicate a different effect of the seasonal factor
on cells with differing degrees of activity of the synthetic apparatus. The greatest
suppression of the activity of the synthetic apparatus of the cell from May to

TABLE VII

Mean Volumes of Nucleoli, Range of Volumes, and Production of Hormones in Different Seasons in Adrenal Cortex of *Citellus erythrogenus*[a]

Season	Logarithms of volumes ($\mu m^2 \times 10^{-4}$)		Range ($\times 10^{-4}$)		Hormone production ($\mu g/100\ \mu g$ tissue/hour)	
	Glomerular	Fascicular	Glomerular	Fascicular	Aldosterone	Hydrocortisone
May	0.153 ± 0.6	0.12 ± 0.6	256	250	1.08 ± 0.16	1.58 ± 0.18
July	-0.17 ± 0.6	-0.19 ± 0.7	190	166	0.59 ± 0.09	0.77 ± 0.007
January	-0.25 ± 0.5	-0.23 ± 0.5	145	112	0.13 ± 0.03	0.29 ± 0.006

[a] From Lutsenko et al. (1973).

January occurs in the cells with the highest level of activity of the synthetic apparatus (i.e., the cells with the largest nucleoli). Further explanation of the decrease in the range of nucleolus volume from May to January comes from the fact of partial synchronization of cells under a short-term effect of low temperatures on the culture (Mazia, 1961). Repeating such effects increases the degree of synchronization. For the animals studied, short periodic awakenings during hibernation are quite characteristic. In this case, the body temperature increases from 3–4°C to 38°C. These periodic temperature changes in the organism may, naturally, lead to a certain synchronization of synthetic processes and, hence, to a decrease in the range of the nucleolus volume.

The study of heterogeneity of volumes of nuclei and nucleoli in the different zones of adrenal cortex has shown that the difference in heterogeneity of the volume of nucleoli is significant only in January. This is a very interesting fact since it draws attention to the different activity of the synthetic apparatus of cells of the two zones of the same organ during hibernation. A certain relationship between the level of protein synthesis (Nussdorfer and Mazzocchi, 1972) and steroidogenesis in the cells of the adrenal cortex permits us to assume different levels of readiness of cells of the different zones to fulfill specific functions. This hypothesis is supported by the data of Kolpakov and Samsonenko (1970) who showed that the production of aldosterone *in vitro* in adrenal glands derived from hibernators at 8°C and placed in conditions of rapid arousal ($t = 37°C$) corresponds to the July level of production of this hormone (i.e., on arousal). The same result is not observed for hydrocortisone. Maintenance of ionic homeostasis in hibernators is believed to be due to the maintenance of low activity levels of the mineralocorticoid systems during hibernation (Hoffman, 1968).

It is noteworthy that on entering into hibernation an undulating character of changes in the volume of nucleoli in the cells of glomerular and fascicular zones of the adrenal gland (Kolaeva *et al.*, 1974) can be observed within the month of October. The height of each new rise in the activity (increase in the size of nucleoli) is much lower than the previous one (Table VIII). Noteworthy here is an almost complete analogy with the "test drop"—a phenomenon occurring in the nervous system (Hoffman. 1968). The undulating character of changes in the size of nucleoli on entering hibernation reflects, probably, the change in the functional state of the cells of the adrenal glands in *Citellus erythrogenus*. The cells of adrenal cortex in adult organisms are classified as a tissue with a low level of proliferative activity. However, in hibernating rodents, the adrenal gland probably possesses greater proliferative potentialities, since the massive destruction of tissue in winter can be restored only by high mitotic activity.

It is likely that during short periodic awakenings a small number of cells becomes involved in mitoses, just as in the small intestine (Suomalainen and Oja, 1967).

It also might be assumed that in other tissues (uninvestigated so far), there is a

TABLE VIII

MEANS VOLUMES OF NUCLEOLI OF ADRENAL CORTEX OF *Citellus erythrogenus* IN THE PREHIBERNATION PERIOD[a]

	Glomerular zone		Fascicular zone	
Month	M ± m	P (as compared with previous one)	M ± m	P (as compared with previous one)
August	0.65 ± 0.62		0.89 ± 0.01	
September	0.70 ± 0.02	0.05	0.83 ± 0.02	0.05
Early October	0.61 ± 0.01	0.85	0.77 ± 0.02	0.05
Middle October	0.74 ± 0.01	0.002	0.92 ± 0.01	0.01
Late October	0.50 ± 0.01	0.001	0.73 ± 0.02	0.001

[a] From Kolaeva *et al.* (1974).

difference in the metabolism of cells, although all of them are in G_1 phase. The phenomena observed are another example of the differences in the character and selectivity of changes in many physiological systems that take place during hibernation.

It is quite possible that metabolic alterations also reflect kinetic peculiarities of cells during hibernation. It was shown (Petrovichev and Yakovlev, 1975) that in liver cells of nonhibernating animals, two populations of cells with different prereplicative periods exist.

In winter, the number of mitochondria in liver cells of hibernants is markedly decreased. At the same time, changes occur in the structure of mitochondria: (1) the cristae are shortened; and (2) the mitochondrial matrix becomes more dense. All these alterations indicate the low metabolic activity in the mitochondria, as well as the reduction in the demand and use of energy. However, other investigators (Cossel and Wohlrab, 1964) observed an increase in the number of mitochondria in the liver of bats during hibernation.

In the cells of nephrons, the number of both microvilli and recesses between the bases of microvilli is sharply increased in winter; this suggests an activation of transport in winter (Brandt, 1973). These data are in agreement with the fact that in winter the activities of alkaline phosphatase and aminopeptidase (localized in the brush border) do not fall. The same data are obtained with other hibernating animals (Zimny and Bourgeois, 1960).

Recently more and more information has been obtained on the unusual properties of membranes in these animals during hibernation. It is known that the cells of hibernating animals are relatively insensitive to the action of toxic factors (Hoffman, 1968); this may be due to the change in the properties of the superficial membrane. The absence of cell swelling in hibernating animals at low temperatures (3–4°C) points to the possibility that these cells maintain normal ion concentrations at low temperatures (Willis and Baudysowa, 1977; Willis, 1972; Zeidler and Willis, 1976). It is known that the cells of different zones of the adrenal cortex are controlled by two different systems: ACTH and angiotenson II. During the active period, the cells of different zones exhibit a differentiated response to these controlling factors, whereas during the hibernation period the response is confused (Kolpakov and Samsonenko, 1970).

Considering that ACTH acts through the adenylate cyclase of the surface (Klegg and Klegg, 1969), it can be believed that specific changes take place on the superficial membrane of these cells.

It is believed (Willis, 1968) that the properties of the membranes in hibernating animals are more important than low metabolism in the survival of cells at low temperature.

A. Structural and Metabolic Properties of Cells during Hibernation Indicative of a Quiescent State

The published data cited in earlier sections indicate that the cells of hibernating animals possess a number of specific kinetic and metabolic peculiarities.

The absence of mitoses (even in tissues with high proliferative activity), the limited ability of cells during this period to utilize [^{3}H]thymidine as a precursor of DNA, and, finally, the time of the appearance of the first cells in the S phase and of the first mitoses on arousal all point to accumulation of cells in the early G_1 period of the cell cycle. Flow microfluorometry data confirmed these results and showed that such an accumulation of cells is observed throughout the 6 months of hibernation. Two explanations for the prolonged period of the G_1 phase in the cell cycle of hibernating animals are (1) slow progression of the cell cycle or (2) transition of cells from the normal cycle into the resting period (G_0). The second hypothesis is favored by (a) the unusually long-term residence of cells in the G_1 phase and (b) the metabolic peculiarities of cells during this period. During the resting period (G_0), the cells show the following metabolic changes: permeability of the plasma membrane to simple substances in lowered and chromatin template activity and the production of macromolecules decrease (Epifanova, 1977).

In other hibernating animals, cells may accumulate in the G_2 phase, which can also correspond to R_2 (resting II) (Epifanova and Terskikh, 1969). Unfortunately, phase R_2 of the cell cycle in hibernators has not been studied with respect to cellular metabolic properties.

Of special interest are those hibernating animals (mainly of the squirrel family) whose ability to utilize [³H]thymidine as a precursor of DNA is markedly decreased during the active period. From this point of view, the data of Adelstein and Lyman (1968) are particularly interesting. They have shown experimentally on the liver cells of two representatives of hibernating animals (with and without [³H]thymidine incorporation during the active period) that the difference in the ability to utilize [³H]thymidine as a precursor of DNA does not apply during hibernation to the nonproliferating liver cells, which possess a high level of catabolic enzymes that take part in DNA synthesis.

Unfortunately, besides the paper of Adelstein and Lyman (1968), we have no other data on the properties of the catabolic processes in the resting cells of hibernating animals. In this respect, not the liver cells, which represent a low-proliferative-activity system, but the cells of tissues with high proliferative activity would be of far greater interest.

Having analyzed a large amount of literature on the metabolic properties of the proliferating and resting cells in numerous biological objects in tissue culture, Epifanova in 1977 came to the conclusion that the factor common to all the resting cells is the high level of activity of catabolic enzymes (mainly lysosome hydrolases); this activity has been confirmed by many ultrastructural and biochemical studies. It was found that activation of enzymes other than the lysosome catabolic enzymes also takes place in cells in the state of quiescence (Epifanova, 1977). According to the author (Epifanova, 1977), the high level of activity of catabolic enzymes in the resting cells is one of the main intracellular control mechanisms maintaining the cells in the state of rest and preserving their vitality.

The number of high-electron-density inclusions in liver and kidneys in hibernators is markedly increased in winter. These inclusions can be divided into several groups: (1) peroxisomes; (2) lyosome-like elements with heterogeneous content; and (3) fatty granules (Brandt, 1973). The increase in the content of the lysosome-like elements in liver is related to the decrease in the output of biliary pigment in the small intestine; this results in an enhanced vacuolization of biliary pigments in liver during hibernation. However, accumulation of these inclusions in winter is not accompanied by an increase in the activity of acid phosphatase. This may indicate that only a part of these lysosomes might be identified as typical lysosomes. It is also quite possible that the decrease in the activity of acid phosphatase may be due to the reduction of secretory activity of hepatocytes in winter (Smith and Farguhar, 1966).

An increase in the amount of peroxisomes is found only in liver and kidneys and in only one representative of hibernators (hedgehog). According to Legg and Wood (1970), the increase in the amount of peroxisomes is observed under conditions characterized by specific metabolic parameters, namely, a considerable glyconeogenesis from lipids. During hibernation, the process of glyconeogenesis appears especially important since the activity of mitochondria is lowered. A large number of peroxisomes suggests that in the winter period carbohydrate metabolism deviates from the norm, and that this deviation is in some way caused by an oxygen deficiency. Probably the adipose tissue is the site of lipid and glycogen production during hibernation, and this finally leads to lipemia (Biorck *et al.*, 1956). The data of Burlington and Klein 1967) show that slices of kidney cortex from hibernating animals (*Citellus tridecemlineatus*) possess greater potential for glyconeogenesis from α-ketoglutarate, α-aspartate, α-glutamate, glycerol, oxaloacetate, and pyruvate than tissues of active animals. These data confirm the data obtained earlier by Rebel *et al.* (1960) who found that the lipids of hibernating animals may be involved in glyconeogenesis pathways.

Published data indicate that the extended G_1 period in the cell cycle of hibernating animals is heterogeneous in its structural and metabolic properties during the 5 or 6 months of hibernation. During the last months of hibernation (February–March), in different areas of brain (Semeshina, 1971) and in the cells of the adrenal cortex (Kolpakov *et al.*, 1974; Kolaeva *et al.*, 1977b), an increase of cytoplasmic RNA occurs and changes appear in the nucleoli that are seen with enhancement of functional activity of the cell, and the type of endoplasmic reticulum is modified, although the body temperature of the animals remains low (Figs. 6 and 7).

There is a simultaneous activation of protein biosynthesis in adrenal cortical cells (Lutsenko *et al.*, 1975). Structural rearrangements typical of the increased functional activity of cells take place in the thyroid gland (Mikhnevich and Kostyrev, 1971) and in the hypothalamo-hypophyseal–neurosecretory system (Yurisova, 1970) well in advance of arousal.

Before arousal, the cells begin to prepare to perform their specific functions. Thus, in *in vitro* experiments with adrenal cortical cells derived from hibernating animals in March, it was shown that the production of aldosterone increases by 150%, that of cortisone by 108%, and that of hydrocortisone by 41% (Kolpakov *et al.*, 1974). The reactivity to controlling factors changes simultaneously (compared to January) in the cells of the adrenals. During hibernation the adrenal glands reacted to ACTH (in incubates) with a decreased production of corticosteroids compared with the period preceding arousal (Kolpakos *et al.*, 1974).

Thus, the G_0 (early G_1) period in the cell cycle of hibernating animals during hibernation is apparently nonhomogeneous in the metabolic and structural properties of cells. The changes occurring during this time show (a) an increase of all elements of the protein-synthesizing system of the cell; (b) an increase in biosynthesis of RNA and protein; (c) changes in the properties of the superficial membrane; (d) changes in the metabolism directed to perform specific function. All these changes slowly increase in intensity as the awakening draws nearer and reach a maximum at the time of awakening. Thus, in the quiescent cells,

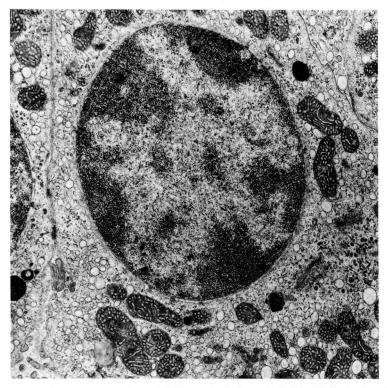

Fig. 6.   Electron micrograph of an adrenal cortical cell from hibernating *Citellus suslicus* (December). ×980. (From Kolaeva *et al.*, 1977b.)

metabolic processes are never fully inhibited; they only acquire new features. A long time before awakening, the specific rearrangement in metabolism begins and is controlled, apparently, by endogenous rhythms.

B. Mechanisms Regulating the Transition of Cells into the State of Quiescence during Hibernation

Mechanisms of the transition from the normal cell cycle in cells of hibernating animals during hibernation remain unknown. When considering this question, one should pay attention to the character of changes in the physiological functions during hibernation—in particular, to the state of endocrine control. One should recall the importance of the endocrine system in maintaining the hibernation state itself (Kayser *et al.*, 1964; Kayser, 1961; Kolpakov *et al.*, 1974) and the role of the endocrine system in the regulation of mitotic homeostasis (Epifanova, 1965; Lagutchev, 1975).

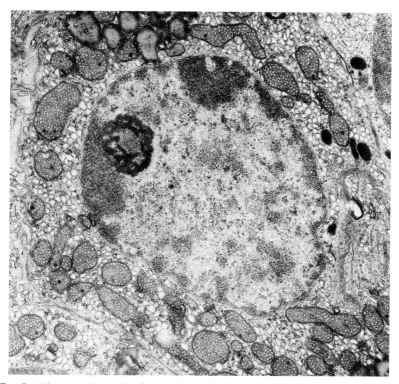

FIG. 7. Electron micrograph of adrenal cortical cell from *Citellus suslicus* in active state (March). ×980. (From Kolaeva *et al.*, 1977b.)

In winter, with the inhibition of the central mechanisms of neuroendocrine control, there is a peculiar "discrepancy" in the system of the control hierarchy when the role of the "metronome" can be taken over by the peripheral links in the endocrine system. The role of such a peripheral metronome can be played by the cortical and medullar substances of the adrenals, by the insular apparatus of the pancreas, and by the parathyroid glands, which retain their activity during hibernation (Kolpakov *et al.*, 1972; Riedesel and Folk, 1954; Mussachia and Jellineck, 1962). It is also known that during hibernation the adrenaline content is high in the adrenal tissue and in a number of other organs (Mussachia and Jellinek, 1962). On awakening, adrenaline is released into the blood and apparently plays a rather important part in the triggering of the awakening (Mussachia and Jellinek, 1962).

During hibernation, there is a selective increase of cortisone production in the adrenals (Kolpakov *et al.*, 1972). It can be assumed that adrenaline and cortisone play a definite role in mitotic homeostasis during hibernation. It is known that adrenaline acts as a cofactor in activating chalones and that cortisone prolongs adrenaline–chalone activity (Bullough, 1973).

The independent functioning of a different apparatus inside each of the peripheral links of the endocrine system is undoubtedly the result of mechanisms elaborated during the process of evolution to maintain homeostasis in the hibernating animals. In 1965, Manasek *et al.* suggested that accumulation of cells in the G_1 phase is most probably due to some peculiarities of the mitotic homeostasis during this period of the cell cycle in the life of the hibernating animals and is not due to the low temperature. It is possible that the selective increase in the level of production of some hormones with the decrease in others is one of the peculiarities of the control of mitotic homeostasis during hibernation.

IV. Conclusion

The period of quiescence in the cell cycle of hibernating animals is an example of the evolutionary adaptation at the cellular level for survival during unfavorable environmental conditions in these organisms. A peculiar type of metabolism in the quiescent cells helps them to survive successfully during this unfavorable period.

The very fact of transition during hibernation from a normal cell cycle to a cell cycle with a prolonged early G_1 period is apparently one of the most interesting phenomena in biology.

The study of the resting period in the cell cycle is at present one of the paramount problems in the further development of conceptions of the cell cycle, and the main attention of investigators is directed at the search for criteria for and identification of the resting cells and the study of mechanisms responsible for the transition from the normal cell cycle and reentry into it.

In the tissues of animals, the cells in the resting phase are distributed among the proliferating cells. The absence of morphological criteria enabling us to distinguish the resting cells from the whole mass of the cellular population makes it impossible so far to study their structure and function *in vivo*. One of the possible approaches is the search for definite physiological states; e.g., under certain conditions, the cells can become synchronized in one phase of cycle G_1 or G_2. In this respect, hibernation seems to be a promising model.

In vivo study of the resting period of hibernation has a number of advantages as compared to the model system of the resting cell now widely used—the stationary phase of cell growth in tissue culture—since in the former instance, the humoral factors that control mitotic homeostasis are retained.

In different periods of hibernation, the quiescent cells serve as a convenient model system for the study of the peculiarities of intracellular metabolism during the period G_0, since events develop at a slow rate and are easy to record.

The slow rate at which the cells of hibernating animals enter the normal cell cycle (on arousal) and depart from it (on entering hibernation) can also be a good model for studying the dynamics of structure in relation to the changing function of the cell, i.e., for the study of the integrative function of the cell.

Finally, the possibility of comparing the dynamics of the physiological functions of the organism with the events of the cell cycle will apparently enable us to find approaches to one additional, little studied, problem—the regulation of mitotic homeostasis.

In 1968, Roger Hoffman, in his paper "Hibernation—as a tool for biological studies," described a wide spectrum of biological problems where the most convenient model could be the hibernation.

The study of the kinetics and metabolism of cells of hibernating animals apparently opens up yet another field where this unique phenomenon in nature can serve to study the structure and function of the resting cells *in vivo*—a very important field not only for theoretical biology, but for applied branches of biology and medicine.

ACKNOWLEDGMENTS

The authors express their deep gratitude to Dr. Olga Epifanova and her colleagues for valuable seminars, which to a great extent determined the problem and plans of the studies on resting periods in the cell cycle of hibernants. We also thank Olga Golovchenko for technical assistance. The authors express their deep gratitude to Professor Charles P. Lyman of the Harvard Medical School for reviewing and valuable criticism of the manuscript.

REFERENCES

Adelstein, S., and Lyman, C. (1968). *Exp. Cell Res.* **50**, 104.
Adelstein, S., Lyman, C., and O'Brian, R. (1964). *J. Comp. Biochem. Physiol.* **12**, 223.
Adelstein, S., Lyman, C., O'Brian, R., and Ito, S. (1966). *In* "Excerpta Medica Monographs on Nuclear Medicin and Biology," No. I. pp. 111–119.
Bianchi, P. (1962). *Biochim. Biophys. Acta* **55**, 547–549.
Biorck, G., Johansson, B., and Veige, S. (1956). *Acta Physiol. Scand.* **37**, 281–294.
Brace, K. (1952). *Science* **116**, 570.
Brandt, M. (1973). *Z. Wiss. Zool.* **185**, 292.
Bullough, W. (1973). *Natl. Cancer Inst. Mongr.* **38**, 5–16.
Burlington, R., and Klein, J. (1967). *J. Comp. Biochem. Physiol.* **22**, 701.
Cossel, L., and Wohlrab, F. (1964). *Z. Zellforsch.* **62**, 608.
Epifanova, O. I. (1965). "Hormones and Cell Proliferation." Academic Press, New York.
Epifanova, O. I. (i977). *Int. Rev. Cytol.* **49**, Suppl. 5, 303.
Epifanova, O. I., and Terskikh, V. V. (1969). *Cell Tissue Kinet.* **2**, 75.
Fisher, K. C. (ed.). (1967). "Mammalian Hibernation," Vol. III. Oliver & Boyd, Edinburgh.
Hoffman, R. (1968). *Fed. Proc.* **27**, 999.
Jaroslow, B., Fry, R., Suhrbier, K., and Sallese, A. (1976). *Radiat. Res.* **66**, 556.

170   S. G. KOLAEVA ET AL.

Kayser, Ch. (1961). "The Physiology of Natural Hibernation," p. 325. Pergamon, New York.
Kayser, Ch. (1962). *New Sci.* **16,** 677.
Kayser, Ch., Vincedon, G., Frank, R., and Porte, A. (1964). *Ann. Acad. Sci. Fenn. Ser. A.* **4,** 269.
Klegg, P., and Klegg, A. (1969). "Horm. Cells and Organisms," p. 161. Stanford University Press, Stanford, California.
Kolaeva, S., and Lutsenko, N. (1972). *Dokl. Acad. Sci. USSR* **203,** 1405.
Kolaeva, S., Kolpakov, M., and Robinson, M. (1974). *Dokl. Acad. Sci. USSR* **216,** 199.
Kolaeva, S., Kramarov, V., and Kramarova, L. (1977a). *In* "Fifth All-Union Conference of Ecological Physiology," Abstr., pp. 217–218. Academic Press of the USSR, Leningrad.
Kolaeva, S., Zubrikhina, G., Ilyasov, F., Ilyasova, E., and Hachko, V. (1977b). *In* "Fifth All-Union Conference of Ecological Physiology," (Frunze, ed.). Abstr., pp. 218–219. Academic Press of the USSR, Leningrad.
Kolpakov, M., and Samsonenko, R. (1970). *Dokl. Acad. Sci. USSR* **191,** 1424.
Kolpakov, M., Kolaeva, S., and Shaburova, G. (1972). *Successes Physiol. Sci. USSR* **3,** 52–67.
Kolpakov, M., Kolaeva, S., Krass, P., Polak, M., and Sokolova, G. (1974). *In* "Mechanism of Seasonal Rhythms of Corticosteroid Regulation in Hibernants" (M. Kolpakov, ed.), pp. 1–43. Nauka, Novosibirsk, USSR.
Lagutchev, S. S. (1975). "Hormones and Cellular Mitotic Cycle." Medicine, Moscow.
Legg, P., and Wood, R. (1970). *J. Cell Biol.* **45,** 118.
Littlefield, S. (1965). *Biochim. Biophys. Acta* **95,** 14.
Lutsenko, N., Ginsburg, G., Kolaeva, S., and Kolpakov, M. (1973). *Cytology USSR* **16,** 198–202.
Lutsenko, N., Kolaeva, S., and Kolpakov, M. (1975). *Probl. Endocrinol.* **4,** 83.
Lyman, C., and Dempsey, E. (1951). *Endocrinology* **49,** 647.
Lyman, C., Weiss, L., O'Brian, R., and Barbeau, A. (1957). *J. Exp. Zool.* **136,** 471.
Manasek, F., Adelstein, S., and Lyman, C. (1965). *J. Cell. Comp. Physiol.* **65,** 319.
Mayer, W., and Bernick, S. (1958). *Anat. Rec.* **130,** 747.
Mazia, D. (1961). "Mitosis and Physiology of Cell Division." Academic Press, New York.
Mikhnevich, O., and Kostyrev, O. (1971). *In* "Hibernation and Seasonal Rhythms of Physiological Functions" (A. Slonim, ed.), p. 33. Nauka, Novosibirsk, USSR.
Musachia, S., Jellinek, M., and Cooper, T. (1962). *Proc. Soc. Exp. Biol. Med.* **110,** 856.
Nussdorfer, G., and Mazzocchi, G. (1972). *Lab. Invest.* **26,** 45.
Petrovichev, N., and Yakovlev, A. (1975). *Cytology USSR* **17,** 1087.
Potter, V. (1962). *In* "Molecular Basis of Neoplasia," p. 367. Univ. of Texas, Austin.
Rebel, G., Weill, S., and Mandel, P., and Kayser, Ch. (1960). *C. R. Soc. Biol.* **154,** 2118.
Riedesel, M., and Folk I. (1954). *Am. J. Physiol.* **179,** 665.
Semeshina, T. (1971). *Physiol. J. USSR* **57,** 1616.
Smith, R., and Farguhar, M. (1966). *J. Cell Biol.* **31,** 319.
Suomalainen, P., and Oja, H. (1967). *Comment. Biol. Soc. Sci. Fenn.* **30,** 1.
Unker, V., and Alekseeva, G. (1974). *J. Evol. Biochem. Physiol. USSR* **2,** 193.
Willis, J. (1967). *In* "Mammalian Hibernation III" (K. Fisher, A. Dane, Ch. Lyman, E. Schonbaum, and F. South, eds.), pp. 356–381. Oliver & Boyd, Edinburgh.
Willis, J. (1972). *Cryobiology* **14,** 511.
Willis, J., and Li, N. (1968). *Am. J. Physiol.* **217,** 321.
Willis, J., and Baudysowa, M. (1977). *Cryobiology* **14,** 511.
Willis, J., Fang, S., and Foster, R. (1972). *In* "Hibernation and Hypothermia: Perspectives and Challenges" (F. South, ed.), pp. 123–147. Elsevier, Amsterdam.
Yurisova, M. (1970). *J. Evol. Biochem. Physiol. USSR.* **6,** 516.
Zeidler, R., and Willis, J. (1976). *Biochim. Biophys. Acta* **436,** 628.
Zimny, M., and Bourgeois, C. (1960). *J. Cell. Comp. Physiol.* **56,** 93.

CELLSIM: Cell Cycle Simulation Made Easy

Charles E. Donaghey

Industrial Engineering Department, University of Houston, Houston, Texas

I. Introduction

The field of cell kinetics comprises studies of the movement and proliferation of cells through their generative cycle and perturbation of this cycle by various treatments. Cells may be destroyed, their movement slowed or blocked, or other kinetics parameters altered by the treatment. The use of digital simulation in this area is not new (Barret, 1966; Kember, 1969; Tibaux *et al.*, 1969). However, simulation has not had wide usage by cell kinetics researchers because of several basic problems:

1. Most of the people in the forefront of the field have come from a medical or cell biology background and have little experience in using computers in their work.

2. The development of computer simulation models can be frustrating, time consuming, and expensive. The researcher has to communicate his concept of the model to a programmer. This can cause a communication problem at this interface between two entirely different disciplines. The programming can require a significant amount of time, since normally a great deal of adjustment must be performed before a satisfactory model is obtained.

171

3. In programming these types of models, cell growth is difficult to handle. Just as a tumor can completely overrun the host in which it resides, computer memory can be quickly exhausted when the tumor is modeled.

It was decided that a special purpose digital simulation language would help to overcome these problems, and the name given to the language was CELLSIM. Our objectives with CELLSIM were as follows:

1. The language should be easy to learn and use, even for those with little or no previous experience with computers and modeling.
2. The language should make the logic of the model apparent to other cell kinetics researchers who examine it.
3. The language should be capable of being utilized on virtually any medium-to large-scale computer system, and it should not require any special hardware.
4. The language should be easily expandable. As new requirements and desires of researchers are discovered, they should be incorporated into the language with a mimimum of difficulty.
5. Models should be handled by the language regardless of the number of cells and the length of the simulation.

It was decided that a FORTRAN interpreter would be the most convenient way of implementing CELLSIM. FORTRAN is the most standard computer higher level language, and this would allow close to machine independence. The group involved in the development of the language also had experience in developing interpretive problem-oriented languages using FORTRAN (Donaghey *et al.*, 1970; Donaghey and Ozkul, 1971).

The sketches in Fig. 1 will be used to give a brief background of some of the concepts in cell kinetics research. Growing cell populations repeatedly pass through four phases. The names traditionally given to these phases or states are G_1, S, G_2, and M. Cell division occurs in state M or mitosis. The state where DNA synthesis takes place is termed S, and the periods before and after S, which represent *gaps* in understanding are called G_1 and G_2. Thus, a newly divided cell leaves M and enters G_1. At some later time, it may pass into the S phase as it begins to make DNA, and then it goes on to G_2 when DNA replication is completed, and eventually enters M to divide again. Figure 1a depicts this flow. The time it takes a cell to complete this cycle is dependent upon the cell type. Some cell types may have a cycle time of several hours; other may take months. Even with the same cell type, the times individual cells spend in each phase will vary from cell to cell and cycle to cycle. One way of describing the phase times is to use probability functions with expected values and variances. Drugs usually act on cells when they are in specific phases and do not discriminate between populations. A drug that destroys tumor cells that are in state S, will also destroy

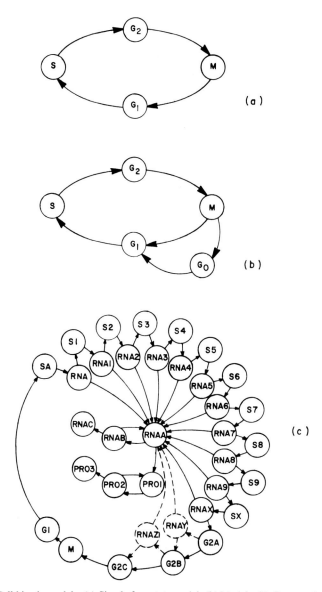

FIG. 1. Cell kinetic models. (a) Simple four-state model. (b) Model with G_0 or resting state. (c) Model with substates to include enzyme activity.

normal cells that happen to be in S. One objective of chemotherapy treatment is to manipulate the cells so that there will be a high proportion of tumor cells in some target state, compared to a low proportion of normal cells in the same state. When this is accomplished, a drug is administered that destroys cells in the target state.

Figure 1b shows another concept that seems to hold for some populations. In this model, some cells upon leaving M go into a resting state called G_0. Cells in G_0 may stay for an extended period before they rejoin the circulating cells, if they ever do. Cells in G_0 appear to be impervious to most drug effects. Figure 1c show how states can be subdivided and the complexity of models increased. This model includes enzyme activity in the flow (Stubblefield and Dennis, 1976).

CELLSIM applies Monte Carlo simulation to cell kinetics modeling. The modeler describes the events in the model in a probabilistic manner, and during execution of the model, the program will cause these events to occur according to these specified probabilities. For example, a modeler may describe the times cells spend in some state S as a random value from a normal distribution with an expected time of 20.0 hours and a variance of 9.0 hours. During execution of the model, each time a cell comes to state S, a random number generator is used to develop a random time from the specified normal distribution. The cell will then reside in S for that amount of time.

The Monte Carlo approach recognizes and handles the variance inherent in cell population kinetics. It differs from deterministic solutions where the model is specified as a set of equations, and the equations solved for an exact solution. The equations are usually differential equations that specify cell flow rates and do not include chance variations in the flow.

The simple model shown in Fig. 2 will be used to introduce CELLSIM. The sketch at the top of the figure describes schematically the flow of cells in the model and shows the probability distributions for the time in each state. The equivalent CELLSIM program is listed below the sketch. The first command in the program is the CELL TYPES command. With this command, a user declares how many cell populations are to be included in the model and assigns a name to each population. In the example model, only one population is included, and it is given the name TUMR. If more than one population were used, the names would be separated by commas in the command.

The second command in the example program is the STATES command. Here, all of the states in the population are assigned names. The 'STATES (1)' in the command indicates that the names assigned to the states apply to the first population in the model. If more than one population had been declared in the CELL TYPES command, the population index would be the order in which the population appeared in the command. The names given to the populations and states are arbitrary and may be up to four characters in length.

The next command in the example program is a TIME IN STATES command.

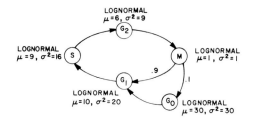

```
CELL TYPES TUMR;
STATES(I) S,G2,M,GI,GO;
TIME IN STATES (I)   S: LOGN(9,16),  G2: LOGN(6,9),
                     M: LOGN (I,I),  GI: LOGN (IO,20),
                     GO: LOGN (30,30);
FLOW (I)   S-G2 (ALL), G2-M (ALL), M-GI(.9), M-GO(.I),
           GI-S(ALL), GO-GI(ALL);
PROLIFERATION (I) M: 2;
INOCULUM (I) IOOOO;
REPORTS I HOUR ;
BLOCK EXIT (I)  BETWEEN 24 AND 48   GI:.9;
GRAPH S(I)/TOTAL * IOO;
SIMULATE IOO HRS;
```

Fig. 2.   Cell model and equivalent CELLSIM program.

This command furnishes the information for the time the cells are to spend in the states. The time in S is to be a random value from a log-normal distribution with an expected value of 9 hours and a variance of 16. The times in each of the other states are also random variables from log-normal distributions. Table I shows the other statistical distributions available in CELLSIM. There are a number of commands that make use of these distributions. If the user had specified the time in state S as S:9, then each cell would spend exactly 9 hours in the state.

The FLOW command appears next in the sample program. The segment of the command, 'S-G2 (ALL),' means that all cells that exit the S state go into state G_2. All cells that leave G_2 go to M. Cells exiting M have a split flow: .9 going to G_1 and .1 going to G_0. All cells that leave G_0 go to G_1 and all cells from G_1 go to S. The flow from a state can be directed to any number of states and may be sent back to reenter the one it just left.

The PROLIFERATION command in the example program shows that state M is a proliferating state with a proliferation rate of 2.0. The proliferation rate of 2.0 indicates that cells divide upon leaving state M, so that twice as many cells will leave M as enter. A proliferation rate less than 1.0 could be used to simulate

TABLE I

CELLSIM Probability Distributions

1.	Normal
2.	Uniform
3.	Erlang
4.	Exponential
5.	Triangular
6.	Trapezoidal
7.	Log-normal
8.	Gamma
9.	Cumulative Distribution Function

cell death in a state. The proliferation rate can also be specified with a statistical distribution.

The INOCULUM command shows that 10,000 cells are to be initially placed in the cell cycle. Table II shows options available in an INOCULUM command. As examples of these options, the command:

INOCULUM (1)   S:5000 AT BEGINNING, M:500 AT END

would place 5000 cells at the beginning of state S and 500 cells at the end of state M. The command:

INOCULUM (1)   S:8000 EVEN (0, 10)

would place 8000 cells evenly spaced in state S with 10 hours as the maximum time remaining in the state, and zero as the minimum time remaining. The command:

INOCULUM (1)   G2:1000 EXP (5.3)

would place 1000 cells in state G_2. The time remaining in the state for each cell would be a random value from an exponential distribution with an expected value of 5.3 hours. The command:

INOCULUM (1)   G2:1000

TABLE II

Options for Inoculum Command

1.	At beginning
2.	At end
3.	Even
4.	Statistical distribution
5.	State default
6.	Population default

would cause default options to be called that would place 1000 cells in state G_2. These default procedures would look to the time in states information for state G_2 and place the cells in the state with the time remaining for each cell a random value from an exponential distribution. The mean will be approximately half of the mean of the time in states.

$$\text{INOCULUM (1)} \qquad 1000$$

This would place 10,000 cells spread throughout population one. The system determines an expected time in each state included in the flow and sums them. The number of cells in each state will be assigned in proportion to that state's expected time to this total expected time.

The REPORTS command in the program specifies that, during execution, the model is to report on its status every hour of simulated time. This report will show for each state in each population the number of cells that reside in that state. Any graphs that are requested will also have a plot point generated at each report time.

The GRAPH command in the example program specifies that a graph of the percentage of the cells in S be displayed. The system recognizes S(1) as the number of cells in state S for population one. The word TOTAL in the expression is a key word and represents the total cells in the model. Therefore:

$$S(1)/\text{TOTAL} *100$$

will give the percentage of cells in state S. Figure 3 shows the graph that is obtained from the model. The graph in the figures is reduced from actual size. The time interval on the graph is set in the REPORT command. The REPORT command in the sample program specifies the reporting interval to be 1 hour. Thus, a point on the graph will be generated every simulated hour. The coordinates of the points are stored in auxiliary storage in a linear file. The graphs are then printed at the end of the simulation. The length of the Y axis is the same for each graph (about 10 in.) with the minimum and maximum values on the Y axis; the minimum and maximum values are obtained from the expression during the simulation. In a single GRAPH command, any number of graphs can be requested; each graph expression is separated by a comma. Table III gives examples of graph expressions. The CELLSIM interpreter is capable of processing GRAPH statements very similar to FORTRAN arithmetic statements. Any level of parentheses may appear in a statement, and the same arithmetic operators are used ($=$, $-$, $*$, $/$, $**$) to represent addition, subtraction, multiplication, division, and exponentiation, respectively. In Table III, TOT(1) refers to the total cells in population one, and TOT(2), the total cells in population two. As stated previously, TOTAL refers to the total cells in the model. RI is the cell rate into a state, and RO is the cell rate out of a state. These CELLSIM rate functions use the reporting interval to calculate an approximate cell rate into and out of states.

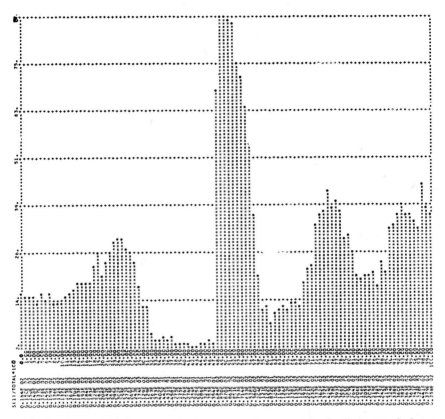

Fig. 3. Output graph from example program showing proportion of cells in state S. Large accumulation is obtained after block removal.

For example, if the reporting interval were 0.25 hours, and during some quarter hour interval 100 cells flowed into state M of population one, the value for RI(M(1)) would be 400 cells/hour. This value is recalculated for each report interval. The LG and LN functions in Table III are $\log_{10}$ and natural log, respectively.

TABLE III

TYPICAL GRAPH EXPRESSIONS IN CELLSIM

M(2)/(M(1) + M(2)) * TOTAL
2.5 * S(1)/TOT(1) + 1.5 * S(2)/TOT(2)
RI(S(1)) + RO(S(2))
LG (TOTAL)
LN (M(2))

The last statement in a CELLSIM program must be a SIMULATE command. This command signals that the model has been completely defined and specifies the length of the simulation. The interpreter then begins the simulation.

II. Monte-Carlo Simulation

Before showing specific examples of CELLSIM programs and describing more of the commands that may be included in a model, it might be well to discuss, in general, Monte Carlo simulation techniques and how they are used in CELLSIM. Monte Carlo simulation is applied to problems that may be too large or too complex to solve by deterministic techniques. It includes the chance variation inherent in most real life problems. The description Monte Carlo is used because of the games of chance that are conducted in the casinos there. As a simple example of Monte Carlo simulation, consider the following problem. Customers arrive at a four-chair barber shop according to the following interarrival time (T) distribution:

T = Interarrival Time	$P(T)$
1	1/6
2	1/3
3	1/3
4	1/6

This shows that 1/6 of the time a customer arrives 1 minute after the previous customer has arrived, and 1/3 of the time he arrives 2 minutes after the previous customer has arrived, etc. The time to get a haircut has the following distribution:

C = Haircut Time	$P(C)$
7	1/2
8	1/6
9	1/6
10	1/6

Haircut time (C) is 7 minutes, 1/2 the time; 8 minutes, 1/6 the time; etc. If this problem were to be simulated, some device would have to be used to get random values from the two distributions. A single die would serve nicely for this purpose. Since the probability of each face turning up on a single roll is 1/6, random interarrival times could be generated by rolling a die and assigning the time depending on the face that turns up as follows:

Face	T
1	1
2, 3	2
4, 5	3
6	4

Likewise, a random haircut time can be generated by rolling the die again and assigning the time C as follows:

Face	C
1,2,3	7
4	8
5	9
6	10

Using this scheme, we could simulate the activity in the barber shop for any length of simulated time. Figure 4 shows graphically the results of the first hour of the simulation. The arrows on the time scale shows the arrivals in the shop. Each arrival is assigned a number, which identifies him when he gets his haircut. This process can continue for any length of simulated time, and the modeler can gather any statistics on the system that he desires. There might be interest in what proportion of time the barbers are busy, or the maximum number of customers waiting for a barber during a simulation, or how the system would perform if

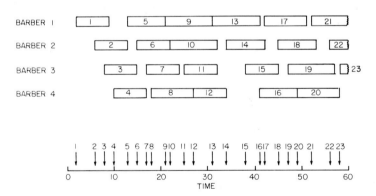

FIG. 4. Simulation of barber shop problem using single die to obtain interarrival times and haircut time. Arrows show customer arrival. Number associated with arrival is used to show the barber that cut his hair.

there were only three barbers. These and other statistics would easily be examined with such a simulation study.

Most simulations are performed on a digital computer. The chance device on the computer is a random number generator. It serves the same purpose as the die in the hand simulation. A random number generator is a computer algorithm that is included with all Monte Carlo simulation programs. Usually the algorithm requires only a few lines of code. Every time this section of code is called, it generates a value uniformly distributed in the interval from 0.0 to 1.0. This value can then be translated to a random value from some probability distribution. Figure 5 shows how values from a random number generator can give T values from the interarrival distribution in the barber shop problem. When the random number is in the interval (0, 0.167), the T value will be 1. When the random number is in the interval (0.167, 0.500), which should occur with a probability of 0.333, the T value will be 2, etc. The T values are assigned by using segments of the (0, 1) interval in proportion to the probability of each value occurring. Many techniques have been developed to get random variables from the classic probability distributions, e.g., normal and exponential. Most of them use the basic random number generator that gives uniformly distributed random values in the (0, 1) interval and performs manipulations on these values to transform them to other distributions. Naylor *et al.* (1966) gives a good disucssion of some of these techniques.

A random number generator is included in the CELLSIM package along with the computer codes for transforming these values to the probability distributions shown in Table I. Suppose a modeler specifies that the time cells are to spend in some state is to be a random value from a log-normal distribution with a given mean and variance. The random number generator and the code to transform random numbers to log-normal distributions are called on every time a cell enters that state. The log-normal value that is generated is then the time the cell will spend in that state. Just as in the barber shop problem a haircut time could be generated as soon as the customer arrived, a state residence time is generated as soon as a cell arrives at a state. Thus, the system can schedule the exit of the cell from the state as soon as it comes into the state. CELLSIM keeps a master schedule of the events that are to occur in the simulation. When a cell leaves one

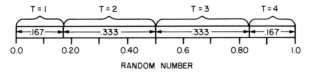

FIG. 5.   Transformation of values from random number generation to random interarrival times in barber shop problem.

state and enters another, the exit time for the state that it is entering is determined and that event is placed on the schedule. At any point in the simulation the master schedule may contain several hundred events. Each of these future events has the scheduled occurrence time of the event and other information describing the event contained in a single list. These lists are stored in the master schedule in chronological order of their scheduled occurrence time. During the simulation, the program takes the earliest event from the master schedule and processes that event. If the event is a cell leaving some state, the program finds which state the cell will enter. The cell is placed in this new state and its time of exit from that state is determined. This exit event is placed on the master schedule and the program then moves to the event that is next in chronological order on the master schedule. This is the way simulation time advances during the model. Event processing may cause one or more future events to be added to the schedule. After these events are added to the schedule, the event now in the earliest position in the schedule is examined and the process repeats. Thus, simulation time during a CELLSIM model execution does not advance at a constant rate. The model progresses from one event to the next in nonuniform time increments.

In the example problem of Fig. 2, the initial inoculum of cells in the model was 10,000. The cells proliferate, so that in a short time there will be many more cells in the model; and if the model is run long enough, it could contain 10^{10} cells or more. The question arises on how CELLSIM keeps track of the tremendous number of cells that might be in a model without using excessive amounts of computer storage and time. The answer is that CELLSIM does not follow individual cells. Cells are assigned to a model in groups. Each groups will then enter, move through, and leave a state together. When a group enters a state, a residence time is determined using the random number generator and the information given in the TIME IN STATES command. This residence time is the time the entire group will spend in the state. The representation of a group of cells requires only a few words of memory in the computer. One word contains the number of cells currently in the group, another contains the time the group is scheduled to exit the state, and another contains the time the group entered the state. A few more words are used as linkages to the master schedule and other parts of the model. The total number of words for each group of cells in the model comes to seven. These seven words can represent any number of cells. The modeler can control the number of groups initially in a CELLSIM model. This is done with a NUMBER OF GROUPS command. The example model in Fig. 2 does not contain such a command, so some built-in default algorithms will control the initial number of groups in the model. The default algorithm will assign between 80 and 90 groups to each population in the model. The number of groups initially assigned to a state within a population is based on the ratio of the time within that state to the total cell cycle time for all states included in the initial flow. Each group initially assigned to a state will represent the same number of cells. For example, if 30 groups were used to initially represent 1200 cells in some state, each group will represent 40 cells.

The number of groups in a model may grow as the model executes. The model in Fig. 2 shows a split in the flow as cells exit state M. Therefore, when a group exits state M, it will be split into two groups: one group going to state G_1 and the other group going to state G_0. It is necessary in CELLSIM then to have collection algorithms that will keep the number of groups within the bounds of the computer's memory. These collection routines oversee the number of groups, and will go through a collection and reassignment of groups whenever the total number reaches some limit. The modeler can specify this limit with a MAXIMUM NUMBER OF GROUPS command or, again, it can be left to default. The default option causes the collection and reassignment to take place whenever the number of groups becomes approximately 25% greater than the number initially assigned to the model. Figure 6 shows graphically how the reassignment algorithm works. In this example, three groups are being combined into a single one. The three original groups had 46, 30, and 50 cells, respectively. Each of the original groups has its own entry and exit time. The resulting group contains the sum of the cells in the original groups and has an entry and exit time that is the weighted average of the entry and exit time of the original groups. The space that the original groups required is then made available for continued execution of the model.

The concept of grouping cells and then collecting and reassigning the groups during execution appears to cause little or no bias or disturbance in CELLSIM models. We have conducted tests in which each cell was individually followed during a model, have compared this model to the same model with grouping, and

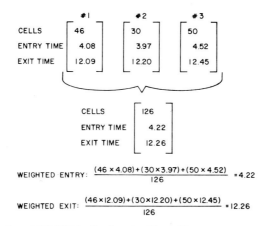

FIG. 6. Operation of CELLSIM collection algorithms. Three groups are combined into a single group. Collected space in memory is then available for further execution of the model.

have found the results from both models to be statistically the same. Following each individual cell in a model would prove to be prohibitively expensive, both in computer time and memory, for realistic models.

III. Example Problems

The example programs presented in this section will be used to demonstrate the capabilities of the CELLSIM language. Each program will be used to introduce different commands and options available in the language. A detailed discussion of all the commands is contained in the CELLSIM II Users Manual (Donaghey, 1975).

A. EXAMPLE I—TUMOR MODEL WITH ASORBING G_0 STATE

The CELLSIM program in Fig. 7 contains a G_0 state. Once cells enter this state, they remain there for the rest of the simulation. This model differs from the one in Fig. 2 where cells were allowed to exit from G_0. States from which cells never exit are called "absorbing states" in CELLSIM, and G_0 is declared to be such a state in the third command of the program. The program would also work by giving G_0 an extremely long time in the TIME IN STATES command. However, the program runs more efficiently by using the absorbing states command, since the system does not attempt to follow cells once they enter such a state.

This program also demonstrates the use of the CHANGE command. The modeler feels that a higher proportion of cells will enter the G_0 state as the total

```
CELL TYPES TEST;
STATES(1) S,G2,M,G1,G0;
ABSORBING STATES(1) G0;
TIME IN STATES(1) S:LOGN(9,6), G2:LCGN(6,4), M:1, G1:LOGN(12,10);
FLOW(1) G1-S(ALL), S-G2(ALL), G2-M(ALL), M-G1(.95), M-GC(.05);
PROLIFERATICN(1) M:1.8;
INOCULUM(1) 1C00;
CHANGE FLOW(1), WHEN LG(TOTAL) > 4, G1-S(ALL), S-G2(ALL),
     G2-M(ALL), M-G1(.80), M-GO(.20);
CHANGE FLOW(1), WHEN LG(TOTAL) > 5, G1-S(ALL), S-G2(ALL),
     G2-M(ALL), M-G1(.50), M-GO(.50);
REPORTS 12 HOURS;
GRAPH TOTAL, LG(TOTAL);
SIMULATE 504 HOURS;
```

FIG. 7.   Example I. Model with absorbing G_0 state. Proportion of cells flowing to G_0 increases as the population grows.

cell population gets larger. The original flow had .05 as the proportion of cells exiting M that go into G_0. The first CHANGE command in the program specifies that when the logarithm (base 10) of the total cells in the model reaches 4.0, the proportion going to G_0 will be .20. The next CHANGE command indicates that when the logarithm reaches 5.0, the proportion will be .50.

There are two graphs specified in this model. The first one is a graph of the total cells in the model, and the second one is a graph of the logarithm of the cells. Figures 8 and 9 show the two graphs that were generated when the model was fun for 504 hours (21 days) with a reporting interval of 24 hours.

Figure 10 shows some of the written reports that were generated during the execution of this model. The first portion of the output shows that the model required 25% of the available memory and could be run with as many as 401 groups. The second portion gives a key for the populations in the model and the states that are in each population. The first, and only, population in the model is population TEST and it is assigned population number one. The five states in the population are also assigned index numbers. These index numbers will be used to refer to the populations and states in the remainder of the reports. The system then reports at $t = 0$. The total cells in the model at that time are 1000, and they are represented by 82 groups. The report then shows the cells and groups in each population and state. At $t = 0$, there are no cells in state G_0, the absorbing state. The default inoculum in this model caused no cells to be initially located there. The number immediately under the time of the report gives a measure of the activity of the model up until that reporting time. The value is the number of change of states divided by 100. A change of state takes place every time a group leaves one state and moves to another. For example, at $t = 12$, the value immediately under the time is 1.03. This means the model went through 103 changes of state as it progressed from the beginning of the simulation until $t = 12$. These reports can be suppressed by the modeler with a NO PRINT command. It should be noticed in this model that the cells in the G_0 state (state 5) are represented by zero groups. This results from declaring state G_0 as an absorbing state.

B. Example II—Fraction Labeled Mitosis Model

This example (Fig. 11) demonstrates how fraction labeled mitosis experiments may be simulated in CELLSIM. The CELL TYPES command indicates that there will be two populations in the model: an unlabeled population called UNLB, and a labeled population called LABL. The information contained in the STATES, TIME IN STATES, FLOW, and PROLIFERATION commands is declared to be the same for both populations by enclosing '(1, 2)' in each command. The initial inoculum of 5000 cells is placed in population one, the unlabeled population. The simulated labeling takes place at $t = 20$ hours. In an

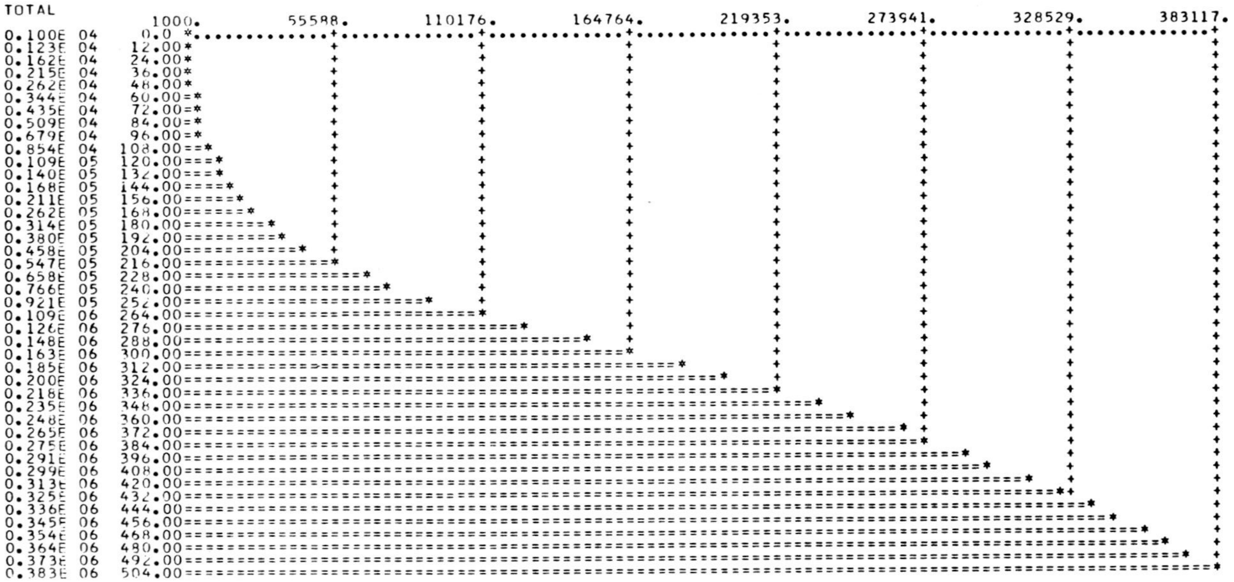

Fig. 8. Example I. Graph of the total cells in the model.

186

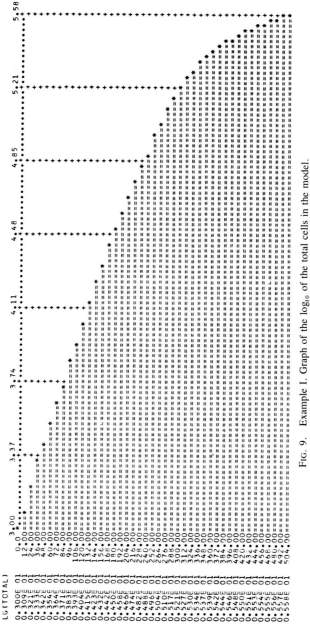

FIG. 9. Example I. Graph of the log₁₀ of the total cells in the model.

187

```
% MEMORY USED 25.0
POSSIBLE GROUPS 401

POPULATIONS      STATES

1 TEST            1 S
                  2 G2
                  3 M
                  4 G1
                  5 G0


TIME      TOTAL            POPULATIONS       STATES
0.0       0.1000E  04  82
0.01
                           1  0.1000E  04  82
                                             1 0.3266E  03  27    2 0.2082E  03  17    3 0.3525E  02  3
                                             4 0.4300E  03  35    5 0.0            0

12.00     0.1233E  04  82
1.03
                           1  0.1233E  04  82
                                             1 0.3842E  03  30    2 0.2545E  03  21    3 0.1209E  02  1
                                             4 0.5563E  03  30    5 0.2625E  02   0

24.00     0.1620E  04  82
2.63
                           1  0.1620E  04  82
                                             1 0.4654E  03  24    2 0.3332E  03  21    3 0.4914E  02  4
                                             4 0.7027E  03  33    5 0.6977E  02   0

36.00     0.2149E  04  82
4.06
                           1  0.2149E  04  82
                                             1 0.6144E  03  28    2 0.2702E  03  13    3 0.4163E  02  2
                                             4 0.1093E  04  39    5 0.1292E  03   0
```

FIG. 10.   Example I. Standard output report. Report is produced at each reporting interval.

actual cell population, a radioactive precursor such as [³H]thymidine is introduced to the population and is taken up by those cells currently making DNA. These are the cells in the S state. The radioactive nuclei of these cells can be readily detected by autoradiographic techniques as the cells progress through the cycle. The MOVE command in the model accomplishes the labeling. At $t = 20$, the cells in the S state of the unlabeled population are moved to the S state of the labeled population. These cells are moved in place. That is, a cell with a specified amount of time remaining in S in the unlabeled population will have exactly the same amount of time remaining in S in the labeled population after the move is made. The fraction labeled mitosis (FLM) curve that is developed in these experiments shows the proportion of labeled mitotic cells. The GRAPH command in the model generates a simulated FLM curve by requesting a plot of:

$$\frac{M(2)}{M(1) + M(2)}$$

Figure 12 shows the resulting graph from this model. The first peak is relatively well defined. There is a decay in successive peaks caused by the variation in the cell cycle times. This graph is very typical of FLM curves obtained from actual data.

This example also demonstrates the use of the NUMBER OF GROUPS command. The modeler declares (see Fig. 11) that there are to be initially 300 groups

```
CELL TYPES UNLB,LABL;
STATES(1,2) G1,S,G2,M;
TIME IN STATES(1,2) G1:LOGNORMAL(9,8), S:LOGNORMAL(8,6),
    G2:LOGNORMAL(5,3), M:LOGNORMAL(1.8,.6);
FLOW(1,2) S-G2(ALL), G2-M(ALL), M-G1(ALL), G1-S(ALL);
PROLIFERATION(1,2) M:2;
INOCULUM(1) 5000;
MOVE, AT TIME=20, S(1) TO S(2);
NUMBER OF GROUPS(1)=300;
REPORTS 1 HOUR;
GRAPH M(2)/(M(1)+M(2));
SIMULATE 85 HOURS;
```

FIG. 11.   Example II. Fraction labeled mitosis experiment.

in population one. Had this command not been placed in the program, the system would have defaulted to approximately 80 groups for population one. Since population two (the labeled population) initially had zero cells, no groups will be allocated for it. At $t = 20$ when the move takes place, all of the groups representing the cells in S(1) will be moved to population two. The use of more groups gives better resolution of CELLSIM graphs. This is especially true of simulated FLM curves, since the mitotic state is usually short in relation to the total cell cycle. With just a few groups in mitosis at any time, the FLM curve will have more abrupt changes and the resulting curve will appear jagged. Of course, putting more groups in a model means there will be more entities that the CELLSIM system must keep track of, and this means longer execution time. Example II with the 300 groups requires about 30% more execution time than it would need if it were run with 80 groups. All of the examples in this paper were run on an IBM 360/44. Execution time is highly dependent on the machine that is used. Example II required about 240 seconds on the 360/44 but might take less than 20 seconds on a more modern computer system.

C. Example III—Drug Effects Model

Example III (Fig. 13) demonstrates some other commands available to a CELLSIM user. This model simulates chemotherapy treatment of a tumor. The first point to be noted in the model is the use of the triangular distribution in the TIME IN STATES command. The time in state M will be a random variable from a probability density function that has the shape shown in Fig. 14. A block of the interface between G_1 and S is put in from $t = 24$ until $t = 48.01$. The block is 90% effective with 10% of the cells moving into S while it is in effect. In addition to blocking, the modeler feels the drug introduced at $t = 24$ also elongates the time in S for cells already there, and for the ones that get through the block. Therefore, the first CHANGE TIME IN STATES command, put into

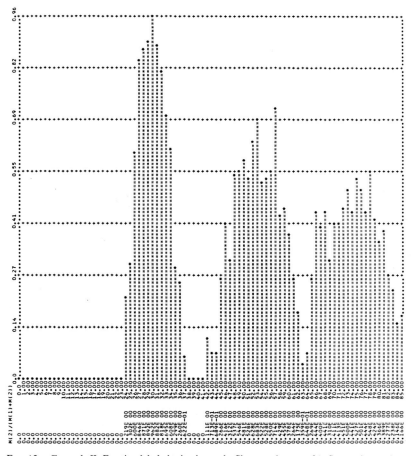

Fig. 12.   Example II. Fraction labeled mitosis graph. Sharp peak at $t = 31$. Successive peaks are lower and less well defined.

effect at $t = 24$, changes the mean time cells will spend in S from 9 hours to 27 hours. Cells that get through the block will obtain their assigned time in S from this revised log-normal distribution. The ACCELERATE command that is also introduced at $t = 24$ affects the cells that are already in S. The acceleration rate is 0.333, which, since it is less than unity, is actually a deceleration of the cells. It causes a change in the scheduled exit time for cells in S. For example, suppose at $t = 24$, a group of cells resides in S and this group is scheduled to exit S at $t = 27$ (3 hours in the future). The ACCELERATION command with a rate of 0.333 would change the exit time to $t = 33$ (9 hours in the future). The system recalculates a new time remaining in state by taking the current time remaining

```
CELL TYPES TUMR;
STATES(1) G1,S,G2,M,GO;
TIME IN STATES(1) G1:LCGN(12,9), S:LOGN(9,6), G2:LOGN(6,3),
      M:TRI(0,.5,1.4);
ABSORBING STATES(1) GO;
FLOW(1) G1-S(ALL), S-G2(ALL), G2-M(ALL), M-G1(.95), M-GO(.05);
PROLIFERATION(1) M:1.9;
INOCULUM(1) 10C000;
BLOCK EXIT(1) BETWEEN 24 AND 48.01 G1:.90;
CHANGE TIME IN STATES(1), AT TIME =24, S:LOGN(27,6);
ACCELERATE(1), AT TIME=24, S:.333;
CHANGE TIME IN STATES(1), AT TIME=48, S:LOGN(9,6);
ACCELERATE(1), AT TIME=48, S:3.0;
KILL(1) AT TIME = 54 S:.95;
KILL(1) AT TIME=75 S:.95;
KILL(1) AT TIME=92 S:.95;
REPORTS 1 HOUR;
GRAPH S(1)/TOTAL, TOTAL;
NO PRINT*;
SIMULATE 96 HOURS;
```

FIG. 13.   Example III. Drug effects model.

(3.0 hours) and dividing it by the acceleration rate (0.333). The group's scheduled exit time in the state is then changed to 9 hours in the future. Of course, anytime an ACCELERATION command contains a value greater than 1.0, the exit time would be moved up.

The slowing effect is removed 24 hours later. Another CHANGE TIME IN STATES command at $t = 48$ puts the time in state S back to its original log-normal distribution, and another ACCELERATE command with an acceleration rate of 3.0 speeds the exit time of the cells residing in S at $t = 48$. The BLOCK EXIT command has its termination time as 48.01 to insure that the wave of cells that will move into S at that time will get their time in S from the original distribution and avoid the acceleration that takes place instantaneously at $t = 48$.

In this model, three kills are scheduled at $t = 54$, $t = 75$, and $t = 92$ hours. The kills destroy 95% of the cells in S. Two graphs are requested: one of the proportion of cells in S and the other of the total cells in the model. Figure 15 shows the graph of the total cells in the tumor. It can be seen that the number of cells has been reduced from the initial inoculum of 10,000 to 5,800 with this schedule of treatment. The NO PRINT* command in the model suppresses the written reports.

The researcher or clinician, using this type of model, could manipulate the scheduled kills and observe which gave the best results. One approach would be

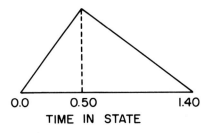

FIG. 14.   Example III. The triangular distribution specified for the time cells are to spend in state M.

to run the model with no kills and determine from the graph of the proportion of cells in S the optimum time for the first administration of the killing drug. The model would be run then with the single kill scheduled at that time. The results from this run would show the best time for a second kill. The best time for a third kill could then be determined and this process could be continued indefinitely. The modeler may also wish to include a normal cell population and observe the effect that the treatment has on these cells.

D. EXAMPLE IV—MODEL WITH DIURNAL VARIATION

This program (Fig. 16) demonstrates how a diurnal effect can be included in a CELLSIM model. Two populations are included: the cell population named CLLS and a timer population called TIMR. The timer population is an artificial one, and it provides the mechanism for including a cyclic action in population CLLS. Population TIMR has only two states, A and B. States A and B both have a constant time in state of 12 hours. The single "cell" in TIMR moves back and forth between these two states. The position of this timer cell determines the times in states for population CLLS. This is accomplished by the WHILE clauses included in the TIME IN STATES commands for population one. When the timer cell is in state B [A(2) < B(2)], the state times, especially for G_1, are longer than when the timer cell is in state A [A(2) > B (2)]. There is another TIME IN STATES command included in the model for population one, that does not contain a WHILE clause. It is necessary in CELLSIM that the time in states be explicitly specified at $t = 0$. This is not accomplished if only TIME IN STATES with WHILE clauses are used. FLOW and PROLIFERATION commands may all include WHILE clauses.

A DATA statement is included in this program. The usual purpose of a DATA statement is to superimpose some actual data on an output graph. For example, suppose a model included the following DATA statement:

DATA 1 : 5, .19: 12, .22: 36, .07;

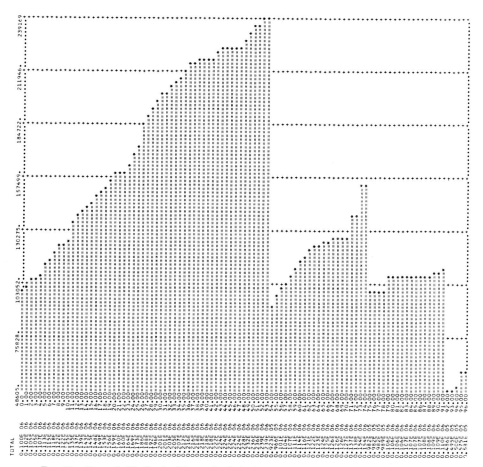

FIG. 15.   Example III. Graph of the total cells in the tumor. The scheduled treatment has reduced the number of cells from 10,000 at $t = 0$ to 5,800 at $t = 96$.

This statement gives three data points for some variable at $t = 5$, 12, and 36. If a GRAPH command included in the same program with this DATA statement were:

GRAPH 1 M(1)/TOTAL;

This would cause, on the resulting graph, the three data points to be superimposed on the model output. The data points are merely indicated with a "0" at the appropriate position on the graph, and the modeler can connect them by hand. However, in Example IV, the use of the DATA statement is to allow the modeler to set for the graph his own minimum and maximum of the proportion of cells in

```
CELL TYPES CLLS, TIMR;
STATES(1) G1,S,G2,M;
STATES(2) A,B;
FLOW(1) G1-S(ALL), S-G2(ALL), G2-M(ALL), M-G1(ALL);
FLOW(2) A-B(ALL), B-A(ALL);
TIME IN STATES(1), WHILE A(2) < B(2), G1:LOG(12,10),
     S:LOG(8,3), G2:LOG(5,2), M:1;
TIME IN STATES(1), WHILE A(2) > B(2), G1:LOG(8,3),
     S:LOG(7,3), G2:LOG(4,2), M:1;
TIME IN STATES(1) G1:LOG(8,3), S:LOG(7,3), G2:LOG(4,2), M:1;
TIME IN STATES(2) A:12, B:12;
PROLIFERATION(1) M:2;
INOCULUM(1) 5000;
INOCULUM(2) B:1 AT BEGINNING;
NUMBER OF GROUPS(1)=200;
NUMBER OF GROUPS(2)=1;
REPORTS 1 HOURS;
DATA 1: 151,0: 152,1;
GRAPH $1$ G1(1)/TOT(1);
NO PRINT*;
SIMULATE 150 HOURS;
```

FIG. 16.   Example IV. Model of cell population with diurnal effect. Population TIMR is a dummy population used only for regulating diurnal rhythm.

G_1. The DATA statement in the program specifies two points: a value of 0 at $t = 151$, and a value of 1.0 at $t = 152$. Since the simulation lasts for only 150 hours, neither of these points will be superimposed on the graph produced by the model. However, CELLSIM determines the minimum and maximum for a graph by examining all values generated during the simulation plus any values included with a DATA statement associated with the graph. Therefore, the minimum and maximum for the proportion of cells in G_1 will be 0.0 and 1.0 as furnished by the data statement. The graph of the proportion of cells in G_1 is shown in Fig. 17.

Figure 17 clearly shows the cyclic action that is established in the cell population. The INOCULUM command specified only that 5000 cells were to be placed initially in the population. The default algorithms that placed the cells initially in the cycle did not establish the diurnal rhythm. However, the initialization effects are quickly suppressed after a few cycles. Any testing or perturbations on the model should only take place after the initialization phase has been completed. A modeler has the capability in CELLSIM, by using a POSITION command, to get a complete description of exactly where all the groups in the model are located. He may then rerun the model using this information in an INOCULUM command. The cells will then be placed in the cell cycle as they were located after the initialization phase had been completed.

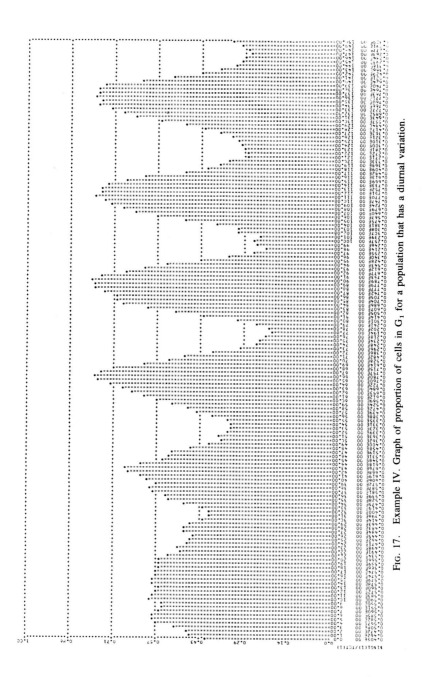

FIG. 17. Example IV. Graph of proportion of cells in G_1, for a population that has a diurnal variation.

```
CELL TYPES TEST;
STATES(1) G1,S,G2,M;
FLOW(1) G1-S(ALL), S-G2(ALL), G2-M(ALL), M-G1(ALL);
TIME IN STATES(1) G1:LOGN(12,9), S:LOGN(7,4), G2:LOGN(6,3), M:1;
BLOCK EXIT(1) BETWEEN 40 AND 64 G1:.95;
PROLIFERATION(1) M:2;
INOCULUM(1) 10000;
REPORTS 1 HOUR;
NUMBER OF GROUPS(1)=300;
NO PRINT*;
FMF: 39, 63.9, 68, 76: G1(1), S(1), G2(1)+M(1): ORIGIN=30: CV=7;
SIMULATE 77 HOURS;
```

FIG. 18.   Example V. CELLSIM model with simulated flow microfluorometry (FMF) output.

E. EXAMPLE V—SIMULATED FLOW MICROFLUOROMETRY OUTPUT

This model (Fig. 18) demonstrates another form of output available to a CELLSIM modeler. The FMF command indicates that the user wishes to obtain simulated flow microfluorometry (FMF) histograms from the model. When a cell population is to be analyzed by FMF equipment, the cells are stained with a fluorescent dye having specific stoichiometric affinity for DNA. The stained cells are then passed, single file, through a laser beam. During the transit across

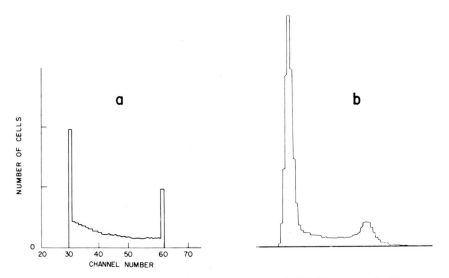

FIG. 19.   Example V. Theoretical FMF histogram (a) and typical FMF histogram (b). Distortion due to light scatter, random noise, etc.

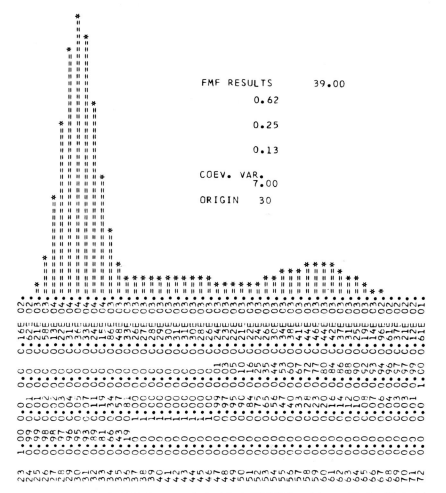

FIG. 20.   Example V. FMF histogram obtained from CELLSIM model at $t = 39$. Specified origin cell is 30. There are 62% of the cells with minimum DNA, 25% with intermediate DNA, and 13% with maximum DNA.

the beam, a fluorescent light flash is emitted from each cell, the intensity of which is proportional to the amount of fluorochrome bound to the DNA. Counters are connected to the equipment and record the number of cells having specific DNA amounts (Dean and Jett, 1974). A theoretical DNA distribution histogram for exponentially growing cells obtained by FMF is shown in Fig. 19a. However, due to instrumentation error, electronic noise, light scattering, etc., the resulting FMF histogram will appear typically as shown in Fig. 19b. The

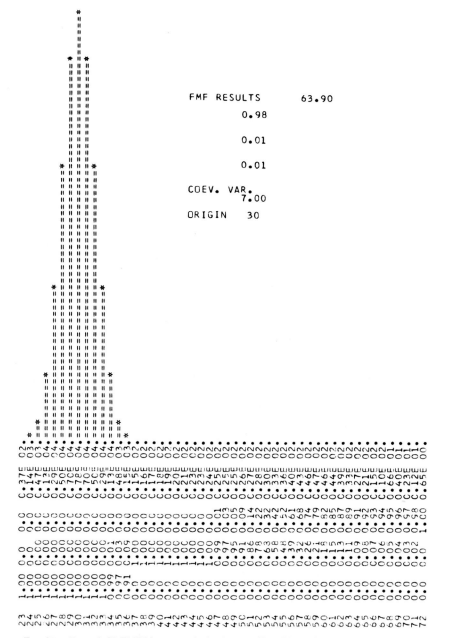

FMF RESULTS   63.90

0.98

0.01

0.01

COEV. VAR.
7.00

ORIGIN   30

FIG. 21.   Example V. FMF histogram obtained at $t = 63.9$. Block of exit from G_1 state has been in effect for 23.9 hours. Cells with minimum DNA make up 98% of population.

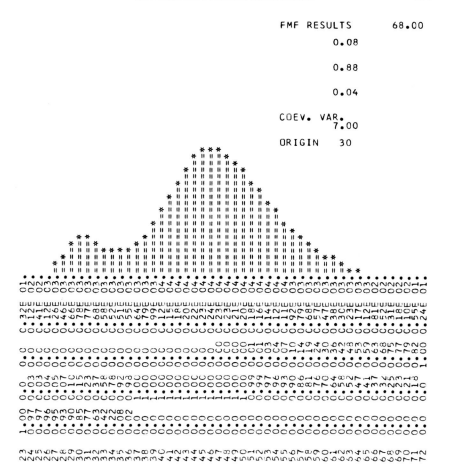

FIG. 22.   Example V. FMF histogram obtained at $t = 68$. Cells were released from G_1 exit block 4 hours earlier. Eighty-eight percent of cells are in the S state (intermediate DNA).

left-hand peak on the histogram represents the cells in G_1 phase (minimum amount of DNA) and the peak to the right represents cells in $G_2 + M$ (maximum amount of DNA). The segment between the peaks represents cells in S phase.

The FMF command in the example specifies that FMF histograms are to be generated at $t = 39, 63.9, 68,$ and 76 hours. The three segments of the histograms are to be: G_1 cells in the left-hand segment with the minimum amount of DNA, S cells in the middle segment with the variable amount of DNA, and cells in G_2 or M in the right-hand segment with the maximum DNA. The origin channel is to be channel 30, and this is where the G_1 cells (minimum DNA) will be concentrated. Since the left-hand peak is specified at channel 30, the right-

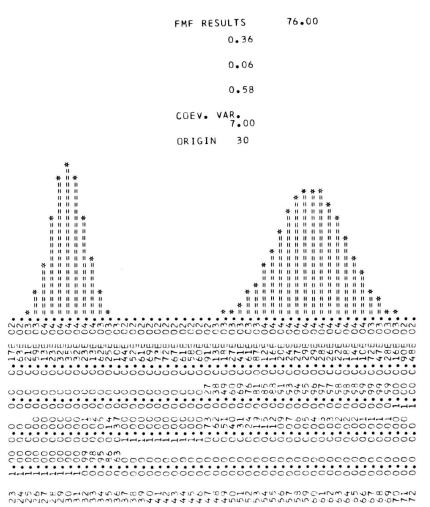

FIG. 23. Example V. FMF histogram obtained at $t = 76$, 12 hours after block removal. Fifty-eight percent of cells have maximum DNA (G_2 + M).

hand peak for cells with 2-fold amounts of DNA will be set at 60. The modeler also specifies in the command that the coefficient of variation (CV) for the histograms is to be 7.0. A coefficient of variation spreads the cells around the channels to which they were originally assigned, and it simulates the instrumentation error, noise, and light scatter present in the actual FMF equipment. The larger the CV value that is used, the greater the spread. If the command set the CV to zero, the resulting histogram would look similar to the one in Fig. 19a.

Figures 20 to 23 show the four simulated FMF histograms generated by this

```
CELL TYPES EXV1;
STATES(1) G1,S,G2,M;
TIME IN STATES(1) G1:LOGN(12,7), S:LOGN(8,4), G2:LOGN(5,3), M:1;
FLOW(1) G1-S(ALL), S-G2(ALL), G2-M(ALL), M-G1(ALL);
INOCULUM(1) 10000;
PROLIFERATION(1) M:2;
BLOCK EXIT(1) BETWEEN 63 AND 75 G1;
NUMBER OF GROUPS(1)=350;
REPORTS 1 HOURS;
NO PRINT*;
HISTOGRAM:63, 75: G1(1), S(1), G2(1), M(1);
DATA 1: 81,0: 82,100;
GRAPH $1$ S(1)/TOTAL*100;
POSITION :80: G1(1),S(1),G2(1),M(1);
SIMULATE 100 HOURS;
```

FIG. 24.   Example VI. CELLSIM model containing HISTOGRAM and POSITION commands.

model. The model has a BLOCK EXIT command that causes a 95% block of cells moving from G_1 to S. This block is in effect from $t = 40$ until $t = 64$. The first histogram was asked for at $t = 39$, and it is shown in Fig. 20. The values along the abscissa gives the channel number, the proportion of cells in each of the segments in that channel, and the total cells in the channel. For example, in channel 23, the proportion of cells with the minimum amount of DNA (G_1) is 1.0, and zero for the proportion in the other two segments. The total number of cells in channel 23 is 16 (0.16×10^2). This histogram was generated at $t = 39$ (before the block was put into effect) and reflects a signature histogram for the unpertrubed cell population. The output also shows the total proportions for the three segments to be 0.62 for the G_1 cells, 0.25 for the S cells, and 0.13 for the G_2 + M cells.

Figure 21 shows the histogram at $t = 63.9$, just before the 24-hour block was removed. At that time, 98% of the cells are concentrated in G_i. The histogram at $t = 68$ is shown in Fig. 22. This is 4 hours after block removal, and the wave of cells that accumulated at the G_1/S interface are now moving through S and beginning to enter G_2 + M. Figure 23 shows the histogram at $t = 76$, 12 hours after block removal. The effect of the block is still evident, with 58% of the cells in G_2 + M.

The use of simulated DNA histograms provides the researcher with an opportunity to validate CELLSIM models. If the simulated histograms are significantly different than those obtained from an actual cell population, one can conclude very quickly that the assumptions in the model are wrong. However, if the simulated histograms match actual ones under a variety of perturbations, the

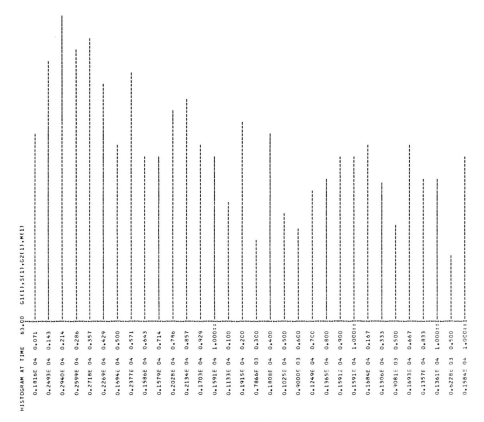

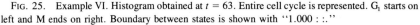

Fig. 25. Example VI. Histogram obtained at $t = 63$. Entire cell cycle is represented. G_1 starts on left and M ends on right. Boundary between states is shown with "1.000 : :."

modeler can have some confidence that the model is correct. There are several defaults and options available in FMF commands. For example, the origin channel and the CV will default to 14 and 5.0, respectively, if not mentioned in the command. The modeler can also specify a DNA rate. The default DNA rate is assigned so that the amount of DNA is exactly the same as the proportion of state S that has been completed. For example, cells that have spent 10% of their scheduled time in S will have manufactured 10% of their DNA. When they have completed 20% of their time in S, they will have produced 20% of their DNA, etc. However, it may be felt for some cell populations that most of the DNA is produced in the early phase of S. In that situation, cells that have spent 10% of their time in S may already have 40% of their DNA. A user may include information on this nonlinear DNA rate by specifying points on a DNA-TIME curve. These points are given in the FMF command. The methods used to process FMF commands are described in Donaghey *et al.* (1978).

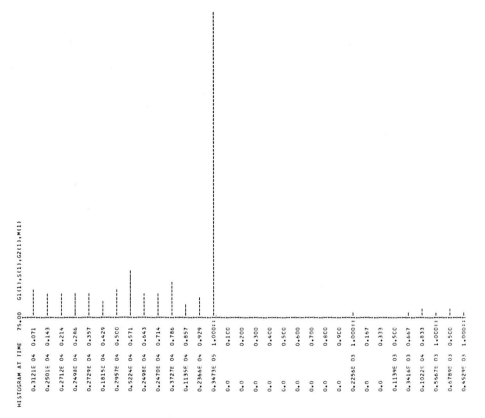

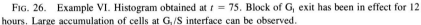

Fig. 26.   Example VI. Histogram obtained at $t = 75$. Block of G_1 exit has been in effect for 12 hours. Large accumulation of cells at G_1/S interface can be observed.

F. Example VI—Use of HISTOGRAM and POSITION Commands

This model (Fig. 24) demonstrates the use of the HISTOGRAM and POSITION commands. The model is a simple four-state model with a 12-hour block of G_1 between $t = 63$ and $t = 75$ hours. Two histograms are requested, one at $t = 63$ when the block is initiated, and one at $t = 75$ at the end of the blocking period. The four states to be included in the histogram are listed in the HISTOGRAM command in the order in which they are to appear. The histogram produced by a HISTOGRAM command gives more detail than one generated by an FMF command. With an FMF histogram, cells with the minimum amount of DNA (G_1/G_0) are concentrated in a single channel. Cells with the maximum DNA ($G_2 + M$) are also placed in a single channel. The histograms that result from a HISTOGRAM command show the distribution of the cells in all states that are included.

```
CELLS=   1
CELLS=  0.2854E 05   GROUPS=  113   TIME=  80.00
```

0.2269E 03	61.688	85.421	0.772
0.2277E 03	62.261	80.348	0.981
0.2269E 03	64.247	80.791	0.952
0.2269E 03	64.893	82.884	0.840
0.2277E 03	65.544	81.418	0.911
0.2269E 03	65.803	82.788	0.836
0.2269E 03	66.164	80.555	0.961
0.2269E 03	66.636	82.896	0.822
0.2269E 03	66.747	82.785	0.826
0.2269E 03	66.790	81.864	0.880
0.2277E 03	67.170	82.805	0.821
0.2277E 03	67.414	82.350	0.843
0.2277E 03	67.506	81.767	0.875
0.2277E 03	67.845	80.250	0.980
0.2277E 03	68.056	81.486	0.889
0.2277E 03	68.066	80.901	0.930
0.2269E 03	68.101	81.189	0.909
0.2269E 03	68.158	80.334	0.973
0.2269E 03	68.165	83.711	0.761
0.2269E 03	68.270	81.680	0.875
0.2269E 03	68.463	80.020	0.998
0.2269E 03	68.690	84.670	0.708
0.2277E 03	68.696	80.332	0.971
0.2277E 03	68.726	82.270	0.832
0.2277E 03	68.730	81.109	0.910
0.2277E 03	68.780	90.970	0.506
0.2277E 03	68.798	80.395	0.966
0.2277E 03	68.884	81.149	0.906
0.2277E 03	69.118	81.035	0.913
0.2256E 03	69.135	85.002	0.685
0.2256E 03	69.220	85.120	0.678
0.2277E 03	69.263	84.697	0.696
0.2269E 03	69.325	81.732	0.860
0.2277E 03	69.544	81.371	0.884
0.2269E 03	69.783	80.487	0.954
0.2277E 03	69.848	80.552	0.945
0.2269E 03	70.185	80.347	0.966
0.2277E 03	70.425	91.801	0.842
0.2269E 03	70.467	86.034	0.612
0.2277E 03	70.517	85.084	0.651
0.2269E 03	70.549	81.016	0.903
0.2277E 03	70.609	80.066	0.993
0.2277E 03	70.614	83.850	0.709
0.2277E 03	70.630	87.736	0.548
0.2277E 03	70.643	84.614	0.670
0.2269E 03	70.775	85.940	0.608
0.2277E 03	70.829	81.400	0.868
0.2269E 03	70.954	85.255	0.633
0.2277E 03	70.963	86.774	0.572
0.2269E 03	71.024	81.445	0.861
0.2269E 03	71.157	83.550	0.714
0.2269E 03	71.190	88.031	0.523
0.4347E 03	71.341	88.759	0.497
0.2277E 03	71.374	83.721	0.699
0.4512E 03	71.521	87.557	0.529
0.2269E 03	71.539	83.075	0.733
0.2277E 03	71.694	82.441	0.773
0.2277E 03	71.742	84.132	0.666
0.2277E 03	71.818	86.654	0.551
0.2277E 03	71.846	82.095	0.796
0.2277E 03	72.071	85.240	0.602
0.2277E 03	72.127	91.174	0.413
0.2277E 03	72.154	87.753	0.503
0.2277E 03	72.220	81.449	0.843
0.2277E 03	72.435	80.693	0.916
0.2277E 03	72.490	85.742	0.567
0.2277E 03	72.792	86.226	0.537
0.2277E 03	72.867	91.802	0.377
0.2277E 03	72.934	82.988	0.703
0.2277E 03	73.105	87.909	0.466
0.4512E 03	73.125	90.191	0.403
0.2277E 03	73.138	82.108	0.765
0.2277E 03	73.146	90.238	0.401
0.2277E 03	73.229	83.552	0.656
0.2277E 03	73.329	81.752	0.792
0.2277E 03	73.354	82.774	0.706
0.2277E 03	73.438	84.848	0.575
0.2277E 03	73.525	82.700	0.706
0.2277E 03	73.617	84.413	0.591
0.2277E 03	73.658	85.782	0.523
0.2277E 03	73.834	90.540	0.369
0.2277E 03	73.926	88.225	0.425
0.2277E 03	74.017	88.802	0.405
0.2277E 03	74.204	85.146	0.530
0.2277E 03	74.259	83.929	0.594
0.2277E 03	74.285	84.516	0.559
0.4347E 03	74.334	85.962	0.487
0.2277E 03	74.400	91.003	0.337
0.4512E 03	74.511	84.855	0.531
0.2269E 03	74.552	86.981	0.439
0.4347E 03	74.733	87.184	0.431
0.4347E 03	74.733	87.866	0.401
0.4512E 03	74.930	85.718	0.470
0.2269E 03	75.234	86.473	0.424
0.2277E 03	75.370	86.966	0.399
0.2277E 03	75.440	90.741	0.298
0.4512E 03	75.580	86.427	0.407
0.4512E 03	75.685	91.728	0.269
0.2269E 03	75.709	85.267	0.449
0.2277E 03	76.000	89.663	0.293
0.4512E 03	76.072	88.174	0.325
0.4537E 03	76.371	85.685	0.390
0.2277E 03	76.420	88.234	0.303
0.2277E 03	76.626	87.721	0.304
0.4512E 03	76.639	91.265	0.230
0.2277E 03	77.294	92.585	0.177
0.4512E 03	77.403	87.181	0.266
0.2277E 03	77.471	91.671	0.178
0.2277E 03	77.503	96.026	0.135
0.2277E 03	77.512	95.071	0.142
0.2277E 03	77.631	85.737	0.292
0.2277E 03	78.926	91.654	0.084
0.2277E 03	79.171	86.964	0.106

```
TIME REMAINING DISTRIBUTION
    0.0        0.0
    1.870      0.25
    3.739      0.39
    5.609      0.54
    7.479      0.71
    9.349      0.84
   11.218      0.93
   13.088      0.98
   14.958      0.98
   16.827      1.00
```

FIG. 27.

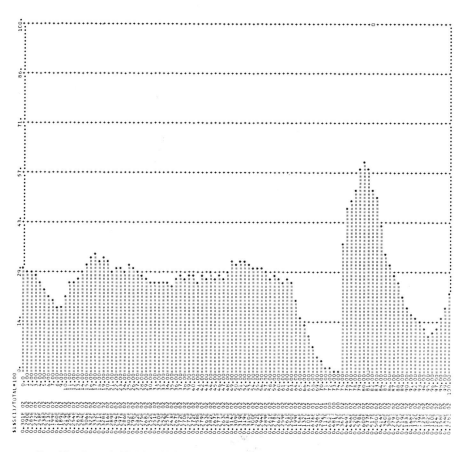

FIG. 28. Example VI. Model initiated with information from POSITION command. The first 20 hours from this model should be similar to the output from the original model from $t = 80$ until $t = 100$.

The histogram obtained at $t = 63$ is shown in Fig. 25. Each state mentioned in the command is assigned a section of the abscissa proportional to the average time spent in that state. Thus, for our model, the abscissa can be thought of as the total cell cycle, starting with G_1 and terminating with M. Each state is then subdivided into segments. The vertical line at each segment represents the number of cells in that segment. Figure 25 shows, for example, that 1816 cells (0.1816×10^4) have completed 0.071 or less of the G_1 state, and the vertical line

FIG. 27. Example VI. Information on G_1 state resulting from POSITION command. Similar tables are prepared for each state mentioned in the command. This table shows that at $t = 80$ the cells in G_1 are represented by 113 groups.

```
CELL TYPES EXV1;
STATES(1) G1,S,G2,M;
TIME IN STATES(1) G1:LOGN(12,7), S:LOGN(8,4), G2:LOGN(5,3), M:1;
FLOW(1) G1-S(ALL), S-G2(ALL), G2-M(ALL), M-G1(ALL);
INOCULUM(1)  G1: 28540 CDF(0,0, 1.870,.25, 3.739,.39, 5.609,.54, 7.479,.71,
                      9.349,.84, 11.218,.93, 13.088,.98, 14.958,.98, 16.827,1.0),
              S: 46130 CDF(0,0, 1.330,.16, 2.661,.33, 3.991,.56, 5.321,.74,
                      6.652,.83, 7.982,.91, 9.312,.97, 10.643,.99, 11.973,1.0),
             G2: 2041 CDF(0,0, 1.142,.17, 2.284,.17, 3.427,.28, 4.569,.39,
                      5.711,.61, 6.853,.83, 7.995,.83, 9.138,.89, 10.280,1.0),
              M: 113.9 CDF(0,0, .029,0, .059,0, .236,0, .265,1.0);
PROLIFERATION(1) M:2;
NUMBER OF GROUPS(1)=350;
REPORTS 1 HOURS;
NO PRINT*;
DATA 1: 81,0: 82,100;
GRAPH $1$ S(1)/TOTAL*100;
SIMULATE 20 HOURS;
```

FIG. 29.   Example VI. Graph of percentage of cells in S from $t = 0$ until $t = 100$.

printed for that segment represents those cells. The next vertical line represents 2493 cells that have completed more than 0.071, but less than 0.143 of their assigned time in G_1. The end of a state is indicated by a 1.000 on the abscissa. Figure 25 shows that the density of cells in early G_1 is about twice as great as the density at the end of the cell cycle. This is to be expected since the mitotic state, M, is the end of the cell cycle, and cells proliferate upon leaving M.

Figure 26 shows the histogram obtained at $t = 75$. At that time, the block of the G_1 exit had been in effect for 12 hours. The large accumulation of cells can be seen at the end of G_1, ready to move into S. If more histograms were requested after the block removal, it could be seen how this accumulation would become dispersed after several cell cycles.

The POSITION command in the model (Fig. 24) causes another form of CELLSIM output. This command will cause a complete description of each group of cells in each of the states mentioned in the command at $t = 80$. This is the most detailed output that can be obtained from a CELLSIM model. Figure 27 shows the information on the G_1 state that resulted from this command. A similar list would be produced for each of the other states. The second line of the list indicates that at $t = 80$ there are 28,540 cells in G_1, and they are represented by 113 groups. Information on each of these groups follows. The number of cells in each group and its state entry and scheduled exit time are displayed. The proportion of the state completed for each group is also shown. The first group in Fig. 27 represents 226.9 cells. It came into G_1 at $t = 61.688$ and its scheduled exit time is 85.421. The proportion of this group's time in G_1 that has been completed at $t = 80$ is .772. This information is given for all 113 groups in G_1.

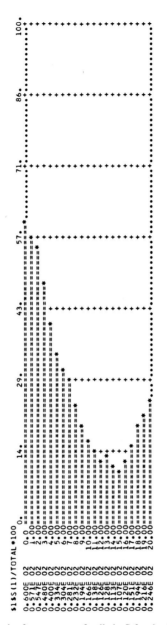

FIG. 30.   Example VI. Graph of percentage of cells in S for the first 20 hours from the restart model.

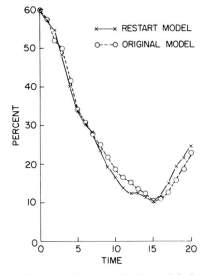

FIG. 31.   Example VI. Comparison of results from original and restart model.

The small table at the bottom of Fig. 27 is also produced by the POSITION command for each state. It summarizes the information on the cells in G_1 and displays it as a cumulative distribution of the time remaining in the state. The table shows that .25 (or one-quarter) of the cells in G_1 have 1.870 hours or less in the state. It also shows that all of the cells have 16.827 hours or less remaining in the state. This distribution can be used in an INOCULUM command, and a model is then initialized very close to the status it was in when the POSITION command was executed. The cumulative distribution function (CDF) option used in an INOCULUM command permits this type of initialization. Figure 28 shows the Example VI model initialized with the data obtained at $t = 80$. Thus the output from this model, from $t = 0$ to $t = 20$, should correspond to the original model from $t = 80$ to $t = 100$. Figure 29 gives the graph of the proportion of cells in S for the original model when it was ran for 100 hours. Figure 30 is the same graph for the model when it ran for 20 hours after being initialized with the information obtained in the POSITION command. Figure 31 superimposes the last 20 hours of the original model on the first 20 hours of the second model. It is obvious that the output is nearly identical. The major use of the POSITION command is to allow a restart of a model.

IV. Discussion

CELLSIM has been in existence for almost eight years. New commands and options have been continuously added to the language since it first appeared. A major concern in the design of CELLSIM was to create a modular system. That is, new features should be capable of being implemented with little or no disruption of previously written code. For the most part, this has been accomplished. Adding a new command to the CELLSIM system requires inserting two instructions to an interpretation routine to recognize the command and adding the necessary subroutines that process the command. Adding new options to an already existing command requires modifying the subroutines that process that particular command. In either case, the additions or modifications are localized. The current FORTRAN program that processes CELLSIM is made up of over 125 subroutines, requiring about 5500 statements.

The system is now in use at over 20 installations. It has been implemented on a variety of computer systems, among them IBM, UNIVAC, Honeywell, CDC, and DEC. The normal distribution mode is to send the FORTRAN source code on magnetic tape to those institutions requesting it. A new master tape that has the latest commands and features is prepared at least twice a year. Copies of the program that are distributed are made from the latest master. A CELLSIM newsletter, which describes the latest version and any new applications, is sent to all users two or three times a year. CELLSIM has recently been added to the PROPHET time-sharing network. PROPHET is a time-shared computer system developed under sponsorship of the United States National Institutes of Health. It has been operational since 1972 and is used for the analysis and visualization of biomedical data and information. The system is resident on a large PDP-10 computer in Massachusettes and is used by researchers at a number of institutions throughout the country. With PROPHET, a user accesses CELLSIM through an interactive graphics terminal. The output is displayed on a television-like screen, and the user may receive a hard copy of the display, if he desires, by utilizing an attached graphics thermal printer. The usual output device for non-PROPHET users is a line printer. Again, in the interest of having the system as universal and machine independent as possible, it was designed with no special input/output requirements. Any computer system that has at least a card reader for input and a line printer for output can process CELLSIM models.

Most of the CELLSIM users are utilizing the language in their research efforts. A model will be prepared containing the assumptions that a researcher has about the behavior of a cell population. It may include the effects of drugs and treatments on the population. The results of the model are then compared with those obtained from the actual population. As with any simulation model, a CELLSIM program does not show directly the true characteristics of a cell population. However, the modeler can very quickly and definitely tell if the assumptions he used to develop his model were correct, by matching simulated FMF histograms or labeling graphs with those obtained experimentally. The ease by which a CELLSIM model can be modified is, perhaps, it greatest advantage over other modeling techniques. A model can be run, the results examined, the model easily modified, and then run again. A significant amount of insight seems to be gained by this iterative process. Some models are run over 30 times before acceptable results are obtained. Each iteration of the model usually generates discussions among the researchers about the fundamental behavior of the population and very often provides insight about potential avenues for experimental verification. (Donaghey *et al.*, 1978).

CELLSIM is also being used to develop schedules for chemotherapy treatment for cancer patients. (Tejada *et al.*, 1978). Fraction labeled mitosis curves are developed from tumor biopsies. Using these actual FLM curves to obtain information on the times in states, a CELLSIM model is prepared and then modified until its simulated FLM curves match the actual ones. The CELLSIM model is then treated with different schedules of drug treatment. The schedule that gives the most promising results is then used for the patient. These *in vivo* applications of CELLSIM are just beginning, but the results are encouraging. CELLSIM models are understandable by clinicians and researchers who have had no previous computer experience, and it provides a communication medium for comparing models and assumptions.

Another area in which CELLSIM seems to be growing in popularity is education. Students can be taught the language in a short amount of time, and then are able to utilize it to obtain cell biology laboratory experience without having to resort to the expense and space–time requirements for actually growing and maintaining cell cultures. The effects of changing various parameters of a population can be demonstrated and investigated quickly and inexpensively. It has been used at both the graduate and undergraduate level.

ACKNOWLEDGMENTS

This work was supported by the following grants: NSF GJ–189, NSF GJ–37786, NSF MCS 76-11459, NIH RO1–CA1–19995–01, and NIH RO1–CA1–19995–02.

REFERENCES

Barret, J. C. (1966). *J. Natl. Cancer Inst.* **37**, 443.
Dean, P. N., and Jett, J. H. (1974). *J. Cell Biol.* **60**, 523.
Donaghey, C. E. (1975). "CELLSIM II User's Manual," 2nd ed. Department of Industrial Engineering, University of Houston, Houston, Texas.
Donaghey, C. E., and Ozkul, O. S. (1971). *Eng. Educ.* **61**, 359.
Donaghey, C. E., Dewan, P., and Singh, D. (1970). *Ind. Eng.* **2**, (12).
Donaghey, C. E., Drewinko, B., Barlogie, B., and Stubblefield, E. (1978). *BioSystems* **10**, 339.
Kember, K. F. (1969). *Cell Tissue Kinet.* **2**, 11.
Naylor, T. H., Balintfy, J., Burdick, D., and Chu, K. (1966). "Computer Simulation Techniques." Wiley, New York.
Stubblefield, E., and Dennis, C. M. (1976). *J. Theor. Biol.* **61**, 171.
Tejada, F., Leung, I., and Zubrod, C. G. (1978). *Proc. Cell Kinet. Soc.* p. 35.
Tibaux, G., Firket, H., and Hopper, A. F. (1969). *Cell Tissue Kinet.* **2**, 333.

INTERNATIONAL REVIEW OF CYTOLOGY, VOL. 66

The Formation of Axonal Sprouts in Organ Culture and Their Relationship to Sprouting *in Vivo*

I. R. DUCE

Department of Zoology, University of Nottingham, Nottingham, England

P. KEEN

Department of Pharmacology, University of Bristol, Bristol, England

This article attempts to discuss the phenomenon of axonal sprouting in ganglion–nerve preparations maintained in organ culture and the relationship of this sprouting to axonal regeneration and the reaction of neurons to injury. No attempt will be made to cover developmental processes, the growth of dissociated neurons in tissue culture or tissue transplantation studies. Since regeneration in nonmammalian species differs markedly from that in mammals, only the latter will be considered.

Particularly when dealing with pseudo-unipolar sensory neurons, some topographical definitions are necessary: (a) central and peripheral denote proximity to the central nervous system; (b) proximal and distal denote proximity to the cell body; and (c) anterograde and retrograde denote movement away from and toward the cell body, respectively.

I. What Is Axonal Sprouting?

The term "axonal sprouting" has been used to describe a variety of phenomena.

1. *Sprouting from Interrupted Axons*

When an axon is interrupted by cutting or crushing, parts of the axon distal to the injury degenerate and sprouting occurs from the axons on the proximal side of the injury (Cajal, 1928). Usually in the peripheral nervous system, but only rarely in the central nervous system, do some of these sprouts grow to reestablish functional synaptic connections. Such sprouts may be *terminal,* i.e., arising from the cut tip of an axon, or *collateral* i.e., arising as a side branch of the axon on the proximal side of the cut.

2. *Reactive Sprouting*

a. *In Adjacent Neurons.* When the normal innervation of an area is interrupted in any of several ways, adjacent axons sprout to occupy the vacated synapses. This phenomenon is called "reactive sprouting" and occurs in both the central (Cotman, 1978) and peripheral (Edds, 1953) nervous sytems. The sprouts formed may be terminal or collateral.

b. *In Axonal Collaterals of a Damaged Neuron.* When two axonal branches of a single neuron project to different areas, damage to one branch can initiate sprouting in its remote, uninjured axonal collateral (Pickel *et al.,* 1974).

3. *Sprouting from Cell Bodies of Axotomized Neurons*

When sympathetic neurons are axotomized, finger-like extensions of neuronal membrane containing cytoplasm form on the surface of their cell bodies (Matthews and Raisman, 1972; Purves, 1975). These have been termed "axonal sprouts" by several authors, presumably because they are supposed to be analogous to axons. Since they arise from the cell body, however, they should be clearly distinguished from true axonal sprouts.

II. Changes following Nerve Section *in Vivo*

To assess the significance of the changes that occur in severed nerves isolated with their ganglia in organ culture, it is necessary first to review the changes that follow section of a nerve *in vivo*.

A. Cell Body

The cell bodies of peripheral neurons pass through three stages following interruption of their axons (Brattgard *et al.*, 1957). There is first a latent stage during which neurotransmitter-related molecules continue to be produced and the levels of these in the cell body may even rise. This is succeeded by a period during which a set of retrograde changes take place in the cell body. These changes have been termed ''chromatolytic'' and are thoroughly reviewed by Lieberman (1971, 1974), Grafstein (1975), and Watson (1974). In summary, these changes are as follows:

1. In light microscope preparations, the Nissl substance, which is identified by its ability to take up basic dyes, disappears, giving rise to the term ''chromatolysis.'' Electron microscopy shows that this is due to a rearrangement of the rough endoplasmic reticulum and an increase of free polyribosomes.

2. Cell bodies become swollen.

3. Nuclei migrate to the periphery of the cell body and the nucleolus increases in size.

4. Synthesis and turnover of nucleolar RNA are increased.

5. The number of lysosomal bodies increases, and there is a concomitant increase in acid phosphatase activity.

6. There is a diversion of synthesis away from molecules required for synaptic function toward those structural elements required for growth and regeneration of the injured axon.

If the axon grows to reestablish functional connections, a third recovery or maturation phase ensues during which the neuron reverts to its original state (Brattgard *et al.*, 1957; Watson, 1970). A number of factors govern the length of the latent period between axonal injury and the onset of chromatolysis and these have been reviewed by Lieberman (1974). In particular, the closer the injury to the cell body, the more rapid the onset of the reaction (Watson, 1968; Kristensson and Olsson, 1975). The latter authors crushed sciatic nerves at various distances from the dorsal root ganglia and showed that there was a correspondence between the time taken for a marker to be transported from the crush to the ganglia and the time for the onset of chromatolysis. These experiments suggest that retrograde transport of some ''message'' (or cessation of transport of a normal constituent) might be the trigger for chromatolysis. The possible nature of the stimulus has been critically reviewed by Cragg (1970) and Grafstein and McQuarrie (1978).

Changes of intermediary metabolism in the cell body do not follow any clear pattern (Watson, 1966a; Härkönen and Kauffman, 1973a,b). Incorporation of radiolabeled amino acids into the cell bodies of axotomized neurons is variously reported as being either increased (Miani *et al.*, 1961) or decreased (Engh *et al.*,

1971). The complexity of the situation was underlined by Kung (1971) who showed that at time points up to 4 hours following administration of radiolabel, axotomized neurons contained less radioactivity than controls, but that from 16 hours onward the situation was reversed. Rather than any consistent change in the rate of protein synthesis, axotomy appears to bring about a qualitative change in the pattern of proteins produced (Hall et al., 1978).

Changes in neurotransmitter-related processes in axotomized cells bodies have been studied most thoroughly in adrenergic neurons. Dense-core vesicles are present in the perikaryon of these neurons; these vesicles contain noradrenaline (NA) (Richards and Tranzer, 1975) and the enzyme that forms it, dopamine-β-hydroxylase (DBH) (Lagercrantz, 1976). The number of vesicles in the cell bodies is reduced by one-half 12–38 hours following axotomy, and concurrently, vesicles accumulate in the interrupted postganglionic axons (Matthews and Raisman, 1972).

Three days after ligating the sciatic nerve, NA levels in lumbar sympathetic ganglia are normal; they then decline to 50% of control values at 14 days, and at 30 days, have returned to normal (Cheah and Geffen, 1973). Levels of tyrosine hydroxylase (TH), which is the enzyme catalyzing the first step in the synthesis of NA from tyrosine, are 76% of control levels at 3 days and 70% at 14 days (Cheah and Geffen, 1973). In the superior cervical ganglion, postganglionic axotomy causes an immediate decline in levels of DBH to 20% of normal at 8 days (Kopin and Silberstein, 1972). When postganglionic axons are interrupted close to the ganglion, the NA content of the cell bodies may initially *increase* due to a pileup of transmitter at the point of transection extending back into the perikaryon (Jacobowitz and Woodward, 1968; Dahlström, 1971).

Matthews and Raisman (1972) describe two types of sprouting from the cell bodies of axotomized adrenergic neurons:

1. During the first two days, finger-like extensions of axonal membrane protrude into the satellite-cell. The cytoplasm of these sprouts contains ribosomes and granular endoplasmic reticulum.

2. Three days after axotomy, a second type of sprout appears. These contain organelle-free cytoplasm and extend circumferentially from the body of the neuron into the space that has, at this stage, been left by separation of the satellite cell from the neuron. These reach a maximum size 6–10 days after axotomy and then regress.

Purves (1975), on the other hand, described numerous abnormal profiles 5–10 μm in diameter containing an unusual complement of tubular and vesicular organelles, mitochondria. and dense-core vesicles. He considered that at least some profiles corresponded to large varicosities of dendrites not seen in normal neurons.

In cholinergic neurons, axotomy causes a reduction of histochemically demonstrable acetylcholinesterase (AcChE) to a minimum at 5 days (Flummerfelt and
Lewis, 1975). Biochemical analysis of individual neurons shows that AcChE
levels remain normal for the first 3 days following a nerve crush, then drop to a
minimum level at day 10, and as regeneration proceeds, recover to normal at day
20 (Watson, 1966b).

B. CHANGES IN AXON PROXIMAL TO INJURY

Following axotomy, there is a reduction in the overall diameter of axons
proximal to the injury, with a corresponding reduction in conduction velocity
(Cragg and Thomas, 1961; Aitken and Thomas, 1962). Nerve section has a
greater effect on conduction velocity than nerve crush (Horch, 1976).

The greater part of neuronal protein synthesis takes place in the cell body.
Macromolecules are carried from the cell body to the axon and nerve terminals
by axonal transport processes, which hence play a major role in the regeneration
process. Three classes of axonal transport have been recognized: (a) Fast anterograde, carrying mainly particulate material from the cell body at velocities of
up to 400 mm/day; (b) Slow anterograde, carrying mainly soluble material at
velocities of the order of 1 mm/day by a process of axoplasmic flow; and (c)
Retrograde, returning material to the cell body at a velocity approximately half
that of the fast anterograde system (Heslop, 1975; Lubinska, 1975; Kristensson,
1978).

Changes in axonal transport following axotomy have been measured in two
ways. First, by injecting radiolabeled precursors into the vicinity of the cell
bodies and following the progress of the incorporated material down the axon,
the *velocity* of the fastest moving component (Ochs, 1976) can be determined.
Second, by measuring the accumulation of the transported substance at an occlusion placed on the axon, the average *rate* of transport (in amount of material per
unit time) can be determined. If a number of assumptions are made, an apparent
average velocity may be calculated from the average rate (Dahlström, 1971).
Following axotomy, there is no change in the velocity of transport of the fastest
transported proteins (Griffin *et al.*, 1976; Ochs, 1976) or glycoproteins (Frizell
and Sjöstrand, 1974a) in the axon proximal to the injury. This is perhaps not
surprising as this maximal velocity is remarkably constant in different nerves and
in different species both *in vivo* and *in vitro* (Ochs, 1974). The amount of protein
transported by the fast system has, on the other hand, been variously reported to
be either increased (Griffin *et al.*, 1976), unchanged (Ochs, 1976), reduced
(Bulger and Bisby, 1978), or initially reduced and then increased (Frizell and
Sjöstrand, 1974b). If, as suggested above, nerve section causes a change in the
type of protein exported from the cell body, one might expect a selective increase
in the transport of glycoproteins that are required for the growth of axon mem-

brane (Bennett *et al.*, 1973). Frizell and Sjöstrand (1974a) found the expected increase in the transport of glycoproteins in hypoglossal motoneurons, but transport in vagal motoneurons was reduced. This could represent a difference between myelinated and unmyelinated axons since the hypoglossal motoneurons are mostly myelinated, whereas those of the vagus are mostly unmyelinated. During the later stages of regeneration, fast transported radiolabeled proteins are carried past the original injury into the regenerating axons. The rate of transport in regenerating axons is similar to that in mature ones and rabiolabel accumulates at the tip of the growing sprouts (Griffin *et al.*, 1976) where it can be used to chart the progress of regenerative growth (Black and Lasek, 1976).

In axotomized neurons, increased amounts of endogenous protein and glycoprotein are returned to the cell body (Frizell *et al.*, 1976; Bulger and Bisby, 1978). This retrograde transport returns to normal following regeneration but remains elevated if regeneration is prevented (Bulger and Bisby, 1978). In particular, it seems that in injured axons, proteins that would normally be distributed to distal parts of the axon are instead prematurely reflected from the site of injury so that they are returned to the cell body a shorter time after leaving it. They are also returned in greater quantities (since less has been incorporated into distal structures) and possibly in a different form (since there has been less distal degradation) (Bisby and Bulger, 1977; Bulger and Bisby, 1978).

Exogenous proteins may be taken up by nerve terminals and transported back to the cell bodies (Kristensson, 1978). Tracers are also taken up into damaged axons immediately following injury and have been proposed as a marker for retrograde transport of the chromatolytic message (Kristensson and Olsson, 1975). Retrograde transport of exogenous molecules cannot be measured during regeneration because the tracers are not taken up by growing axons (Kristensson and Olsson, 1976). However, retrograde transport of tracers injected either locally (Kristensson and Sjöstrand, 1972) or systemically (Olsson *et al.*, 1978) is increased in recently regenerated neurons.

Frizell and Sjöstrand (1974c) showed that slow anterograde transport of radiolabeled proteins in regenerating hypoglossal and vagal motoneurons was, respectively, increased and diminished in amount. This inconsistency parallels the same authors' unexplained findings with regard to fast anterograde transport of glycoproteins in these neurons. Lasek and his co-workers have described two peaks of slowly transported protein in motoneurons: one (SCa) moves at 1 mm/day and includes two tubulin peptides and a triplet of peptides thought to be derived from neurofilaments; the second, smaller, peak (SCb) moves at 3–4 mm/day and includes tubulin together with several other peptides. In regenerating neurons, the composition of SCa is unchanged whereas the amount of material in SCb is increased, largely due to a selective increase of tubulin in this fraction (Lasek and Hoffman, 1976). The velocity of each peak is unchanged, but since the amount of material in the faster moving peak is increased, the apparent rate of transport of the slow component as a whole will increase. This may explain the increase in apparent average velocity of slow transport in axotomized neurons reported by Frizell and Sjöstrand (1974c). It is suggested that the cytoskeletal elements in SCb move at the same velocity as the growing tip and thus provide both the impetus and the raw material for axonal growth. Lasek and Black (1977) have further shown that, in mature neurons, the column of cytoskeletal components does not turnover in passage down the axon, but on reaching the terminal, it is disassembled and the components rapidly broken down. It is postulated that this breakdown is brought about by a Ca^{2+}-sensitive protease (Pant et al., 1979) that is activated by the relatively high Ca^{2+} levels in the terminals. According to this hypothesis, Ca^{2+} levels in growing axonal sprouts would be low; the cytoskeletal column would not be disassembled, and so growth would continue until synaptic contact was made; Ca^{2+} levels in the tip would rise; disassembly of the column would commence; and growth would be arrested. Thus, when nerve is prevented from regenerating by ligature and resection, the proteins that turn over rapidly in normal nerve terminals are unable to do so and instead accumulate in the occluded tips of the axons, which become swollen with neurofilaments (Lasek and Black, 1977).

When a nerve is occluded, the initial rise in the amount of NA at the occlusion ceases at 48 hours (Dahlström and Haggendal, 1966). It seems possible that this is due to failure of axonal transport of NA. Boyle and Gillespie (1970) found that a reduced amount of NA accumulates at a more proximal ligature placed on the nerve either 8 or 25 days after nerve section. Cheah and Geffen (1973) showed that axonal transport of NA continues normally for the first 48 hours after nerve section and then falls to a minimum at 7 days. Following a crush injury, recovery occurs by day 17, but if nerves are prevented from regenerating, no recovery of axonal transport has occurred at 23 days (Karlström and Dahlström, 1973). In cholinergic neurons, the axonal transport of acetylcholinesterase is unchanged 1 day after crushing the sciatic nerve, has dropped to 40% of normal at 2 days, and to 20% at 6 days (Schmidt and McDougal, 1978). Transport is later restored in frozen or crushed nerves, but not in resected ones (Heiwall et al., 1978).

C. Changes at the Transected Stump

Within hours of injury to either a myelinated or unmyelinated axon, that part of the axon just proximal to the injury becomes swollen and contains increased numbers of organelles (Wettstein and Sotelo, 1963; Zelena et al., 1968; Kapeller and Mayor, 1969; Martinez and Friede, 1970; Morris et al., 1972a; Matthews, 1973). The length of the pellet of accumulated organelles increases until, at 24 hours after injury, it is some 200 μm long (Zelena et al., 1968). In myelinated axons, the swelling of the axoplasm is accommodated by a slipping of the myelin sheath (Friede and Martinez, 1970). The accumulated organelles include:

1. Mitochondria, which are most numerous in that part of the pellet farthest from the injury. Mitochondrial enzymes also have been shown to accumulate at a crush injury (Banks *et al.*, 1969).
2. Neurofilaments
3. Multivesicular bodies and autophagic vaculoes. There is a concomitant increase in the activity of lysosomal enzymes (Holtzman and Novikoff, 1965; Johnson, 1970).
4. Tubulovesicular smooth endoplasmic reticulum (SER). A network of tightly packed tubular and vesicular membrane profiles 40–110 nm in diameter collects close to the injury. Some of these profiles have an electron-dense core.

The nomenclature of this material is reviewed by Martinez and Friede (1970). The profiles appear similar to the much sparser SER membranes seen in normal neurons (Droz *et al.*, 1975), but it is uncertain whether they are transported from more proximal parts of the axon or are formed locally.

This raises the question: to what extent can protein synthesis take place in the stump of a transected nerve? Although a limited amount of protein synthesis occurs in normal axons (Koenig, 1970), the great preponderance of neuronal macromolecules originate in the perikaryon. Following axotomy, however, there is an increase of more than 20-fold in protein-synthesizing activity in axons in the proximal stump region. After a latent period of 12 hours, this activity rises to a peak at 18 hours and then falls to twice control values at 96 hours (Tobias and Koenig, 1975). The increased protein synthesis, which is cytoribosomal and not mitochondrial, entails synthesis of a novel component with an apparent molecular weight similar to that of actin. It is suggested that this component may be required for axonal growth (Koenig *et al.*, 1977).

Since it is known (see Section II,B) that neurotransmitters and related molecules continue to be produced in the cell body and transported down the axon for some time following axotomy, it is not surprising to find that neuro-transmitter vesicles accumulate proximal to an axonal injury. Using staining techniques specific for catecholamines, an accumulation of dense-core vesicles can be demonstrated proximal to a constriction (Tomlinson, 1975; Till and Banks, 1976). NA accumulates linearly with time proximal to a crush for at least 48 hours (Dahlström and Haggendal, 1966) and DBH, which is another major constituent of adrenergic vesicles, also accumulates (Wooten and Coyle, 1973; Brimijoin, 1972). It appears that there is initially no net synthesis of NA prox-imal to an injury (Dahlström, 1967), although the NA is turning over (McLean and Keen, 1972). At a later stage, however, local synthesis of NA contributes to the accumulation (Wooten *et al.*, 1977).

D. SPROUTING AND REGENERATION

Following injury to a nerve, sprouts form at the proximal stump and recanalize the endoneurial tubes in the peripheral stump. Following a crush injury, which leaves the endoneurial tube intact, sprouting takes place within the endoneurial tube, and there is usually complete restitution of fiber numbers and caliber spectrum distal to the injury, although maturation of fiber diameter depends on reestablishing functional synaptic connections (Aitken *et al.*, 1947). Damage to the endoneurial tube, however (for instance, when the nerve is cut), is followed by a more profuse sprouting. These bare sprouts traverse the scar in an attempt to restore connection with the peripheral endoneurial tubes. Not all are successful, and so, following this type of injury, there is not complete restitution of fiber numbers or caliber spectrum (Sunderland, 1968).

1. Myelinated Axons

Both terminal and collateral sprouts are formed by interrupted myelinated axons. These sprouts are initially nonmyelinated.

a. *Terminal Sprouts.* Terminal sprouts arise from the tip of the smaller myelinated axons during the first day (Cajal, 1928). They have clublike ends and some are branched (Ranson, 1912). Zelena *et al.* (1968) show, in a nerve crushed 4 hours previously, several terminal sprouts growing from a single myelinated axon and contained within one endoneurial tube. Duce and Keen (1976) show a free terminal sprout in a cut nerve that can be traced back to a myelinated axon.

b. *Collateral Sprouts.* Collateral sprouts arise from the body of an axon some distance proximal to the transected tip. At their most proximal point, they are seen as a subdivision of the axoplasm (Morris *et al.*, 1972b); the two compartments often have very different cytoplasmic inclusions (Duce and Keen, 1976). More distally, they penetrate the myelin sheath, usually at a node of Ranvier, and run along the outside of the myelinated fiber but beneath the basement membrane. Cajal (1928, Fig. 62) shows multiple collateral sprouts coursing along the degenerating tip of a single myelinated axon. Morris *et al.* (1972a) describe these as "type I regenerating units" in which the degenerating myelinated process and its Schwann cell are accompanied by numerous nonmyelinated sprouts, which are at this stage only loosely associated with the Schwann cell. The whole unit is enclosed in a single basement membrane. These "regenerating units" first appear between 24 and 36 hours after nerve section.

"Type II regenerating units" are similar to their type I counterparts but contain no degenerating myelinated process and probably represent a bunch of collateral sprouts that have extended beyond the tip of their parent myelinated axon (Morris et al., 1972a).

2. Unmyelinated Axons

Cajal (1928) reported that cut unmyelinated axons gave rise to numerous branched sprouts with terminal clubs. Ranson (1912) found that, unlike their counterparts from myelinated axons, these sprouts died back on the third day, and so he termed this "abortive regeneration." These observations are complicated by the fact that they were made on mixed nerve in which it is not always possible to distinguish between nonmyelinated sprouts and unmyelinated axons. Aguayo et al. (1973) cut nerves consisting almost entirely of unmyelinated fibers and 2 days later found bundles of up to 200 sprouts loosely associated with a Schwann cell all in a single basal lamina. These coexisted with degenerating axons and so were probably collateral sprouts arising from a point proximal to the tip. Dyck and Hopkins (1972) described similar regenerating units following crush injury to an unmyelinated nerve.

3. Loss of Redundant Sprouts

Although each interrupted axon may initially give rise to as many as 50 sprouts (Ranson, 1912; Weddell, 1942), the number of regenerated axons that finally becomes established in the portion of nerve distal to the injury does not exceed the number of parent axons (Sunderland, 1968). Hence there must be considerable loss of redundant sprouts along the way. It appears that this loss takes place in two phases.

a. *Phase I.* Following an injury that leaves the endoneurial tube intact, multiple sprouts are formed within it (see earlier), and these traverse the injured region within the intact tube. When the endoneurial tube is severed, however, a greater number of sprouts is formed (Shawe, 1955), and these are free of any investment. These free sprouts ramify through the scar (Cajal, 1928), and those that contact an endoneurial tube in the distal stump continue to grow along it. Other sprouts persist in the scar and some grow retrogradely along the proximal stump (Aitken and Thomas, 1962). Between 24 and 48 hours following injury, many sprouts degenerate (Ranson, 1912; Grafstein and McQuarrie, 1978; Duce et al., 1976). Growing sprouts are guided by contact with a substratum rather than by humoral factors (Weiss, 1945), and neurites of neuronal cells in tissue culture require surface contact for growth (Letorneau, 1975). Thus, it seems possible that in order to survive an axonal sprout needs to make contact with a suitable substratum. Ideally, this would be an endoneurial tube in the peripheral stump, but it appears that tissues of the proximal stump or elements in the scar itself (as when sprouts form a neuroma) can also form a suitable surface. If this is so, the initial wave of sprout degeneration would be a local phenomenon resulting from failure to contact a substratum.

b. *Phase II.*   Only sprouts that have colonized an endoneurial tube grow along the distal portion of the nerve, and free sprouts are not seen beyond the scar. In the case of myelinated axons, several sprouts colonize a single endoneurial tube (Ranson, 1912), and in regenerating unmyelinated fibers, the number of axonal processes per Schwann cell is higher than usual (Aguayo *et al.*, 1973). Hence, distal to the injury in a regenerating nerve, both the total number of axons and the axon: Schwann cell ratio exceed normal values. During regeneration, these redundant sprouts are resorbed so that the total number of axons and the axon: Schwann cell ratio are restored to normal (Bray and Aguayo, 1974; Ranson, 1912; Sunderland, 1968; Weddell, 1942). The origin of this second phase of degeneration is obscure, but it seems likely that, in contrast to the first phase, the phenomenon is integrated at the level of the neuron and entails a selective degeneration of all but one of the sprouts arising from any one axon. Wall and Devor (1978) discuss the alternative possibility that one axon may maintain multiple distal branches.

4. Myelination

A wave of myelination follows the advance of the regenerating axon tip with a latency of some weeks (Quilliam, 1958). Cross-anastomosis experiments between myelinated and unmyelinated fibers indicate that it is the nature of the regenerating neuron that determines whether it becomes myelinated. When the proximal stump of an unmyelinated nerve is anastomosed to the distal stump of a myelinated nerve, the majority of the regenerating axons do not become myelinated. Conversely, when the proximal stump of a myelinated nerve is anastomosed to the distal stump of an unmyelinated nerve, the regenerating axons become myelinated by the indigenous Schwann cell population (Weinberg and Spencer, 1975, 1976; Aguayo *et al.*, 1976a). Schwann cells do not migrate from the proximal to the distal stump (Aguayo *et al.*, 1976b), and hence the Schwann cells must be multipotent. Following nerve section, the Schwann cells of the distal stump proliferate, and it is suggested that these cells may need to undergo a mitotic division before acquiring the potential to respond to axonal influences (Aguayo *et al.*, 1976a) [The possible nature of these axonal influences is discussed by Weinberg and Spencer (1978).]

5. Rates of Growth of Axonal Sprouts

The methods that have been used to measure the rate of axonal growth include (a) identification of growing axon tips by histological methods; (b) testing for the presence of the regenerating tips of sensory axons by determining the most distal part of the nerve at which a direct pinch will elicit a reflex response; (c) testing for axonal conductivity by electrophysiological methods; (d) injection of radiolabeled tracers into cell bodies and measurement of the distance along the nerve at which axonally transported radioactivity accumulates in the axon tips; (e) assessment of recovery of function by physiological tests; and (f) testing for regenerating adrenergic neurons by their ability to take up [^{3}H]NA or by fluorescence histochemistry. Rates of regeneration differ between species. The rat has been the most common source of the ganglia studied in organ culture, and so, data for this species will be given where available.

a. *Latent Period.* Most studies report a latent period between injury to a nerve and the start of regeneration, which is more protracted following a cut than a crush injury (Gutmann *et al.,* 1942). Weiss and Taylor (1943), using histological methods, reported a latent period of 4 days between cutting and splicing the sciatic nerve and the start of regeneration. Estimates of the latent period following crush injuries to this nerve are rather shorter; i.e., 1.6 and 2.0 days for sensory neurons (McQuarrie *et al.,* 1977; Berenberg *et al.,* 1977), 1.3 days for adrenergic neurons (McQuarrie *et al.,* 1978), and for motoneurons, an average of 3.2 days and a minimum of 2.1 days (Forman and Berenberg, 1978). Whether there is a true latent period is clearly of prime importance in determining whether the sprouts that form within hours of injury are in fact those that eventually grow to reinnervate the distal stump. Cajal (1928) and other workers quoted above described axonal sprouting within hours of injury, followed by a slow rate of regeneration across the scar. Measurement of regeneration beyond the injury is likely to overestimate any latent period because (a) it does not take into account the time needed for sprouts to traverse the scar and (b) collateral sprouts arise some distance proximal to the injury and so have some distance to travel before reaching the scar. Thus Jacobson and Guth (1965), using electrophysiological recording, found that, immediately following injury, conduction failed some 2 mm proximal to the lesion and that regeneration from this point commenced immediately, albeit at a very slow initial rate.

b. *Rate.* Estimates of the rate of growth of regenerating sprouts following crush injury in the rat are in fair concordance. Using radiolabeled tracers in motoneurons, Forman and Berenberg (1978) found an average rate of 3.0 mm/day and a maximal rate of 4.4 mm/day, whereas Black and Lasek (1976) reported an average rate of 3.6 mm/day. (The latter authors report a slower average rate over the first 4 days, but their technique did not determine whether this included an initial latent period.) Rate of growth in sensory neurons is approximately 4.4 mm/day (McQuarrie *et al.,* 1977; Berenberg *et al.,* 1977). In adrenergic neurons, [^{3}H]NA uptake gave a rate of 3.9 mm/day (McQuarrie *et al.*. 1978), whereas the more subjective fluorescence histochemical method gave rates of 1.4 and 2.9 mm/day during the first and second weeks, respectively (Olson, 1969). In the electrophysiological studies quoted earlier, Jacobson and Guth (1965) found that the rate of growth increased exponentially from an initial rate of 0.3 to 3.0 mm/day at 18 days.

6. *Factors Affecting Growth of Axonal Sprouts*

The rate of regeneration is affected by the nature of the lesion, regeneration being slower after a cut than a crush injury (Gutmann *et al.*, 1942). Isenschmid (1932) reported that thyroidectomy delayed functional recovery from nerve injury and that this delay was abolished by administration of thyroid extract. Thyroid extract had no effect on nerve regeneration in normothyroid animals. Kiernan and co-workers, however, have found that daily administration of triiodothyronine (T3) to normal rats with a crush injury of the sciatic nerve accelerates both growth of axonal sprouts (Cockett and Kiernan, 1973) and reinnervation of muscle (McIsaac and Kiernan, 1975a) as judged by histological criteria. T3 also accelerates recovery of function (McIsaac and Kiernan, 1975b). The latter finding has been confirmed by Berenberg *et al.* (1977), but these authors found no acceleration of the rate of regeneration of axonal sprouts by the pinch test. They therefore suggest that the effect of T3 in their experiments was on maturation of synapses rather than on axonal growth.

Cyclic AMP, which stimulates growth of neurites in cultured neurons (Roisen *et al.*, 1972), accumulates proximal to a nerve crush (Bray *et al.*, 1971; Appenzeller and Palmer, 1972). Hence it seems reasonable to test whether this cyclic nucleotide stimulates growth of axonal sprouts in regenerating nerves. Pichichero *et al.* (1973) found that intramuscular dibutyryl cyclic AMP (dbcAMP) increased the rate of functional recovery following a crush injury to the sciatic nerve. However systemically administered dbcAMP did not affect the velocity of axonal growth in either sensory (McQuarrie *et al.*, 1977) or motor (Black and Lasek, 1976) neurons in this situation. This may indicate that, like T3 above, dbcAMP affects maturation of functional synapses rather than velocity of axonal growth. Although these experiments are pertinent to the question of whether regeneration can be accelerated by dbcAMP, they are not a critical test of whether endogenous cAMP has a role in the initiation or maintenance of axonal sprouting.

McQuarrie *et al.* (1977) showed that the veloctiy of growth of sensory axonal sprouts beyond a crush injury of the sciatic nerve was accelerated if the nerve had been sectioned at a more distal point 2 weeks previously. The latent period was unaffected. Wells (1977) showed qualitatively similar results by ^{3}H labeling. Since the effect is also seen if the nerve is resected at the second operation (McQuarrie and Grafstein, 1973), it is unlikely to result from a change in the distal segment. Possibly the retrograde changes in the cell body caused by the conditioning lesion (see Section II,A) prime the neuron to respond more adequately to the second lesion. Paradoxically, outgrowth of adrenergic axons (as assessed by [3H]NA uptake) was *slowed* by a conditioning lesion (McQuarrie *et al.,* 1978) although the latent period was reduced. This may indicate an important difference between sensory and sympathetic neurons. The authors argue against the alternative explanation that the conditioning lesion, by bringing about retrograde changes in the neuron resulting in a diversion of synthesis away from transmitter-related functions (see Section II,A), may have simply reduced the ability of the sprouts to take up [^{3}H]NA.

Nerve growth factor (NGF) stimulates axonal outgrowth from sympathetic neurons and from immature dorsal root ganglia (Levi-Montalcini and Angeletti, 1968; Hendry, 1976). Hence it increases the regeneration rate in short adrenergic axons in mature animals (Bjerre and Rosengren, 1974) and prevents the chromatolytic response in axotomized superior cervical ganglion neurons (West and Bunge, 1976). NGF has also been reported to promote regeneration of spinal neurons in kittens (Scott and Liu, 1964).

7. *Relationship of Regenerative to Reactive Sprouting*

Reactive sprouting, that is, sprouting of axons to widen their synaptic field following interference with other input to that area, occurs in both the peripheral and central nervous systems of mammals. In the peripheral nervous system, it has been described in the nerves supplying skeletal muscle (Hoffman, 1950; Edds, 1953), superior cervical ganglia (Murray and Thompson, 1957), and skin (Weddell *et al.,* 1941).

Partial section of the motor nerve supply to a skeletal muscle gives rise to reactive sprouting in the remaining intact fibers. Two types of sprout are formed: (a) terminal sprouts from the end-plate region and (b) collateral sprouts from nodes of Ranvier in preterminal parts of the axon (Hoffman, 1950). Although both types of sprouting arise in response to partial denervation, they are apparently distinct phenomena (Ironton *et al.,* 1978). Terminal sprouting can be selectively induced by a number of manipulations including local injection of botulinum toxin (Duchen and Strich, 1968), deafferentation (Brown *et al.,* 1978), and blockade of conduction by applying tetrodotoxin to the nerve at a point distant from the muscle (Brown and Ironton, 1977). The factor common to these procedures is that they all produce a denervation type of supersensitivity of the postjunctional muscle membrane. There is a close correlation between the extent of sprouting and the number of extrajunctional acetylcholine receptors. Blockade of the postsynaptic receptors with α-bungarotoxin prevents the sprouting (Pestronk and Drachman, 1978) although it does not block sprouting at a crush injury. A further distinction between terminal and collateral sprouting in response to partial denervation is that development of only the former is prevented by direct stimulation of the muscles, a procedure that prevents denervation-like changes (Ironton *et al.,* 1978). Thus it appears that terminal sprouting, but not collateral sprouting, is associated with changes in the postjunctional membrane, although whether there is a causal relationship remians to be determined.

Two theories have been advanced for the causation of collateral reactive sprouting. First, nerve terminals may normally release an inhibitory substance that suppresses collateral sprouting in adjacent terminal axons. Diamond *et al.* (1976) showed that if axonal transport (but not nerve conduction) in a motor nerve was blocked by local application of colchicine, collateral sprouting occurred in adjacent motoneurones. This suggests that axoplasmic transport might normally carry some factor to the terminals that prevents adjacent terminals from enlarging their synaptic field. This experiment was carried out on the salamander, but Cotman and Nadler (1978) have obtained similar results in the mammalian CNS. The second possibility is that the breakdown products of degenerating nerves may act as a stimulus to collateral sprouting (Hoffman, 1950; Ironton *et al.*, 1978). This is of particular interest since it could constitute a common mechanism for the induction of collateral sprouts in both reactive sprouting and sprouting in response to a nerve injury. Support is given to this suggestion by the report of Tweedle and Kabara (1977) who isolated a factor from denervated muscle that, when injected into normal muscles, stimulated collateral sprouting in the motor nerves. Extract of normal muscle was without effect, although a chemically pure lipid also produced collateral sprouting.

Cotman and Nadler (1978) showed that, when a freezing probe or colchicine is applied to fimbrial axons, they develop collateral branches at the point of application. They suggest that this collateral branching may be a direct result of blocking axonal flow. Alternatively, since local application of colchicine mimics axotomy by producing chromatolytic changes in the cell body (Purves, 1976), the sprouting could be mediated through changes in the perikaryon.

8. *Properties of Axonal Sprouts in Vivo*

Regenerating sprouts conduct electrical impulses, and this has been used as a means of measuring regeneration rates. Sprouts from regenerating myelinated axons, being initially nonmyelinated, conduct like normal unmyelinated axons (Sanders and Whitteridge, 1946). Following myelination, caliber spectrum and conduction velocity may or may not be eventually restored to normal (Guth, 1956; Sunderland, 1968). In regenerating unmyelinated fibers, the conduction velocity is initially half-normal, but returns to normal once synaptic contact has been made (Hopkins and Lambert, 1972).

The rate of fast axonal transport of radiolabeled proteins in regenerating sprouts is similar to that in normal axons (Griffin *et al.*, 1976), and the radiolabel accumulates in the tips of the sprouts. It is an intriguing but as yet unexplained finding that, whereas radiolabeled proteins are transported beyond a crush into regenerating axons *in vivo*, when the nerves are isolated *in vitro*, proteins are transported only as far as the crush. For some reason, *in vitro* incubation conditions will support transport in mature fibers but not in regenerating axons (McLean *et al.*, 1976b).

Horseradish peroxidase (HRP) is taken up by nerve terminals and transported retrogradely to the cell body (Kirstensson, 1978). Although HRP can be taken up into axons immediately following injury, it is not taken up by regenerating sprouts (Kirstensson and Olsson, 1976). Uptake into nerve terminals is thought to be associated with membrane turnover (Heuser and Reese, 1973), which presumably does not occur in regenerating sprouts. In their inability to take up HRP, axonal sprouts differ from the growth cones of cultured neurons with which they are often assumed to be homologous.

Following axotomy, the neuronal cell body continues to synthesize neurotransmitter-related materials (Section II,A) that accumulate in the transected axon (Section II,C). Hence axonal sprouts have some similarities to nerve terminals. Acetylcholine (AcCh) is released from transected ventral roots (Evans and Saunders, 1974). The stumps of neurosecretory axons in the transected pituitary stalk can form an "ectopic posterior pituitary gland" and release vasopressin in response to a physiological stimulus (Billenstien and Leveque, 1955). The axonal sprouts of regenerating adrenergic neurons contain NA, which can be visualized by fluorescence histochemistry (Olson, 1969) and is probably contained in dense-core vesicles (Kapeller and Mayor, 1966). These sprouts also resemble nerve terminals in accumulating [^{3}H]NA by a high-affinity uptake process (McQuarrie *et al.*, 1978). Thus, in these neurons at least, growth and neurotransmitter-related function are apparently not mutually exclusive states (cf. Watson, 1974).

The axonal sprouts of sensory neurons may resemble peripheral sensory terminals in responding to various stimuli. Thus the tips of regenerating sensory axons are excited by mechanical stimulation (Guth, 1956) and by acetylcholine (Diamond, 1959), whereas the axonal sprouts in a neuroma are excited by adrenaline and NA (Wall and Gutnick, 1974a). Axonal sprouts in a regenerating carotid sinus nerve become sensitive to chemical stimuli before they contact the glomus cells of the carotid body (Bingmann *et al.*, 1977), and the sprouts in a neuroma on the same nerve also show chemoreceptor activity (Mitchell *et al.*, 1972).

Little is known of the molecular composition of the membrane of growing sprouts. It has been suggested that the outer surface of the sprouts bears specific molecules (possibly glycoproteins), enabling mutual recognition between the growing axon tip and the membrane with which it eventually synapses (Barondes and Rosen, 1976). Support is given to this suggestion by the report that the growth cones of cultured neuronal cells (Pfenninger and Rees, 1976) and the axonal sprouts of developing cerebellar parallel fibers (Zanetta *et al.*, 1978) both bind large amounts of concanavalin A, which is a lectin that selectively binds to glycosyl/mannosyl residues. It has been suggested that other recognition molecules are also present on growing sprouts (Varon and Bunge, 1978), including those that specify Schwann cell function and myelinogenesis (Weinberg and Spencer, 1978).

III. Sympathetic Ganglia in Organ Culture

A. General

Superior cervical ganglia (SCG) from mature rats retain their ability to respond to acetylcholine after some 2 weeks in organ culture (Larrabee, 1970). Dolivo (1974) has reviewed the metabolism of sympathetic ganglia in culture. Dibutyryl cyclic AMP or depolarizing concentrations of K^+ induce the synthesis of DBH and TH in cultured SCG (Mackay and Iversen, 1972; Silberstein et al., 1972b; Keen and McLean, 1972). It has been suggested that this may provide a model for the transsynaptic induction of these enzymes that occurs following increased sympathetic activity in vivo. Banks et al. (1971a) studied axonal transport of NA in vitro. Since these studies required a length of postganglionic nerve, recourse was made to the cat inferior mesenteric ganglion–hypogastric nerve preparation. As in vivo, the accumulation of NA at a ligature on the nerve is linear over a 48-hour period, and dense-core vesicles accumulate proximal to the ligature. These authors also used a two-compartment incubation chamber to apply drugs selectively to either the ganglion or the nerve. Colchicine inhibits axoplasmic NA transport and the degree of inhibition exerted by various concentrations of the drug correlates well with the extent to which it depolymerizes axonal microtubules (Banks et al., 1971b). This preparation has also been used to study the metabolic and ionic requirements for NA transport (Banks et al., 1973; Kirpekar et al., 1973).

B. Formation of Axonal Sprouts from the Perikaryon

Silberstein et al. (1972c) reported that when rat SCG are maintained in organ culture, multiple "axonal sprouts" bud from the perikaryal plasma membrane of the sympathetic neurons. These sprouts, which are 10–25 μm in length, are easily visible 48 hours after explantation and are quite extensive by 96 hours. Their cytoplasm contains microtubules, mitochondria, and small numbers of dense-core vesicles (Fig. 1). Do these sprouts form only under culture conditions or are they a normal reaction to injury? Three types of perikaryal sprout have been described in SCG neurons following axotomy (Section II,A) and these differ in the cytoplasmic organelles that they contain. In this respect, the sprouts that develop in organ culture resemble those containing mitochondria and large

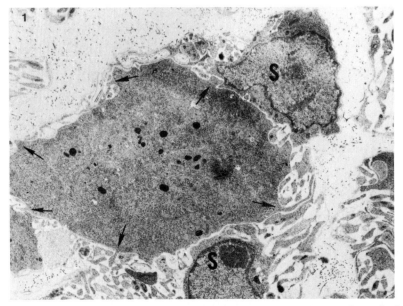

Fɪɢ. 1.   Electron micrograph of a neuronal cell body in a superior cervical ganglion maintained for 48 hours in organ culture. Numerous small axonal sprouts (arrows) project from the perikaryal plasma membrane. Note that most of the neuronal surface is devoid of coverage by the adjacent satellite cells (S), which are surrounded by multiple extensions of the axonal sprouts. (From Silberstein *et al.*, 1972c, by kind permission of Dr. I. J. Kopin.)

dense-core vesicles described by Purves (1975) in axotomized guinea pig SCG and so may not be a phenomenon that occurs only in culture. When ganglia are cultured for 48 hours in the presence of colchicine or vinblastine, few sprouts are seen and those that are seen are small, devoid of microtubules or dense-core vesicles, and contain only amorphous material. Cell bodies contain characteristic sheets of neurofilaments (Silberstein *et al.*, 1972c).

Perikaryal "axonal sprouts" have the ability to take up [^{3}H]NA. Fresh SCG incubated with [^{3}H]NA take up a limited amount of the labeled amine, but in cultured ganglia, [^{3}H]NA accumulation is markedly increased (Silberstein *et al.*, 1972c). Autoradiography shows that in ganglia cultured for 48 or 96 hours, the accumulated radiolabel is localized almost entirely in the sprouts. Colchicine and vinblastine, which inhibit development of sprouts, also inhibit the development of enhanced uptake, thus providing further evidence that [^{3}H]NA uptake is largely a function of the sprouts. However, whereas sprouts are evident at 48 hours and extensive at 96 hours, [^{3}H]NA uptake reaches a maximum after 48 hours in culture (when it is 6-fold that in fresh ganglia) and then declines. Presumably the increased growth of sprouts between 48 and 96 hours is offset by a functional deterioration. Added NGF does not increase [^{3}H]NA uptake in cultured ganglia, but antibody to NGF consistently reduces the enhancement of uptake (and by implication, the development of sprouts) by 25%, suggesting that endogenous NGF plays a role in (but is not essential for) sprout development (Silberstein *et al.*, 1972c).

Uptake of [³H]NA into sprouts is blocked by reserpine, which inhibits uptake into amine storage granules (Carlsson, 1966). This finding indicates that the [³H]NA accumulated by the sprouts is subsequently sequestered in storage granules where it is protected from monoamine oxidase (MAO). These storage sites are presumably the large dense-core vesicles that are seen in most but not all of the sprouts. [³H]Metaraminol uptake also is increased during culture, but to a lesser extent than that of [³H]NA. This increase is prevented by colchicine, indicating that the [³H]metaraminol is entering sprouts, too. Metaraminol is an analog of NA that is taken up by the membrane but is not broken down by MAO; hence its uptake does not depend on subsequent sequestration in storage granules and the enhanced uptake is not blocked by reserpine. Thus, whereas the enhanced [³H]NA uptake indicates increased storage capacity, the enhanced [³H]metaraminol uptake denotes increased membrane uptake capacity, probably due to an increase in membrane surface.

What are the implications of these changes for NA metabolism in cultured SCG? First, there is an increase in the number of storage granules as shown by an increase in reserpine-sensitive [³H]NA uptake discussed earlier. There is also an increased amount of DBH and NA in the position of heavy granules when a ganglion homogenate is placed on a sucrose gradient (Brown *et al.*, 1977). The increased number of storage granules leads to an increase in levels of DBH and NA in cultured ganglia and reaches a maximum at 24 hours (Webb *et al.*, 1975; Brown *et al.*, 1977). At 24 hours, NA levels are 4-fold those in fresh ganglia and are still elevated at 96 hours (Webb *et al.*, 1975). There is a concomitant slowing of NA turnover (Brown *et al.*, 1977). Levels of the cytoplasmic enzyme TH do not increase in cultured ganglia (Webb *et al.*, 1975; Brown *et al.*, 1977). Addition of NGF to the medium does not affect the changes in levels of NA, DBH, and TH that occur during culture.

Some of the increased number of storage granules are accommodated in the newly formed perikaryal sprouts, while others accumulate at the stumps of the transected postganglionic nerves. Webb *et al.* (1975) describe an increased histofluorescence in the postganglionic axons; this increase is prevented by colchicine, which blocks axoplasmic transport. When ganglia were divided into rostral and caudal portions before assay, there was a relatively greater increase in the NA and DBH content of the rostral portion, which contains most of the stumps of the transected postganglionic axons.

To what extent do the changes that take place during culture resemble those that follow axotomy *in vivo*? The increase in NA levels to a maximum at 1 day that is seen in culture has not been reported following axotomy *in vivo*. Cheah and Geffen (1973) found normal NA levels 3 days after axotomy *in vivo* but made no measurements at earlier time points. Whereas DBH levels rise in cultured ganglia, they fall following axotomy *in vivo* (Kopin and Silberstein, 1972). This discrepancy would, however, be expected if the axotomy was carried out at some distance from the ganglion so that the transected stump was not included when the ganglion was assayed. Differences between axotomized and cultured ganglia would also be expected if the chromatolytic reaction, which follows axotomy and involves a diversion of synthesis away from transmitter-related functions, did not occur in culture. This seems not to be the case, however, because typical chromatolytic changes have been reported in the neuronal cell bodies of SCG kept in organ culture for 72 hours (Brown *et al.*, 1977). A further difference between axotomized and cultured neurons is that, in the latter, the preganglionic nerves are also cut. However, preganglionic section does not change the NA levels of SCG *in vivo* (Brown *et al.*, 1977).

1. *Similarities between "Axonal Sprouts" and Nerve Terminals*

Although the sprouts that develop in cultured SCG arise from the perikaryon, they have several features characteristic of sympathetic nerve terminals rather than cell bodies.

a. *Uptake of [³H]NA.* Although a limited amount of NA is taken up by ganglion cell bodies, NA uptake is primarily a function of nerve terminals, where the uptake process acts to recapture released NA, thus limiting NA action and conserving neurotransmitter stores. It is a feature of nerve terminal uptake processes that they have a relatively high affinity for the neurotransmitter concerned. The K_m for the uptake of NA into axonal sprouts in cultured ganglia is similar to that for uptake into nerve terminals, a value some 10-fold lower than that for uptake into fresh ganglia (Hanbauer *et al.*, 1971). A further similarity between the uptake processes in axonal sprouts and in nerve terminals is that they are inhibited to the same extent by cocaine, metaraminol, and phenoxybenzamine, whereas uptake into fresh ganglia is less sensitive to inhibition by these drugs (Hanbauer *et al.*, 1971).

b. *Release of [³H]NA.* The [³H]NA taken up by axonal sprouts can be released by electrical stimulation or by depolarizing concentrations of K^+. These procedures produce a much greater release of [³H]NA from ganglia that have been maintained for 48 hours in organ culture than from fresh ganglia (Vogel *et al.*, 1972). The release by electrical stimulation resembles natural release of neurotransmitter from nerve terminals, for both are Ca^{2+}-dependent. Bretylium, which blocks NA release from nerve terminals, diminishes the stimulated release of [³H]NA from cultured ganglia, and conversely, phenoxybenzamine enhances release from cultured ganglia, as well as from nerve terminals (Vogel *et al.*, 1972). It is likely that phenoxybenzamine increases NA release from nerve terminals by blocking the presynaptic α-receptors through which NA feeds back to reduce its own release (Kirpekar, 1975). The observation that phenoxybenzamine also enhances [³H]NA release from axonal sprouts suggests that the latter further resemble nerve terminals in having α-adrenoceptor sites. It is possible, however, that in these experiments phenoxybenzamine could have increased release by blocking re-uptake of [³H]NA. The presynaptic membrane of sympathetic neurons also carries β-receptors, activation of which *increases* NA release (Stjarne and Brundin, 1975). Isoprenaline, which stimulates β-adrenoceptors, increases the K^+-stimulated release of [³H]NA from cultured SCG (Weinstock, *et al.*, 1978). Isoprenaline continues to exert its effect when NA re-uptake is blocked by desmethylimipramine and so isoprenaline cannot be acting by inhibiting uptake. The effect of isoprenaline on the release of [³H]NA from axonal sprouts is blocked by propranolol (a β_1 and β_2 antagonist) and by butoxamine (a β_2 antagonist) but not by practolol (a β_1 antagonist). Although it thus appears that the receptors are of the β_2 subtype, they are not blocked by another β_2 antagonist, sotalol (Weinstock *et al.*, 1978). Isoprenaline does not enhance the K^+-stimulated release of [³H]NA from freshly excised ganglia, and so it appears that the sites upon which it acts develop with the axonal sprouts.

One point of dissimilarity between sprouts and nerve terminals is that, whereas terminals contain predominantly small vesicles, only large vesicles have been described in sprouts. Nonetheless, it appears that the sprouts that form from the perikaryon of SCG neurons in cell culture provide an opportunity to study the function of sympathetic nerve terminals in the absence of postsynaptic elements.

C. Innervation of the Iris by Axonal Sprouts

When a SCG from an adult rat is placed in contact with an homologous iris in organ culture in the presence of NGF, sprouts grow from the pole of the ganglion to reinnervate the iris (Silberstein *et al.*, 1971). These sprouts can be visualized by fluorescence histochemistry, and it seems clear that they originate from the severed ends of the postganglionic axons and so are true axonal sprouts and quite distinct from the perikaryal sprouts discussed in the preceding section. Heterologous reinnervation also occurs; thus, the rat SCG will reinnervate a guinea pig or mouse iris whereas the rat iris can in turn be reinnervated by a mouse SCG (Silberstein *et al.*, 1972a). Fluorescence histochemistry shows that these adrenergic sprouts ramify in both the sphincter and dilator regions (Silberstein *et al.*, 1971) in contrast to the situation *in vivo* where the adrenergic innervation is concentrated in the dilator region and the sphincter muscle receives a parasympathetic, cholinergic innervation. It is not clear whether the innervation of either of these muscles is a functional one. This apparently adrenergic innervation of both muscles is in contrast to the situation when the same experiment is carried out with tissues from animals 3–5 days old, which is an age at which normal innervation of the iris has not occurred. In this case, the dilator becomes innervated by adrenergic sprouts, whereas, like *in vivo,* the sphincter develops a cholinergic innervation. The latter is a functional one; stimulation of the ganglion causes a contraction of the iris (Hill *et al.,* 1976). These cholinergic fibers, like the adrenergic ones, must have originated from the SCG, whereas *in vivo,* the cholinergic innervation is derived from the ciliary ganglion. It appears that in young animals, before synaptogenesis has taken place, SCG neurons retain the potential to form either adrenergic or cholinergic junctions as dictated by the tissue with which they synapse.

The adrenergic sprouts that colonize the adult iris in organ culture are able to accumulate [³H]NA, and this property can be used to monitor the progress of the reinnervation (Silberstein *et al.,* 1971). The adrenergic terminals in the freshly excised iris accumulate [³H]NA, but after 24 hours in culture, the intrinsic nerves degenerate and are no longer visible by fluorescence histochemistry, and the ability of the tissue to accumulate [³H]NA is lost. When, but only when, the iris is in contact with a ganglion, there is then a gradual return of [³H]NA-concentrating ability. After 48 hours, [³H]NA uptake is approximately one-third that in the freshly excised iris, and after 6 days is restored to normal. This recovery is paralleled by an increase in the density of the adrenergic innervation as seen by fluorescence histochemistry. Sprouts spread from the point of contact between ganglion and iris and do not evenly reinnervate the whole of the tissue; hence at 6 days, in those parts of the iris that are reinnervated, the neuronal network is much denser than *in vivo.* This, together with the fact that fibers are seen in both dilator and sphincter regions, raises the question of whether the sprouts are truly replacing the original fibers or simply ramifying in the tissue. That there is a degree of specificity is shown by the finding that sprouts from a SCG will not colonize a tissue that does not normally possess a sympathetic innervation (Silberstein *et al.,* 1972a). Return of the sprouts is accompanied by a restoration of the DBH content of the iris (Silberstein *et al.,* 1971). The growth of sprouts to reinnervate the iris is much more dependent upon NGF than the growth of perikaryal sprouts (Section III,B). Thus, if NGF is omitted from the medium, the increase in [³H]NA uptake is halved and antibody to NGF further reduces it to one-third (Johnson *et al.,* 1972). These observations suggest a role for endogenous NGF. What is the source of the NGF in these cultured tissues? [*In vivo,* it is thought, NGF is produced in peripheral tissues and retrogradely transported to exert a trophic effect on the cell body (Hendry, 1976).] Both iris and SCG contain NGF when excised, and their NGF content declines during culture (Johnson *et al.,* 1972). During the first 2 days of culture, both iris and ganglion release into the medium 2–3 times the amount of NGF that they originally contained, showing that synthesis of NGF continues in culture. The total amount released is the same irrespective of whether they are cultured together or separately. If the iris is precultured for 2 days to allow endogenous NGF levels to fall, reinnervation by the ganglion, as assessed by [³H]NA upatake, is much reduced. This reduction can largely be prevented by adding NGF to the medium during the preculture period. It can also be shown that when fresh tissues are cocultured, full reinnervation can be obtained by adding NGF for the first 2 days of culture only. Hence it appears that NGF is required for early events only (Johnson *et al.*, 1972).

IV. Changes following Section of Sensory Axons *in Vivo*

A. Special Characteristics of Dorsal Sensory Neurons

The cell bodies of the pseudo-unipolar dorsal sensory neurons lie in the dorsal root ganglia (DRG), the morphology of which has been thoroughly reviewed by Lieberman (1976). Two main types of ganglion cells, "large light" and "small dark," have been described and these may be further classified into subtypes on morphological grounds (Andres, 1961; Duce and Keen, 1977b). The peptides substance P and somatostatin have been localized to separate populations of small dark cells (Hökfelt *et al.*, 1976), and it has been suggested that substance P may function as a neurotransmitter in these neurons (Otsuka and Konishi, 1976). Small dark cells exhibit fluoride-resistant acid phosphatase (FRAP) activity, as do their terminals in the substantia gelatinosa of the spinal cord (Colmant, 1959; Gerebtzoff and Maeda, 1968; Knyihar and Gerebtzoff, 1973).

The axons of the pseudo-unipolar DRG cells bifurcate within the ganglion to give a peripheral branch and a centrally directed branch that runs in the dorsal root to enter the spinal cord. Fast axoplasmic transport has the same velocity in the two branches. However, the peripheral branches are in general longer and receive approximately 4 times as much material by the fast transport system as do their central counterparts (Lasek, 1968; Ochs, 1972; Ochs *et al.*, 1978). By contrast, the slow transport system carries similar amounts of material centrally and peripherally but has a lower velocity in the central branch (Komiya and Kurokawa, 1978). It might be supposed that transmitter-related proteins would be selectively diverted down the central branch to the terminals in the spinal cord. Reports differ, however, as to whether there is a qualitative difference between the proteins transported centrally and peripherally following injection of radiolabeled amino acids into the DRG. Anderson and McClure (1973) reported a difference between the radiolabeled proteins within the spinal cord and those in the sciatic nerve. Other workers have detected no difference between labeling patterns in dorsal root and sciatic nerve (Sabri and Ochs, 1972; White and White, 1977; Mori *et al.*, 1977). The putative neurotransmitter substance P is transported centrally, as indicated by its accumulation at an occlusion on the dorsal root (Takahashi and Otsuka, 1975). It is also found in the peripheral terminals (Olgart *et al.*, 1977a) from which it may be released by nerve stimulation (Olgart *et al.*, 1977b), but whether its presence here is incidental or denotes a peripheral function is a matter of conjecture.

Both HRP and NGF are transported retrogradely along the peripheral branch to the DRG. Twenty hours after injecting NGF into peripheral tissues only the large light cells are labeled (Stöckel et al., 1975). HRP, on the other hand, is mainly concentrated in the small dark cells 22 hours after application and in the large light cells at 70 hours (Neuhuber et al., 1977). There is controversy as to whether HRP is transported transganglionically from the peripheral branch to the dorsal root (cf. Furstman et al., 1975; Neuhuber et al., 1977). HRP is also transported retrogradely in dorsal roots from the spinal cord to both types of DRG cell simultaneously (Neuhuber et al., 1977).

The dorsal roots are typical peripheral nerves except that they lack an epineurial sheath (Gamble, 1964). The small caliber fibers of the dorsal root become segregated in the lateral part of the rootlet just before entering the spinal cord in man (Sindou et al., 1974) and monkey (Snyder, 1977), but possibly to a lesser extent in the cat (cf. Berthold and Carlstedt, 1977; Snyder, 1977). On entering the spinal cord, there is a transition zone in which the dorsal root axons assume the properties of CNS axons, that is, (a) they become invested by oligodendroglia instead of Schwann cells, and (b) the larger unmyelinated fibers gain a myelin sheath (Gamble, 1976; Berthold and Carlstedt, 1977).

B. Section of Peripheral Branches

Section of the peripheral branches of sensory neurons results in classic chromatolytic changes in their cell bodies in the DRG (Lieberman, 1971). A proportion of the axotomized cells die, and it seems that this neuronal loss involves chiefly the small dark cells (Ranson, 1906). The small dark cells also show chromatolysis earlier than do the large light cells (Lieberman, 1974); possibly this may be correlated with the finding that retrograde transport from an injury is more rapid in dark than in light cells (Neuhuber et al., 1977). Section of peripheral branches also causes transganglionic degenerative changes in the nerve terminals within the spinal cord. Terminals in the substantia gelatinosa lose their fluoride-resistant acid phosphatase activity (Colmant, 1959) and show ultrastructural signs of degeneration (Knyihar and Csillik, 1976). Interestingly, peripheral axotomy also causes a change in the somatotopic organization in the dorsal horn of the spinal cord. Thus, following chronic section of the nerves to the foot, neurons in the medial part of the dorsal horn, which normally respond to stimulation of the foot and toes and which have thus been deprived of their normal input, gain the ability to respond to stimuli from more proximal parts of the limb. Thus, the intact afferents from the more proximal parts of the limb have enlarged their synaptic field (possibly by reactive sprouting, although other mechanisms are possible) (Devor and Wall, 1978).

The peripheral axons of sensory nerves regenerate at approximately 4 mm/day (Section II,D,5), and their peripheral terminals regain the ability to respond to stimuli. Where sensory nerves connect with nonneural receptors (e.g., Type I cutaneous receptors), section of the nerve is followed by degeneration of the nonneural elements. These receptors return when the skin is reinnervated (Brown and Iggo, 1963).

Electrical Activity following Peripheral Axotomy

Following injury to a sensory nerve, as with other nerves, there is an initial short-lasting "injury discharge" followed by a period of silence during which the nerves can still be excited electrically. Severed sensory (but not motor) nerves may then start to show spontaneous activity that can be registered either in the dorsal root (Wall and Gutnick, 1974b) or in the injured nerve between the stump and the DRG (Govrin-Lippman and Devor, 1978) and may start as early as 3.5 hours after injury (Kirk, 1974). If regeneration is prevented, a neuroma consisting of a tangled mass of unmyelinated sprouts is formed at the severed stump. These sprouts may be the source of spontaneous activity, which reaches a maximum 2 weeks after axotomy (Govrin-Lippman and Devor, 1978) when some 30% of the fibers may show activity. Application of a local anesthetic to the neuroma abolishes the activity. Electrical activity may be increased by applying various stimuli to the sprouts within a neuroma. They are sensitive to mechanical pressure (Kirk, 1974; Wall and Gutnick, 1974a) as are the growing tips of regenerating sensory neurons (Section II,D,8). Neuroma sprouts also respond to NA (Wall and Gutnick, 1974a) and to AcCh (Diamond, 1959). When a neuroma is electrically stimulated, the resultant impulses are carried in small myelinated ($A\delta$) fibers (Devor and Wall, 1976). It seems that either large myelinated fibers do not contribute sprouts to a neuroma or, if they do, stimulation of sprouts arising from large myelinated fibers does not initiate impulses in the parent axons. Devor and Wall (1976) did not include unmyelinated fibers in their study. Following damage, the DRG themselves also become mechanically sensitive (Howe *et al.*, 1977) and emit a train of impulses in response to prolonged pressure.

C. Section of Dorsal Roots

1. General

Section of dorsal roots, in contrast to peripheral section, does not give rise to a chromatolytic response in DRG cell bodies (Cragg, 1970; Lieberman, 1971). This is possibly because the bulk of axoplasm is in the peripheral branch, which, being undamaged, sustains the cell body in its normal state. When a dorsal root is sectioned, its terminals in the spinal cord degenerate and the axons from adjacent roots undergo reactive sprouting to occupy the vacated synapses in the deafferented region (Liu and Chambers, 1958.) There is a corresponding metabolic response in the adjacent DRG (Watson, 1973); thus, DRG neurons show a greater metabolic response to section of adjacent dorsal roots than they do to section of their own central branches. As a result of this reactive sprouting, neurons in a deafferented region of the spinal cord come to respond to inputs from the nearest intact fibers (Wall, 1976). However, reactive sprouting cannot entirely explain this reorganization of the sensory map following dorsal root section. A few acutely deafferented neurons gain the ability to respond to afferent impulses, which, in the normal animal, excite only neurons in adjacent segments. This change occurs too soon for reactive sprouting to have taken place and in the gracile nucleus can even be brought about by temporary deafferentation by ice block and reversed after removal of the blockade. Wall (1976) argues that this may be explained by unmasking of existing but normally ineffective afferents from adjacent areas to these cells.

Section of dorsal roots also differs from peripheral section in causing no change in axonal conduction velocity. When the peripheral branch is sectioned, conduction velocity in the nerve proximal to the injury is reduced, as in other nerves, and conduction velocity in the dorsal roots is also reduced. Dorsal root section on the other hand does not reduce conduction velocity in either branch (Czeh et al., 1977). The parallel between chromatolysis and change in conduction velocity suggests that the latter may be mediated by a change in the cell body.

2. Changes at the Transected Stump

Following section of the dorsal root, there is an accumulation of material in the proximal stump (Duce and Keen, 1976). In myelinated axons, this extends some 750 μm back from the tip and includes tubulovesicular SER, dense-core vesicles, and mitochondria. With increasing distance from the cut tip, there is an increase in the ratio of mitochondria to other accumulated organelles. Unmyelinated fibers in the proximal stump become increasingly distorted as they approach the tip. At a distance of 1 mm from the tip, they appear normal with a group of fibers embedded in the cytoplasm of a Schwann cell that has a central nucleus. At a distance of 500 μm from the tip, the axons are distended but the overall architecture is still recognizable. Closer to the tip, the unmyelinated fibers become grossly swollen and distorted although still embedded in Schwann cell cytoplasm. They are packed with accumulated material including SER, clear and dense-core vesicles, and mitochondria. The nature of the accumulated material may differ between axons in the same group: some, for instance, contain a very high proportion of dense-core vesicles. Some dense bodies occur in both axoplasm and Schwann cell cytoplasm.

3. Sprouting from the Cut Tip

Light microscopy reveals a collection of nonmyelinated neuronal processes at the tip of the proximal stump (Fig. 2). Electron microscopy (EM) shows that

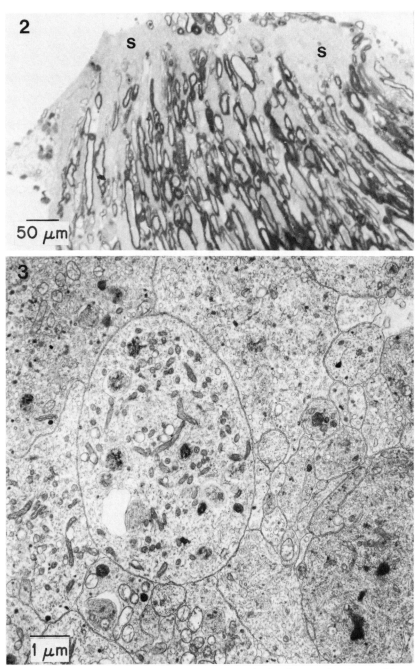

FIGS. 2 and 3. Axonal sprouting at the proximal stump of a dorsal root sectioned *in vivo* 20 hours previously. Fig. 2. Light micrograph showing area of nonmyelinated sprouts (S) that have formed at the cut tip. Fig. 3. Electron micrograph of the cut tip showing numerous bare axonal sprouts that contain various organelles including mitochondria, smooth endoplasmic reticulum, and dense-core vesicles.

these are 1–15 μm in diameter, none of them is myelinated and there are no Schwann cells or other nonneuronal elements in the area (Fig. 3). The organelles contained in the sprouts include clear and dense-core vesicles, tubulovesicular SER, mitochondria, multivesicular bodies, dense bodies, neurofilaments, and neurotubules. In longitudinal section, some of the nonmyelinated sprouts are seen to arise as a terminal extension of a myelinated axon (Duce and Keen, 1976). The cytoplasm at the tip of these sprouts is free of organelles, but more proximally, they contain accumulated material continuous with that in the myelinated part of the axon. There is also evidence of collateral sprouting from myelinated axons (Duce and Keen, 1976). At proximal points, these sprouts appear to arise as invaginations of the axolemma so that in transverse section the sprout is completely surrounded by the cytoplasm of the parent axon. More distally, sprouts are seen running alongside the parent axon but still within its myelin sheath. It is difficult to determine whether any of the sprouts arise from unmyelinated axons because of the small size and tortuous course of these axons. However, some of the smaller sprouts may well have arisen from unmyelinated axons, as they are similar in size and contain a high proportion of dense-core vesicles, which are common in unmyelinated fibers.

Examination of the tip of the cut nerve by scanning electron microscopy (SEM) demonstrates numerous sprouts typically consisting of a stalk 2–7 μm in diameter with a bulbous end 5–20 μm in diameter (Figs. 4–6). A few branching sprouts can be seen. These sprouts are visible as early as 4 hours after cutting and are well-developed by 7 hours (Duce and Keen, 1976). At 20 hours, the sprouts are more profuse but smaller than at 7 hours and have a rougher surface membrane. At 48 hours they appear shrunken with a deeply furrowed membrane.

Regeneration within the dorsal root takes place readily. Four days after cutting a dorsal root, Nathaniel and Pease (1963) reported that regenerating units had formed from myelinated axons as described by Morris et al. (1972a) in the sciatic nerve (Section II,D,1). Remyelination began as early as the seventh day and was complete by the eighth week. Moyer et al. (1953) reported that further regeneration of cut dorsal roots into the spinal cord was prevented by the astrocytic reaction that occurs at the transition zone between central and peripheral nervous tissues (Section IV,A). However, Nathaniel and Nathaniel (1973) showed by EM that thinly myelinated regenerating dorsal root fibers were present in the substantia gelatinosa and posterior funiculi 10 weeks after crushing a dorsal root. Unmyelinated fibers regenerated more rapidly than myelinated ones and were detected in the substantia gelatinosa by the sixth week. Regeneration following injury to the dorsal root has been confirmed by following axonal transport of [³H]leucine from DRG to spinal cord (Ikeda and Campbell, 1971). The regenerating axons in the substantia gelatinosa contain both clear and dense-core vesicles (Nathaniel and Nathaniel, 1973). There is, however, some doubt about whether regenerating dorsal root fibers establish effective synaptic

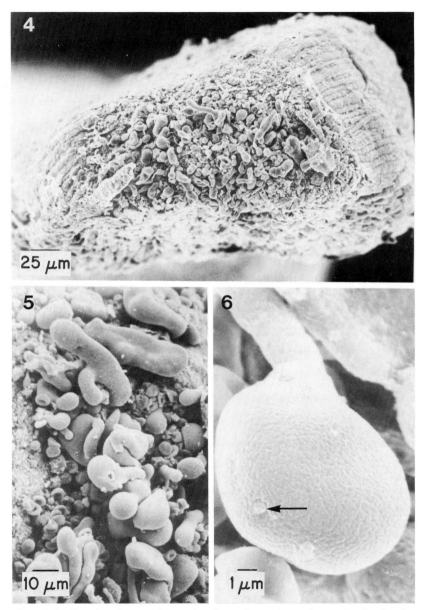

FIGS. 4–6.   Scanning electron micrographs of axonal sprouting at the cut tip of a dorsal root sectioned *in vivo* 20 hours previously. Fig. 4. General view of cut tip. Fig. 5. Higher power view. Note occasional branching sprouts. Fig. 6. Detail of sprout showing stalk and bulbous tip. Note furrowed membrane and dimple (arrow). Such dimples are frequently seen in sprouts and may represent the growth point.

connections (Carlsson and Thulin, 1967). Surprisingly perhaps, ventral root fibers anastomozed to the central stump of a sectioned dorsal root regenerate to form cholinergic synapses within the spinal cord, although the relation of these to the normal synaptic pattern is unclear (Barnes and Worrall, 1968).

V. Dorsal Root Ganglia in Organ Culture

Ochs (1971) established that the cat L7 DRG, with dorsal root and part of the sciatic nerve attached, remained viable when isolated in a moist chamber and that axonal transport of radiolabeled proteins was maintained for 3–4 hours at the same rate as *in vivo*. Longer incubation times were not attempted.

Roberts and Keen (1974) maintained rat DRG in organ culture for 48 hours and studied changes in uptake of [^{14}C]glutamate into the ganglia. The aim of these experiments was to determine whether culture caused an increase in the high-affinity uptake of glutamate, which is postulated to be a neurotransmitter in certain sensory neurons (Johnson, 1978). If this occurred, it might indicate the formation of perikaryal sprouts by analogy with the finding of Silberstein *et al.* (1972c) that [^{3}H]NA was taken up by a high-affinity system into perikaryal sprouts in cultured sympathetic ganglia (Section III,B,1,a). A high-affinity uptake system for [^{14}C]glutamate was found in freshly isolated DRG, but the capacity of this system did not increase during culture, although there was a change in the metabolic fate of the accumulated glutamate (Roberts and Keen, 1974).

A. MORPHOLOGY OF SPROUTS

Duce and Keen (1977a) maintained rat lumbar DRG, with their dorsal roots attached, in organ culture for up to 48 hours and examined the sequence of changes that took place at the cut tip of the dorsal root. There was an accumulation of organelles in both myelinated and unmyelinated axons very similar to that which occurs following dorsal root section *in vivo* (Section IV,C,2). Accumulated material included tubulovesicular SER, mitochondria, bundles of neurofibrils, and densecore vesicles; the latter were more numerous in the unmyelinated axons. After 20 hours in culture, numerous free axonal sprouts form at the cut tip (Figs. 8,11). The majority of these have a bulbous tip, as has long been known from observations *in vivo* by Cajal (1928) and is seen by SEM in dorsal roots *in vivo* (Figs. 4–6). The sprouts differ widely in size: stalks are 1–7 μm in diameter and the bulbous tips are 2–20 μm. Branched forms are seen and also "double" forms in which a sprout appears to emerge from a ruptured membrane (Fig. 11). The main elements contained in the sprouts are tubulovesicular SER, clear and dense-core vesicles, and mitochondria (Figs. 14–16). Occasional dense and

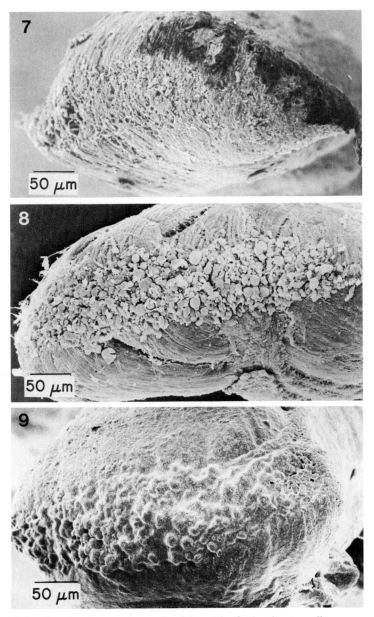

FIGS. 7–9.   Scanning electron micrographs of the cut tip of a dorsal root ganglion–nerve preparation. Fig. 7. Immediately after cutting. Fig. 8. After 20 hours in organ culture. Numerous free sprouts
have formed. Fig. 9. After 20 hours in organ culture in the presence of 3×10^{-6} M demecolcine. No
free sprouts are seen.

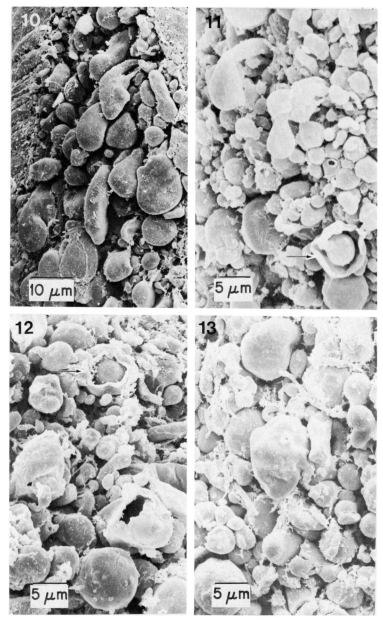

FIGS. 10–13. Scanning electron micrographs of axonal sprouting at the cut tip of a dorsal root ganglion–nerve preparation after various times in organ culture. Fig. 10. After 7 hours in organ culture. Note the preponderance of large sprouts (magnification is half that of other figures on this plate). Fig. 11. After 20 hours in organ culture. More small sprouts are present and also a number of sprouts have ruptured to reveal inner sprouts (arrow). Figs. 12 and 13. After 48 hours in organ culture. Sprouts are more degenerate as shown by their more furrowed and irregular surface. Note ruptured sprout (arrow).

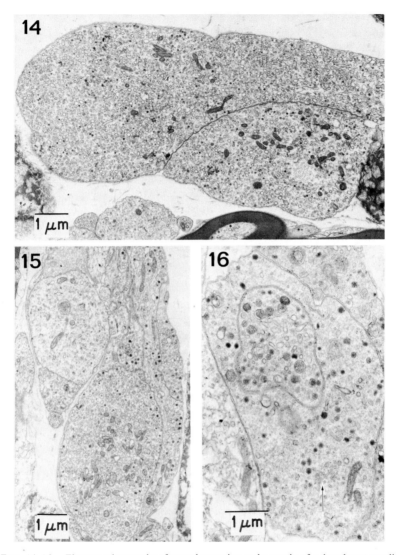

FIGS. 14–16.   Electron micrographs of axonal sprouting at the cut tip of a dorsal root ganglion-nerve preparation maintained for 20 hours in tissue culture. Fig. 14. Two closely apposed sprouts. Fig. 15. A number of closely packed sprouts containing a variety of organelles including mitochondria, smooth endoplasmic reticulum, and dense-core vesicles. Fig. 16. A complex sprout consisting of inner and outer segments. Note cluster of small electron-lucent vesicles (arrow).

lamellated bodies, multi-vesicular bodies, and islands of neurofibrils are also present.

Dense-core vesicles are present in all sprouts; in some, they are associated with packed tubulovesicular SER (Figs. 14, 15) but other sprouts contain numerous dense-core vesicles unaccompanied by any large amount of SER (Fig. 16). These dense-core vesicles can be classified as large (100–150 nm diameter) or small (40–100 nm). The latter can be further subdivided into two types: the electron-dense cores in one being smaller and less dense than in the other (Fig. 16). The large dense-core vesicles are of particular interest since it has been suggested that the peptides substance P and somatostatin may function as neurotransmitters in dorsal sensory neurons (Otsuka and Konishi, 1976). These peptides are present in some DRG neuronal cell bodies (Hökfelt et al., 1976) and substance P is transported down the dorsal root (Takahashi and Otsuka, 1975). Substance P-containing vesicles within nerve terminals in the spinal cord are 60–80 nm in diameter (Pickel et al., 1977) and somatostatin-containing vesicles in the median eminence are 90–110 nm in diameter (Pelletier et al., 1974). Hence, some of the dense-core vesicles in dorsal root sprouts could represent an accumulation of peptide-containing neurotransmitter vesicles analogous to the accumulation of noradrenaline-containing vesicles that occurs in interrupted sympathetic nerves (Section II,C). This is by no means certain, however, because the growth cones of dorsal root neurons in cell culture contain numerous 100 nm dense-core vesicles that are thought to be concerned in the rapid transport of building materials to the growing tip (Yamada et al. 1971). Also, vesicles need not have been transported to the sprouts but may have been formed locally from SER (Stelzner, 1971), although this is less likely to be the case in those sprouts that contain little SER (e.g., Fig. 16). Some sprouts contain flattened dense-core vesicles in a situation where adjacent sprouts contain the more usual round vesicles, suggesting that the flattening cannot have been solely an artifact of fixation (Duce and Keen, 1977a). Flattened vesicles are usually associated with inhibitory nerve terminals, and so it is surprising to find these in dorsal root neurons that are usually assumed to be solely excitatory. Flattening of vesicles may, however, be associated with axonal degeneration (Walberg, 1966). In some sprouts that do not contain large amounts of SER, it is possible to see clumps of small clear vesicles approximately 40 nm in diameter (Fig. 16) similar to those described in presynaptic terminals, particularly those known to be cholinergic. This is of interest because one of the postulated sensory neurotransmitters is glutamate (Curtis and Johnston, 1974; Johnson, 1978), and in the crayfish, those presynaptic terminals that utilize glutamate as a neurotransmitter contain clear vesicles 50 nm in diameter (Atwood and Morin, 1970).

Some sprouts contain secondary sprouts within their axoplasm (Figs. 16, 20). There is often a considerable difference between the contents of the inner and outer segments. In Fig. 16, for instance, the inner segment contains tubulovesicular material, dense-core vesicles, multivesicular bodies, and dense bodies, whereas the outer contains much less tubular material but many clear and dense-core vesicles. This could represent the region proximal to the bifurcation of a sprout, and rupture of the outer membrane could give rise to the "double sprouts" seen in scanning micrographs (Figs. 11, 12).

B. TIME-COURSE OF SPROUTING

Immediately after cutting the dorsal root, no specific structure is discernible at the cut proximal stump (Fig. 7). At 7 hours, the cut surface is packed with sprouts, the majority of which have terminal bulbs greater than 5 μm in diameter (Fig. 10). At 20 hours, these large sprouts have developed longer stalks and thus protrude further from the cut surface, and numerous small sprouts with bulbs less than 5 μm in diameter have appeared proximal to them (Figs. 8, 11). At 48 hours, some of the larger sprouts appear degenerate (Figs. 12, 13). Many of the terminal bulbs have a very irregular shape and their surface has become deeply furrowed. A number of bulbs have ruptured: in some cases, revealing a secondary inner sprout. At 48 hours, many of the larger sprouts contain an increased complement of lysosomes and dense bodies, confirming their degenerate nature (Duce and Keen, 1976).

C. PLASMALEMMA OF SPROUTS

Because axonal sprouts are free of connective tissue or myelin sheaths and of glial or Schwann cells, they afford a rare opportunity to study a bare neuronal membrane. By SEM, the plasmalemma of the sprouts has a furrowed appearance that becomes more marked with time. Lewis (1971) obtained SEM photographs of neuronal processes in the abdominal ganglion of *Aplysia*, and these have a surface structure similar to that of the free sprouts reported here. Hansson (1970) and Angelborg and Engström (1974), on the other hand, examined neuronal cells in the retina and tympanic layer, respectively, and in these cases, the neurons appear to have a smooth surface. The surface of neurons in tissue culture has a rather ridged appearance (Privat *et al.*, 1972). Spencer and Lieberman (1971) examined single myelinated nerve fibers by SEM, but in this preparation, the neuronal membrane is covered by myelin and, at the node of Ranvier, the endoneurial sheath. Gershenbaum and Roisen (1978) used SEM to study the sequence of changes that occur in a crushed sciatic nerve. Although regenerating fibers are seen crossing the injured region 15 days after crushing, no surface structure is clearly visible, possibly because the endoneurial sheaths have remained intact.

The technique of freeze–etching has yielded much information on the structure of neuronal membranes (see Sandri *et al.*, 1977, for review). The membranes of

freeze–etched axonal sprouts (Figs. 17–19) do not show any sign of the furrowing seen in SEM, and since the plasmalemma of sprouts appears substantially
smooth when examined by transmission EM also (Figs. 14, 15), it must be
concluded that the furrowing seen by SEM is partially at least an artifact of
preparation for that technique. It may nonetheless reflect some property of the
sprout membrane as the membrane of erythrocytes appears smooth in the same
preparation (Fig. 6, also Duce *et al.*, 1976), and the membranes of 48 hour
sprouts are consistently more furrowed than those at earlier time-points (Duce *et
al.*, 1976).

It may be of interest to inquire whether the plasmalemma of sprouts bears any
similarity to the membranes either of neuronal cells in culture or of nerve terminals. The P-face of freeze–etched axonal sprouts contains a few small intramembraneous particles (Fig. 17) and in this resembles the membrane of the growth
cone, which is notable for having a low density of intramembraneous particles
when compared with glial cell or perikaryal membranes (Pfenninger and Bunge,
1974). The number of these particles in growth cone membranes increases with
prolonged growth in culture, and Pfenninger and Bunge (1974) suggest that low
particle density is a sign of membrane immaturity. A number of pits occur in the
P-face of dorsal root sprouts (Fig. 17). Since these sprouts are packed with
vesicles, it seemed possible that the pits were analogous to the vesicle attachment
sites that occur in the presynaptic membranes of nerve terminals (Heuser *et al.*,
1974), although the pits in the sprouts are not associated with large membrane
particles as are the vesicle attachment sites in the presynaptic membrane (Venzin
et al., 1977). The vesicle attachment sites are thought to be the sites of exocytotic
discharge of neurotransmitter, and the number of both attachment sites and large
membrane particles is much reduced in anesthetized animals as compared with
conscious ones (Streit *et al.*, 1972; Venzin *et al.*, 1977) and, conversely, is
increased by nerve stimulation (Heuser *et al.*, 1974). Hence, the sparsity of these
structures in axonal sprouts could reflect the absence of electrical activity. Since
4-aminopyridine increases the number of vesicle attachment sites in both anesthetized animals (Tokunaga *et al.*, 1978) and in *in vitro* preparations (Heuser *et
al.*, 1974), we tested the effect of treating sprouts with 10^{-3} M 4-aminopyridine
for 3 minutes before fixation prior to freeze-etching (Duce and Keen, unpublished). No change in the morphology of the sprout membrane could be detected,
and we must conclude that the plasmalemma of the sprout is an immature one
and, as might be expected, resembles that of the growth cone rather than the
nerve terminal.

To determine whether the sprout plasmalemma contained the high density of
Na^+ channels that is characteristic of certain neuronal membranes, notably axon
initial segments and nodes of Ranvier, we stained 20-hour-cultured dorsal roots
by the ferric ion-ferrocyanide technique of Waxman and Quick (1978) and
examined them by transmission EM (Duce and Keen, unpublished). This treatment gave intraaxonal staining at the nodes of Ranvier of myelinated fibers as

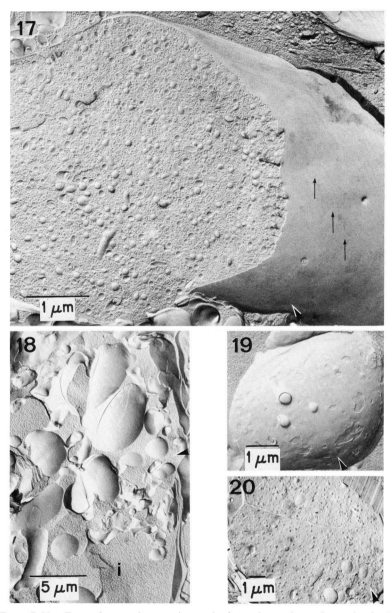

FIGS. 17–20.   Freeze–fracture electron micrograph of axonal sprouting at the cut tip of a dorsal root ganglion–nerve preparation maintained for 20 hours in organ culture. Broad arrows show direction of Pt shadowing. Fig. 17. Cross-fracture of a sprout showing intramembranous particles (arrows) and vesicular contents of the axoplasm. Fig. 18. Surface of sprouts protruding from matrix of ice (i). Fig. 19. Surface of a single free sprout. Fig. 20. Two sprouts originating from a single myelinated axon.

reported by these atuhors, but did not stain the axonal sprouts, suggesting that they are not rich in Na^+ channels.

Since the sprouts are growing rapidly and it has been shown, in growth cones at least, that new membrane is added at the tip (Bray, 1973), it is possible that a point of growth would be identifiable on developing sprouts. Many sprouts show a specialized region on their bulbous tip when examined by SEM (Fig. 6). In freeze-etched preparations, these regions are seen as domes approximately 0.4 μm in diameter (Fig. 19), and the possibility that these could represent growth points is reinforced by their similarity to the domes described on growth cones by Pfenninger and Bunge (1974), which are of very similar size. The domes of axonal sprouts seem not, however, to be surrounded by a circle of plasmalemmal invaginations as is the case with growth cones (Pfenninger and Bunge, 1974).

D. REQUIREMENTS FOR SPROUTING

A dorsal root severed from its ganglion and cultured for 20 hours does not form axonal sprouts whether or not the ganglion is also present in the culture (Duce and Keen, 1977a). Lack of a cell body does not result in degenerative changes in the axoplasm of the nerve over the 20-hour incubation period and hence failure to form axonal sprouts under these conditions may indicate that sprouting is dependent upon material brought to the tip from the cell body. Fast axonal transport, unlike the slow process (McLean et al., 1976a), continues in a nerve severed from its axon (Ochs, 1971). However, if materials were brought to the tip by the fast transport system, the cell body would still be necessary because, at a transport rate of 410 mm/day (Ochs, 1974), the material already present in the nerve would soon be depleted.

To examine further the role of axonal transport in sprouting, nerve–ganglion preparations were cultured for 20 hours in the presence of various concentrations of demecolcine (DMC), which depolymerizes microtubules and blocks axonal transport (Paulson and McClure, 1975). Sprouting was prevented by $3 \times 10^{-7} M$ DMC (Fig. 9), and this reduced the density of microtubules in ummyelinated, small myelinated, and large myelinated axons by 55, 70, and 80%, respectively, suggesting that microtubules may be required for sprouting (Duce and Keen, 1977a). Unfortunately, the apparently all-or-none nature of the sprouting process made it impossible to make a fuller correlation between the effects of DMC on microtubules and on sprouting. Banks et al. (1971b) found a close correlation between NA transport and microtubule density over a range of colchicine concentrations from 10^{-7} to $10^{-5} M$. If the effect of DMC on sprouting is indeed due to its effect on microtubules, this could be either because sprouting requires fast axonal transport, and/or because microtubules are required for sprouting per se as is suggested by the finding that colchicine inhibits extension of axonal processes in cultured dorsal root ganglion cells (Daniels, 1975).

The apparently greater sensitvity of the microtubules of myelinated axons to DMC noted above need not denote any fundamental difference between the microtubules of myelinated and unmyelinated axons but may simply reflect differences in the rate of diffusion into them of oxygen, substrates, drug, or fixative.

E. POSSIBLE STIMULANTS OF AXONAL SPROUTING

The agents that have been suggested to increase sprouting *in vivo* include triiodothyronine, cyclic AMP, and nerve growth factor (Section II,D,6). We have attempted to study the effect of these agents on axonal sprouting in organ culture. To measure axonal sprouting, nerves are fixed and embedded in resin. Transverse sections are taken at 5 μm intervals starting at the tip and examined by light microscopy. The total area of bare axonal sprouts is then measured from photomicrographs (Craven and Keen, unpublished). Preliminary results suggest a significant stimulation of sprouting by triiodothyronine (2.3×10^{-7} M) but no effect of dibutyryl cyclic AMP (5×10^{-3} M), cyclic GMP (1×10^{-3} M), or nerve growth factor (2 units/ml).

VI. Conclusion

As a working hypothesis we would put forward the following explanation of the phenomena of axonal sprouting from mature neurons *in vivo* and in organ culture.

1. Within hours of injury to a nerve, axonal sprouts form at the proximal stump.
2. These sprouts are the initial stage of the regenerative process and have many of the characteristics of the growth cones of neurites formed from neuronal cells in tissue culture.
3. The axonal sprouts formed from a nerve–ganglion preparation in organ culture are, initially at least, indistinguishable from their *in vivo* counterparts and so provide a method for studying axonal sprouting under controlled conditions.
4. In particular, the nerve–ganglion preparation allows a study of the factors that initiate sprouting. Blockade by colchicine suggests that axonal transport is necessary for axonal sprouting, but it is uncertain whether the initial stimulus for sprouting is intraneuronal or a response to an exogenous factor such as a product of tissue damage. It seems possible that collateral reactive sprouting and regenerative sprouting may share a common stimulus.
5. The initial stage of sprouting precedes the onset of chromatolytic changes in the cell body, which divert synthesis away from the molecules required for synaptic function toward those structural elements required for axon growth.

6. Although the sprouts are regenerative, they continue to receive transmitter-related molcules during the initial period before the onset of chromatolysis and so, both *in vivo* and in organ culture, may also show some of the characteristics of nerve terminals such as neurotransmitter uptake and release. They may thus afford an opportunity to study presynaptic processes in the absence of postsynaptic elements.

7. Initially, the sprouts grow slowly because they (a) often follow an erratic course across the scar before contacting an endoneurial tube in the distal stump, and (b) during this initial period the supply of materials is limited because the cell body has not yet undergone the chromatolytic changes that increase the supply of substances required for growth.

8. Initially, a great excess of sprouts is formed (up to 50 per axon) but eventually a 1:1 relationship is established between axons proximal and distal to the original injury. This reduction in the number of sprouts is brought about by two distinct phases of degeneration.

9. Two to three days after the injury, the first phase of degeneration takes place both *in vivo* and *in vitro*. The reason for this degeneration is uncertain, but it seems possible that sprouts from mature neurons resemble the neurites of cultured cells in requiring a suitable substratum on which to grow and that those sprouts that do not find a substratum within the first 24–48 hours regress. The substratum of choice would be an endoneurial tube, but it seems that the scar tissue of a neuroma and, in the case of retrograde sprouts, the tissues or the proximal stump itself are adequate.

10. Not all of the sprouts degenerate; those that colonize an endoneurial tube continue to regenerate beyond the injury. Thus regeneration is a continuous process starting within hours of injury and reports of a latent period before regeneration commences are due to (a) the slow initial growth of sprouts and (b) the fact that sprouts may originate at some distance proximal to the injury.

11. In a crushed or a cut and sutured nerve *in vivo* and in organ culture if a suitable recipient tissue is present, a period of more rapid growth ensues. Growth is accelerated because (a) synthetic mechanisms in the cell body have become adapted to the production of the raw materials for growth and (b) the regenerating sprouts are at this stage directed down endoneurial tubes.

12. At this stage, there is still an excess of axonal sprouts regenerating in the nerve distal to the injury. The 1:1 ratio between axons proximal and distal to the injury is restored by the second phase of degeneration that occurs some weeks after injury. In contrast to the first phase of degeneration, which, we have suggested, is a local phenomenon dependent upon finding a suitable substratum, the second phase is apparently integrated at the neuronal level and possibly consists of a selective degeneration of all but one of the axonal sprouts arising from each parent axon.

13. Once synaptic contact has been established, there is a maturation phase in which axon caliber and conduction velocity return toward normal.

REFERENCES

Aguayo, A. J., Peyronnard, J. M., and Bray, G. M. (1973). *J. Neuropathol. Exp. Neurol.* **32,** 256–270.

Aguayo, A. J., Epps, J., Charron, L., and Bray, G. M. (1976a). *Brain Res.* **104,** 1–20.

Aguayo, A. J., Charron, L., and Bray, G. M. (1976b). *J. Neurocytol.* **5,** 565–573.

Aitken, J. T., and Thomas, P. K. (1962). *J. Anat. (London)* **96,** 121–129.

Aitken, J. T., Sharman, M., and Young, J. Z. (1947). *J. Anat. (London)* **81,** 1–22.

Anderson, L. E., and McClure, W. O. (1973). *Proc. Natl. Acad. Sci. U.S.A.* **70,** 1521–1525.

Andres, K. H. (1961). *Z. Zellforsch.* **55,** 1–48.

Angelborg, C., and Engström, B. (1974). *Acta Oto-Laryngol. (Stockholm)* Suppl. **319,** 43–56.

Appenzeller, O., and Palmer, G. (1972). *Brain Res.* **42,** 521–524.

Atwood, H. L., and Morin, W. A. (1970). *J. Ultrastruct. Res.* **32,** 351–369.

Banks, P., Mangnall, D., and Mayor, D. (1969). *J. Physiol. (London)* **200,** 745–762.

Banks, P., Mayor, D., Mitchell, M., and Tomlinson, D. (1971a). *J. Physiol. (London)* **216,** 625–640.

Banks, P., Mayor, D., and Tomlinson, D. R. (1971b). *J. Physiol. (London)* **219,** 755–761.

Banks, P., Mayor, D., and Mraz, P. (1973). *J. Physiol. (London)* **299,** 383–394.

Barnes, C. D., and Worrall, N. (1968). *J. Neurophysiol.* **31,** 689–694.

Barondes, S. H., and Rosen, S. D. (1976). *In* "Neuronal Recognition" (S. H. Barondes, ed.), pp. 331–356. Chapman & Hall, London.

Bennett, G., di Giamberardino, L., Koenig, H. L., and Droz, B. (1973). *Brain Res.* **60,** 129–146.

Berenberg, R. A., Forman, D. S., Wood, D. K., De Silva, A., and Demaree, J. (1977). *Exp. Neurol.* **57,** 349–363.

Berthold, C.-H., and Carlstedt, T. (1977). *Acta Physiol. Scand.* Suppl. **446,** 5–85.

Billenstien, D. C., and Leveque, T. F. (1955). *Endocrinology* **56,** 704–717.

Binghmann, D., Kienecker, E. W., and Knoche, E. (1977). *In* "Chemoreception in the Carotid Body" (H. Acker *et al.,* eds.), pp. 36–39. Springer-Verlag, Berlin and New York.

Bisby, M. A., and Bulger, V. T. (1977). *J. Neurochem.* **29,** 313–320.

Bjerre, B., and Rosengren, E. (1974). *Cell Tissue Res.* **150,** 299–322.

Black, M. M., and Lasek, R. J. (1976). *Anat. Rec.* **184,** 360–361.

Boyle, F. C., and Gillespie, J. S. (1970). *Eur. J. Pharmacol.* **2,** 77–84.

Brattgard, S. O., Edström, J.-E., and Hyden, H. (1957). *J. Neurochem.* **1,** 316–325.

Bray, D. (1973). *Nature (London)* **244,** 93–95.

Bray, G. M., and Aguayo, A. J. (1974). *J. Anat. (London)* **117,** 517–530.

Bray, J. J., Kon, C. M., and Breckenridge, B. M. (1971). *Brain Res.* **26,** 385–394.

Brimijoin, S. (1972). *J. Neurochem.* **19,** 2183–2193.

Brown, A. G., and Iggo, A. (1963). *J. Physiol. (London)* **165,** 28–29P.

Brown, J. H., Nelson, D. L., and Molinoff, P. B. (1977). *J. Pharmacol. Exp. Ther.* **201,** 298–311.

Brown, M. C., and Ironton, R. (1977). *Nature (London)* **265,** 459–461.

Brown, M. C., Holland, R. L., and Ironton, R. (1978). *Nature (London)* **275,** 652–654.

Bulger, V. T., and Bisby, M. A. (1978). *J. Neurochem.* **31,** 1411–1418.

Cajal, S. R. (1928). "Degeneration and Regeneration of the Nervous System." Oxford Univ. Press, Oxford.

Carlsson, A. (1966). *In* "Handbook of Experimental Pharmacology" (V. Erspamer, ed.), Vol. XIX, pp. 529–592. Springer-Verlag, Heidelberg and New York.
Carlsson, C. A., and Thulin, C. A. (1967). *Experientia* **23**, 125–126.
Cheah, T. B., and Geffen, L. B. (1973). *J. Neurobiol.* **4**, 443–452.
Cockett, S. A., and Kiernan, J. A. (1973). *Exp. Neurol.* **39**, 389–394.
Colmant, H.-J. (1959). *Arch. Psychiat. Z. Ges. Neurol.* **199**, 60–71.
Cotman, C. W., (1978). "Neuronal Plasticity." Raven, New York.
Cotman, C. W., and Nadler, J. V. (1978). *In* "Neuronal Plasticity" (C. W. Cotman, ed.), pp. 227–271. Raven, New York.
Cragg, B. G. (1970). *Brain Res.* **23**, 1–21.
Cragg, B. G., and Thomas, P. K. (1961). *J. Physiol. (London)* **157**, 315–327.
Curtis, D. R., and Johnston, G. A. R. (1974). *Ergeb. Physiol.* **69**, 97–188.
Czeh, G., Kudo, N., and Kuno, M. (1977). *J. Physiol. (London)* **270**, 165–180.
Dahlström, A. (1967). *Acta Physiol. Scand.* **69**, 158–166.
Dahlström, A. (1971). *Philos. Trans. R. Soc. London Ser. B.* **261**, 325–358.
Dahlström, A., and Haggendal, J. (1966). *Acta Physiol. Scand.* **67**, 278–288.
Daniels, M. (1975). *Ann. N.Y. Acad. Sci.* **253**, 535–544.
Devor, M., and Wall, P. D. (1976). *Nature (London)* **262**, 705–708.
Devor, M., and Wall, P. D. (1978). *Nature (London)* **276**, 75–76.
Diamond, J. (1959). *J. Physiol. (London)* **145**, 611–629.
Diamond, J., Cooper, E., Turner, C., and Macintyre, L. (1976). *Science* **193**, 371–377.
Dolivo, M. (1974). *Fed. Proc.* **33**, 1043–1048.
Droz, B., Rambourg, A., and Koenig, H. L. (1975). *Brain Res.* **93**, 1–13.
Duce, I. R., and Keen, P. (1976). *Cell Tissue Res.* **170**, 491–505.
Duce, I. R., and Keen, P. (1977a). *Cell Tissue Res.* **180**, 111–121.
Duce, I. R., and Keen, P. (1977b). *Cell Tissue Res.* **185**, 263–277.
Duce, I. R., and Reeves, J. F., and Keen, P. (1976). *Cell Tissue Res.* **170**, 507–513.
Duchen, C. W., and Strich, S. J. (1968). *Q. J. Exp. Physiol.* **53**, 84–89.
Dyck, P. J., and Hopkins, A. P. (1972). *Brain* **95**, 223–234.
Edds, M. V. (1953). *Q. Rev. Biol.* **28**, 260–276.
Engh, C. A., Schofield, B. H., Dofy, S. B., and Robinson, R. A. (1971). *J. Comp. Neurol.* **142**, 465–480.
Evans, C. A. N., and Saunders, N. R. (1974). *J. Physiol. (London)* **240**, 15–32.
Flumerfelt, B. A., and Lewis, P. R. (1975). *J. Anat.* **119**, 309–331.
Forman, D. S., and Berenberg, R. A. (1978). *Brain Res.* **156**, 213–226.
Friede, R. L., and Martinez, A. J. (1970). *Brain Res.* **19**, 165–182.
Frizell, M., and Sjöstrand, J. (1974a). *Brain Res.* **78**, 109–123.
Frizell, M., and Sjöstrand, J. (1974b). *J. Neurochem.* **22**, 845–850.
Frizell, M., and Sjöstrand, J. (1974c). *Brain Res.* **81**, 267–283.
Frizell, M., McLean, W. G., and Sjöstrand, J. (1976). *J. Neurochem.* **27**, 191–196.
Furstman, L., Saporta, S., and Kruger, L. (1975). *Brain Res.* **84**, 320–324.
Gamble, H. J. (1964). *J. Anat. (London)* **98**, 17–25.
Gamble, H. J. (1976). *In* "The Peripheral Nerve" (D. N. Landon, ed.), pp. 330–354. Chapman & Hall, London.
Gerebtzoff, M. A., and Maeda, T. (1968). *C.R. Soc. Biol. (Paris)* **162**, 2032–2035.
Gershenbaum, M. R., and Roisen, F. J. (1978). *Neuroscience* **3**, 1241–1250.
Govrin-Lipmann, R., and Devor, M. (1978). *Brain Res.* **159**, 406–410.
Grafstein, B. (1975). *Exp. Neurol.* **48**, 32–51.
Grafstein, B., and McQuarrie, I. G. (1978). *In* "Neuronal Plasticity" (C. W. Cotman, ed.), pp. 156–195. Raven, New York.

Griffin, J. W., Drachman, D. B., and Price, D. L. (1976). *J. Neurobiol.* **7,** 355-370.
Guth, L. (1956). *Physiol. Rev.* **36,** 441-478.
Gutmann, E., Guttmann, L., Medawar, P. B., and Young, J. Z. (1942). *J. Exp. Biol.* **19,** 14-44.
Hall, M. E., Wilson, D. L., and Stone, G. C. (1978). *J. Neurobiol.* **9,** 353-366.
Hanbauer, I., Johnson, D. G., Silberstein, S. D., Kopin, I. J., and Bloom, F. E. (1971). *Pharmacologist* **13,** 203.
Hansson, H.-A. (1970). *Z. Zellforsch.* **107,** 23-44.
Härkönen, M. H. A., and Kauffman, F. C. (1973a). *Brain Res.* **65,** 127-139.
Härkönen, M. H. A., and Kauffman, F. C. (1973b). *Brain Res.* **65,** 141-157.
Heiwall, P.-O., Dahlström, A., Larsson, P.-A., and Bööj, S. (1979). *J. Neurobiol.* **10,** 119-136.
Hendry, I. A. (1976). *Rev. Neurosci.* **2,** 149-193.
Heslop, J. P. (1975). *Adv. Comp. Physiol. Biochem.* **6,** 75-163.
Heuser, J. E., and Reese, T. S. (1973). *J. Cell Biol.* **57,** 315-344.
Heuser, J. E., Reese, T. S., and Landis, D. M. D. (1974). *J. Neurocytol.* **3,** 109-131.
Hill, C. E., Purves, R. D., Watanabe, H., and Burnstock, G. (1976). *Pflügers Arch.* **361,** 127-134.
Hoffman, H. (1950). *Aust. J. Exp. Biol. Med. Sci.* **28,** 383-397.
Hökfelt, T., Elde, R., Johansson, O., Luft, R., Nilsson, G., and Arimura, A. (1976). *Neuroscience* **1,** 131-136.
Holtzman, E., and Novikoff, A. B. (1965). *J. Cell Biol.* **27,** 651-669.
Hopkins, A. P., and Lambert, E. H. (1972). *Brain* **95,** 213-222.
Horch, K. W. (1976). *Brain Res.* **117,** 19-32.
Howe, J. F., Loeser, J. D., and Calvin, W. H. (1977). *Pain* **3,** 25-41.
Ikeda, K., and Campbell, J. B. (1971). *Exp. Neurol.* **30,** 379-388.
Ironton, R., Brown, M. C., and Holland, R. L. (1978). *Brain Res.* **156,** 351-354.
Isenschmid, R. (1932). *Schweiz. Med. Wchnschr.* **62,** 785-798.
Jacobowitz, D., and Woodward, J. K. (1968). *J. Pharmacol. Exp. Ther.* **162,** 213-226.
Jacobson, S., and Guth, L. (1965). *Exp. Neurol.* **11,** 48-60.
Johnson, D. G., Silberstein, S. D., Hanbauer, I., and Kopin, I. J. (1972). *J. Neurochem.* **19,** 2025-2029.
Johnson, J. L. (1970). *Brain Res.* **18,** 427-440.
Johnson, J. L. (1978). *Prog. Neurobiol.* **10,** 155-202.
Kapeller, K., and Mayor, D. (1966). *J. Physiol. (London)* **187,** 36-37P.
Kapeller, K., and Mayor, D. (1969). *Proc. R. Soc. London Ser. B.* **172,** 39-51.
Karlström, L., and Dahlström, A. (1973). *J. Neurobiol.* **4,** 191-200.
Keen, P., and McLean, W. G. (1972). *Naunyn-Schmiedeberg's Arch. Pharmacol.* **275,** 465-469.
Kirk, E. J. (1974). *J. Comp. Neurol.* **155,** 165-176.
Kirpekar, S. M. (1975). *Prog. Neurobiol.* **4,** 163-210.
Kirpekar, S. M., Prat, J. C., and Wakade, A. R. (1973). *J. Physiol. (London)* **228,** 173-179.
Knyihár, E., and Csillik, B. (1976). *Exp. Brain Res.* **26,** 73-87.
Knyihár, E., and Gerebtzoff, M. A. (1973). *Exp. Brain Res.* **18,** 383-395.
Koenig, E. (1970). In "Biochemistry of Simple Neuronal Models" (E. Costa and E. Giacobini, eds.) pp. 303-315. Raven, New York.
Koenig, E., Frankel, R. D., and Tobias, G. S. (1977). In "Mechanisms, Regulation and Special Functions of Protein Synthesis in the Brain" (S. Roberts, A. Lajtha and W. H. Gispen, eds.) pp. 107-113. Elsevier/North-Holland, Amsterdam.
Komiya, Y., and Kurokawa, M. (1978). *Brain Res.* **139,** 354-358.
Kopin, I. J., and Silberstein, S. D. (1972). *Pharmacol. Rev.* **24,** 245-254.
Kristensson, K. (1978). *Annu. Rev. Pharmacol. Toxicol.* **18,** 97-110.
Kristensson, K., and Olsson, Y. (1975). *J. Neurocytol.* **4,** 653-661.
Kristensson, K., and Olsson, Y. (1976). *Brain Res.* **115,** 201-213.

Kristensson, K., and Sjöstrand, J. (1972). *Brain Res.* **45,** 175-181.

Kung, S. H. (1971). *Brain Res.* **25,** 656-660.

Lagercrantz, H. (1976). *Neuroscience* **1,** 81-92.

Larrabee, M. G. (1970). *Fed. Proc.* **29,** 1919-1928.

Lasek, R. (1968). *Brain Res.* **7,** 360-377.

Lasek, R. J., and Black, M. M. (1977). *In* "Mechanisms, Regulation and Special Functions of Protein Synthesis in the Brain" (S. Roberts, A. Lajtha and W. H. Gispen, eds.) pp. 161-169. Elsevier/North-Holland, Amsterdam.

Lasek, R. J., and Hoffman, P. N. (1976). *In* "Cell Motility. Book C. Microtubules and Related Proteins" (R. Goldman, T. Pollard and J. Rosenbaum, eds.), pp. 1021-1049. C. S. Harbor Lab., New York.

Letorneau, P. C. (1975). *Dev. Biol.* **44,** 92-101.

Levi-Montalcini, R., and Angeletti, P. U. (1968). *Physiol. Rev.* **48,** 534-569.

Lewis, E. R. (1971). *In* "Scanning Electron Microscopy - 1971" (O. Johari and I. Corvin, eds.), pp. 281-288. ITT Research Institute, Chicago, Illinois.

Lieberman, A. R. (1971). *Int. Rev. Neurobiol.* **14,** 49-124.

Lieberman, A. R. (1974). *In* "Essays on the Nervous System" (R. Bellairs and G. G. Gray, eds.), pp. 71-105. Clarendon, Oxford.

Lieberman, A. R. (1976). *In* "The Peripheral Nerve" (D. N. Landon, ed.), pp. 188-278. Chapman & Hall, London.

Liu, C.-N., and Chambers, W. W. (1958). *Arch. Neurol. Psychiat.* **79,** 46-61.

Lubinska, L. (1975). *Int. Rev. Neurobiol.* **17,** 241-296.

McIsaac, G., and Kiernan, J. A. (1975a). *J. Anat. (London).* **120,** 551-560.

McIsaac, G., and Kiernan, J. A. (1975b). *Exp. Neurol.* **48,** 88-94.

Mackay, A. V. P., and Iversen, L. L. (1972). *Naunyn-Schmiedeberg's Arch. Pharmacol.* **272,** 225-229.

McLean, W. G., and Keen, P. (1972). *Eur. J. Pharmacol.* **18,** 74-78.

McLean, W. G., and Frizell, M., and Sjöstrand, J. (1976a). *J. Neurochem.* **26,** 1213-1216.

McLean, W. G., Frizell, M., and Sjöstrand, J. (1976b). *Exp. Neurol.* **52,** 242-249.

McQuarrie, I. G., and Grafstein, B. (1973). *Arch. Neurol.* **29,** 53-55.

McQuarrie, I. G., Grafstein, B., and Gershon, M. D. (1977). *Brain Res.* **132,** 443-453.

McQuarrie, I. G., Grafstein, B., Dreyfus, C. F., and Gershon, M. D. (1978). *Brain Res.* **141,** 21-34.

Martinez, A. J., and Friede, R. L. (1970). *Brain Res.* **19,** 183-198.

Matthews, M. R. (1973). *Philos. Trans. R. Soc. London Ser. B.* **264,** 479-505.

Matthews, M. R., and Raisman, G. (1972). *Proc. R. Soc. London Ser. B.* **181,** 43-79.

Miani, N., Rizzoli, A., and Bucciante, G. (1961). *J. Neurochem.* **7,** 161-173.

Mitchell, R. A., Sinha, A. K., and McDonald, D. M. (1972). *Brain Res.* **43,** 681-685.

Mori, H., Komiya, Y., and Kurokawa, M. (1977). *Proc. Jpn. Acad. Ser. B.* **53,** 252-256.

Morris, J. H., Hudson, A. R., and Weddell, G. (1972a). *Z. Zellforsch.* **124,** 103-130.

Morris, J. H., Hudson, A. R., and Weddell. G. (1972b). *Z. Zellforsch.* **124,** 131-164.

Moyer, E. K., Kimmel, D. L., and Winborne, L. W. (1953). *J. Comp. Neurol.* **98,** 283-307.

Murray, J. G., and Thompson, J. W. (1957). *J. Physiol. (London)* **135,** 133-162.

Nathaniel, E. J. H., and Nathaniel, D. R. (1973). *Exp. Neurol.* **40,** 333-350.

Nathaniel, E. J. H., and Pease, D. C. (1963). *J. Ultrastruct. Res.* **9,** 533-549.

Neuhuber, W., Niederle, B., and Zenker, W. (1977). *Cell Tissue Res.* **183** 395-402.

Ochs, S. (1971). *Proc. Natl. Acad. Sci. U.S.A.* **68,** 1279-1282.

Ochs, S. (1972). *J. Physiol. (London).* **227,** 627-645.

Ochs, S. (1974). *Fed. Proc.* **33,** 1049-1058.

Ochs, S. (1976). *J. Physiol. (London)* **255,** 249-261.

Ochs, S., Erdman, J., Jersild, R. A., and McAdoo, V. (1978). *J. Neurobiol.* **9**, 465-481.
Olgart, L., Hökfelt, T., Nilsson, G., and Pernow, B. (1977a). *Pain* **4**, 153-159.
Olgart, L., Gazelius, B., Brodin, E., and Nilsson, G. (1977b). *Acta Physiol. Scand.* **101**, 510-512.
Olson, L. (1969). *Histochemie* **17**, 349-367.
Olsson, T. P., Forsberg, I., and Kristensson, K. (1978). *J. Neurocytol.* **7**, 323-336.
Otsuka, M., and Konishi, S. (1976). *Cold Spring Harbor Symp. Quant. Biol.* **40**, 135-143.
Pant, H. C., Terakawa, S., and Gainer, H. (1979). *J. Neurochem.* **32**, 99-102.
Paulson, J. C., and McClure, W. O. (1975). *Ann. N.Y. Acad. Sci.* **253**, 517-527.
Pelletier, G., Labrie, F., Arimura, A., and Schally, A. V. (1974). *Am. J. Anat.* **140**, 445-450.
Pestronk, A., and Drachman, D. B. (1978). *Science* **199**, 1223-1225.
Pfenninger, K. H., and Bunge, R. P. (1974). *J. Cell Biol.* **63**, 180-196.
Pfenninger, K. H., and Rees, R. P. (1976). *In* "Neuronal Recognition" (S. H. Barondes, ed.), pp. 131-178. Chapman & Hall, London.
Pichichero, M., Beer, B., and Clody, D. E. (1973). *Science* **182**, 724-725.
Pickel, V. M., Segal, M., and Bloom, F. E. (1974). *J. Comp. Neurol.* **155**, 43-60.
Pickel, V. M., Reis, D. J., and Leeman, S. E. (1977). *Brain Res.* **122**, 534-540.
Privat, A., Mandon, P., Drian, M. J. (1972). *Exp. Cell Res.* **71**, 232-235.
Purves, D. (1975). *J. Physiol.* (*London*) **252**, 429-463.
Purves, D. (1976). *J. Physiol.* (*London*) **259**, 159-175.
Quilliam, T. A. (1958). *J. Anat.* (*London*) **92**, 383-398.
Ranson, S. W. (1906). *J. Comp. Neurol.* **16**, 265-293.
Ranson, S. W. (1912). *J. Comp. Neurol.* **22**, 487-545.
Richards, J. G., and Tranzer, J. P. (1975). *J. Ultrastruct. Res.* **53**, 204-216.
Roberts, P. J., and Keen, P. (1974). *J. Neurochem.* **23**, 201-209.
Roisen, F. J., Murphy, R. A., Pichichero, M. E., and Braden, W. G. (1972). *Science* **175**, 73-74.
Sabri, M. I., and Ochs, S. (1972). *J. Neurobiol.* **4**, 145-165.
Sanders, F. K., and Whitteridge, D. (1946). *J. Physiol.* **105**, 152-174.
Sandri, C., Van Buren, J. M., and Akert, K. (1977). "Membrane Morphology of the Vertebrate Nervous System: Progress in Brain Research," Vol. 46. Elsevier, Amsterdam.
Schmidt, R. E., and McDougal, D. B. (1978). *J. Neurochem.* **30**, 527-535.
Scott, D., and Liu, C. N. (1964). *In* "Mechanisms of Neural Regeneration" (M. Singer and J. P. Schadé, eds.), pp. 127-148. Progr. Brain Res. Vol. 13. Elsevier, Amsterdam.
Shawe, G. D. H. (1955). *Br. J. Surg.* **42**, 474-488.
Silberstein, S. D., Johnson, D. G., Jacobowitz, D. M. and Kopin, I. J. (1971). *Proc. Natl. Acad. Sci. U.S.A.* **68**, 1121-1124.
Silberstein, S. D., Berv, K. R., and Jacobowitz, D. M. (1972a). *Nature* (*London*) **239**, 466-468.
Silberstein, S. D., Brimijoin, S., Molinoff, P. B., and Lemberger, L. (1972b). *J. Neurochem.* **19**, 919-921.
Silberstein, S. D., Johnson, D. G., Hanbauer, I., Bloom, F. E., and Kopin, I. J. (1972c). *Proc. Natl. Acad. Sci. U.S.A.* **69**, 1450-1454.
Sindou, M., Quoex, C., and Baleydier, C. (1974). *J. Comp. Neurol.* **153**, 15-26.
Snyder, R. (1977). *J. Comp. Neurol.* **174**, 47-70.
Spencer, P. S., and Lieberman, A. R. (1971). *Z. Zellforsch.* **119**, 534-551.
Stelzner, D. J. (1971). *Z. Zellforsch.* **120**, 332-345.
Stjarne, L., and Brundin, J. (1975). *Acta Physiol. Scand.* **94**, 139-141.
Stöckel, K., Schwab, M., and Thoenen, H. (1975). *Brain Res.* **99**, 1-16.
Streit, P., Akert, K., Sandri, C., Livingston, R. B., and Moor, H. (1972). *Brain Res.* **48**, 11-26.
Sunderland, S. (1968). "Nerves and Nerve Injuries." Churchill-Livingstone, London.
Takahashi, T., and Otsuka, M. (1975). *Brain Res.* **87**, 1-11.
Till, R., and Banks, P. (1976). *Neuroscience* **1**, 49-55.

Tobias, G. S., and Koenig, E. (1975). *Exp. Neurol.* **49**, 221-234.

Tokunaga, A., Sandri, C., and Akert, K. (1978). *Neurosci. Lett.* Suppl. **1**, S194.

Tomlinson, D. R. (1975). *J. Physiol. (London)* **245**, 727-735.

Tweedle, C. D., and Kabara, J. J. (1977). *Neurosci. Lett.* **6**, 41-46.

Varon, S. S., and Bunge, R. P. (1978). *Annu. Rev. Neurosci.* **1**, 327-361.

Venzin, M., Sandri, C., Akert, K., and Wyss, U. R. (1977). *Brain Res.* **130**, 393-404.

Vogel, S. A., Silberstein, S. D., Berv, K. R., and Kopin, I. J. (1972). *Eur. J. Pharmacol.* **20**, 308-311.

Walberg, F. (1966). *Exp. Brain Res.* **2**, 107-128.

Wall, P. D. (1976). *In* "Perspectives in Brain Research" (M. A. Corner and D. F. Swaab, eds.), pp. pp. 359-377. Progr. Brain Res. Vol. 45. Elsevier, Amsterdam.

Wall, P. D., and Devor, M. (1978). *In* "Physiology and Pathobiology of Axons" (S. G. Waxman, ed.), pp. 377-378. Raven, New York.

Wall, P. D., and Gutnick, M. (1974a). *Exp. Neurol.* **43**, 580-593.

Wall, P. D., and Gutnick, M. (1974b). *Nature (London)* **248**, 740-743.

Watson, W. E. (1966a). *J. Neurochem.* **13**, 849-856.

Watson, W. E. (1966b). *J. Neurochem.* **13**, 1549-1550.

Watson, W. E. (1968). *J. Physiol. (London)* **196**, 655-676.

Watson, W. E. (1970). *J. Physiol. (London)* **210**, 321-343.

Watson, W. E. (1973). *J. Physiol. (London)* **231**, 41P-42P.

Watson, W. E. (1974). *Br. Med. Bull.* **30**, 112-115.

Waxman, S. G., and Quick, D. C. (1978). *Brain Res.* **144**, 1-10.

Webb, J. G., Moss, J., Kopin, I. J., and Jacobowitz, D. M. (1975). *J. Pharmacol. Exp. Ther.* **193**, 489-502.

Weddell, G. (1942). *J. Anat. (London)* **77**, 49-62.

Weddell, G., Guttmann, L., and Gutmann, E. (1941). *J. Neurol. Neurosurg. Psychiatry* **4**, 206-225.

Weinberg, H. J., and Spencer, P. S. (1975). *J. Neurocytol.* **4**, 395-418.

Weinberg, H. J., and Spencer, P. S. (1976). *Brain Res.* **113**, 363-378.

Weinberg, H. J., and Spencer, P. S. (1978). *J. Neurocytol.* **7**, 555-569.

Weinstock, M., Thoa, N. B., and Kopin, I. J. (1978). *Eur. J. Pharmacol.* **47**, 297-302.

Weiss, P. A. (1945). *J. Exp. Zool.* **100**, 353-368.

Weiss, P., and Taylor, A. C. (1943). *Arch. Surg.* **47**, 419-447.

Wells, M. R. (1977). *Anat. Rec.* **187**, 745-746.

West, N., and Bunge, R. (1976). *Annu. Meet. Soc. Neurosci., 6th* **2**, 1038 (Abstr.).

Wettstein, R., and Sotelo, J. R. (1963). *Z. Zellforsch.* **59**, 708-730.

White, F. P., and White, S. R. (1977). *J. Neurobiol.* **8**, 315-324.

Wooten, G. F., and Coyle, J. T. (1973). *J. Neurochem.* **20**, 1361-1371.

Wooten, G. F., Weise, V. K., and Kopin, I. J. (1977). *Brain Res.* **136**, 174-177.

Yamada, K. M., Spooner, B. S., and Wessells, N. K. (1971). *J. Cell Biol.* **49**, 614-635.

Zanetta, J.-P., Roussel, G., Ghandour, M. S., Vincendon, G., and Gombos, G. (1978). *Brain Res.* **142**, 301-319.

Zelena, J., Lubinska, L., and Gutmann, E. (1968). *Z. Zellforsch.* **91**, 200-219.

INTERNATIONAL REVIEW OF CYTOLOGY, VOL. 66

When Sperm Meets Egg: Biochemical Mechanisms of Gamete Interaction

BENNETT M. SHAPIRO AND E. M. EDDY

Departments of Biochemistry and Biological Structure, University of Washington, Seattle, Washington

> Embryology may seem, at first sight, a kind of Penelope's web.
>
> Dalcq (1938)

I. Introduction

"Shall it be male or female? say the cells, And drop the plum like fire from the flesh" (Thomas, 1957).

"Aristotle and Democritus maintained that women have no sperm, and that it is only a sweat that they discharge in the heat of the pleasure and the movement, which contributes nothing toward generation; Galen and his followers argued, on the contrary, that without the meeting of seeds, generation can not take place" (Montaigne, 1576). Such speculation on the union of male and female reaches back to our neolithic ancestors, whose cave sculpture and drawings attest their interest. Small wonder then that with the dawn of experimental science (the light of which may have come, in good part, from this preoccupation), fertilization became a central focus of biology. This occurred not only because of the lure of unraveling such a complex and mysterious process, but also because the agricultural revolution and the control of epidemic disease gave us, for the first time, the

257

ability to overpopulate our environment. Thus the desire to find means to attenuate human reproduction, or to enhance that of the animals we eat, served as additional impetus to the study of fertilization and probably accounts for much of the support for such research. The problem remains a critical one for exploration, since the event of fertilization is a web made of diverse strands—from genetics, evolution, cell biology, and development—and as such, serves as a metaphor for the living state, as well as a lever for prying forth its secrets.

For this reason, attention has been paid repeatedly to problems of fertilization, each new sortie employing techniques of contemporary science to extract yet more information. Fortunately, unlike Penelope, we do not need to tear apart the whole fabric and start anew, for a rich and useful literature dates back over a century. For the past few years, several laboratories have been addressing some molecular aspects of the problem of sperm–egg interaction, and that is the subject we shall consider in this review. As our understanding of molecular processes in fertilization increases, it is striking to find a remarkable similarity between mechanisms in species whose gametes are dissimilar in certain structural and functional properties. Just as there are not many differences in important biochemical pathways between mice and men, or, for that matter, between prokaryotes and eukaryotes, so biochemical aspects of sperm–egg interaction have strong similarities across phyla. Of course, we are just on the threshold of rigorous study of the biochemistry of fertilization, so such generalizations should not be extended too far. Nonetheless, we shall point out similarities as we see them, while remaining aware of the risks of such simplification.

This review focuses on three principal subjects: (1) the effect of the egg on the sperm; (2) the contact between sperm and egg; and, (3) the effect of the sperm on the egg. We briefly consider a fourth point: the activation of egg metabolism that follows fertilization; this subject will be more extensively covered in another article in this series (Steinhardt and Epel, in preparation). Additionally, although many insights into fundamental mechanisms of fertilization have come from studies of prokaryotes, lower eukaryotes, and plants, this article concerns itself principally with animal fertilization, both because we do not have the space to do justice to these other areas, and because the article relates to our immediate research interests. We intend to focus mainly on recent work that leads to understanding fundamental biochemical mechanisms and to trace the roots of this understanding in the work of earlier embryologists. However, because of the enormous literature on fertilization, we are forced to make many omissions. The material we concentrate on reflects our personal biases, not necessarily the quality or importance of the work in the field. Several other recent reviews discuss aspects of fertilization in general (Longo, 1973; Austin, 1975; Epel, 1977; Shapiro, 1977; Moscona and Monroy, 1978), or focus on sea urchins (Giudice, 1973; Czihak, 1975; Lallier, 1977; Epel and Vacquier, 1978) or mammals (Gwatkin, 1977; Bedford and Cooper, 1978).

II. The Effect of the Egg on the Sperm

The most dramatic morphological response of the sperm to the egg is the acrosome reaction, in which a change that is a prerequisite for fertilization occurs in the head of sperm from many species (see Baccetti and Afzelius, 1976). In mammals, sperm undergo a process known as capacitation before they respond to the egg, so this less well-understood phenomenon will be discussed first.

A. Capacitation

Capacitation is a physiological event: for the internal fertilization of the mammal, sperm must spend a period of time in the female reproductive tract before they can penetrate the egg (Austin, 1951; Chang, 1951). Although the physiological aspects of this process have been extensively studied (reviewed in Rogers, 1978), there is little known at a molecular level. Capacitation involves certain changes in sperm metabolism, as well as alterations in the distribution of intramembranous particles and lectin binding sites over the sperm plasma membrane (reviewed in Gwatkin, 1977; Friend, 1977; Millette, 1977; Koehler, 1978; Talbot and Franklin, 1978). Capacitation may also involve removal of seminal fluid components that suppress the metabolic activity and fertilizing ability of the sperm, as suggested by the discovery in seminal fluid of decapacitation factors that block the ability of capacitated sperm to fertilize eggs (Chang, 1957). Some decapacitation factors appeared to be glycoproteins (e.g., Davis, 1971; Reyes *et al.*, 1975), as assessed by their deleterious effect on fertilization. However, with a complex biological assay, extensive purification and careful analysis of homogeneous decapacitation factors will be required before we can understand the mechanism of such an inhibitory effect. Of some interest, in a comparative sense, is the finding that seminal fluid of the sea urchin *Arbacia punctulata* inhibits the metabolism of its sperm (Hayashi, 1946): when *Arbacia* sperm are diluted, their respiration increases; if diluted in the presence of egg surface material ("jelly"), respiration ultimately decreases and the sperm die (as discussed below, this is related to induction of the acrosome reaction). However, when sperm are added to jelly in the presence of seminal fluid, they do not decrease their respiratory rate nor their viability, and thus it appears that "adsorption of protein substances and their removal from the surface of the cell" regulate the physiology of sea urchin sperm (Hayashi, 1946). This resemblance to capacitation may indicate a general ability of fluids of the male reproductive tract to lower the metabolic rate of sperm. Perhaps capacitation is not seen routinely in invertebrate fertilization due to a rapid dissociation of seminal fluid components from the sperm.

There is more to capacitation in the mammal than removal of inhibitory effects of seminal fluid components, for epididymal sperm also need to be capacitated.

A principal problem with capacitation is that it confuses the interpretation of the mechanism of acrosome reaction triggering. In many mammalian systems where capacitation requires hours, substances that affect either the rate of capacitation or of the acrosome reaction will affect the triggering of the acrosome reaction. However, it has not been definitively established that capacitation is a necessary prelude to the acrosome reaction (Bedford, 1970), and this makes an interpretation of experiments on mechanisms of induction of the mammalian acrosome reaction difficult. In only a few cases have there been attempts made to distinguish between it and capacitation. Given these considerations, it is remarkable how many similarities in metabolic events exist between the acrosome reactions of invertebrates and of mammals.

B. The Acrosome Reaction

1. General

In a study of *Arbacia* sperm by vital staining, Popa (1927) discovered several changes in sperm incubated in sea water in which eggs had previously been present. The midpiece of the sperm was dislocated, and there was "elimination of a substance through the pointed apex of the spermatozoon." Popa presumed this material to be an adhesive substance, since sperm stuck together after this change. However, the first comprehensive study of this reaction was by Jean Dan, in a series of experiments performed in the 1950s. Dan (1952) found that exposure to the so-called "egg water" led to an extension of a process at the head of the sperm, termed the "acrosome reaction," in several sea urchins. She also discovered that raising the pH of the sea water from 8 to 9.2 led to a similar reaction. In addition, Dan found (1954) that Ca^{2+}, previously shown to be required for fertilization, was required for the acrosome reaction. However, isoagglutination of sperm in jelly, previously suggested to be a component of sperm–egg recognition (Lillie, 1919), occurred in the absence of Ca^{2+}, and thus was not related to the acrosome reaction directly. Dan also observed that sperm died soon after undergoing the acrosome reaction. The requirement for Ca^{2+} in the acrosome reaction is absolute; indeed it may be the only role for extracellular calcium in fertilization, since sperm, having undergone the acrosome reaction in the presence of Ca^{2+}, can fertilize eggs in Ca^{2+}-free sea water (Takahashi and Sugiyama, 1973). In addition to studying the mechanism of the reaction, Dan's group also found that one of the consequences was to release an enzyme that aids in penetrating the egg coats—this was the first evidence for an acrosin-like protein having a role in fertilization (Wada *et al.*, 1956). These early experiments with invertebrate sperm provided the foundation for current ideas about the acrosome reaction: it is triggered by contact with egg surface components; there is release of lytic material during the process; and the head of the sperm is rearranged, in the case of invertebrates, with the production of a filament.

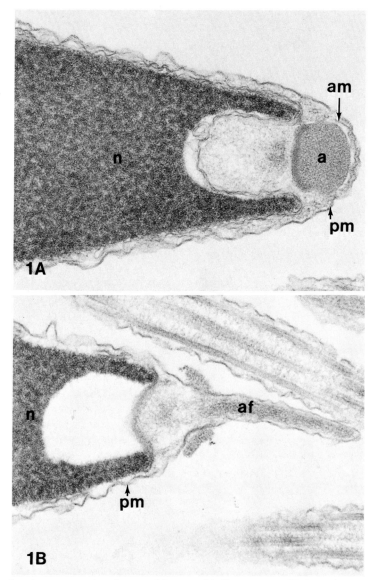

FIG. 1.   Acrosome reaction of sperm of the sea urchin *S. purpuratus*. (A) shows the anterior part of an unreacted sperm. The cup-shaped indentation in the sperm nucleus contains the globular actin and is capped by a dense acrosomal granule. The membrane of the acrosomal granule lies closely apposed to the plasma membrane at the anterior tip of the sperm. (B) shows a sperm that has been triggered by egg jelly to undergo the acrosome reaction. The contents of the acrosomal granule have been exposed by exocytosis and the actin has polymerized to extend the acrosomal filament. a, Acrosome; am, acrosomal membrane; pm, plasma membrane; n, nucleus; af, acrosomal filament.

An example of the acrosome reaction of the sea urchin, *Strongylocentrotus purpuratus*, is shown in Fig. 1. The acrosomal filament, which is quite short in the sea urchin, can be prodigious in other invertebrates. Actin was shown to be a principal component of the acrosomal filament in species with long filaments (*Asterias* and *Thyone*) both by its similarity in molecular weight (MW) to muscle actin and by the fact that it was specifically stained by heavy meromyosin (Tilney *et al.*, 1973). In unreacted sperm, actin is found in a monomeric form that polymerizes rapidly during the acrosome reaction. The actin from acrosomal filaments is packed in paracrystalline arrays arranged hexagonally with periodic banding (DeRosier *et al.*, 1977), and, perhaps, is held together by a second protein of 55,000 MW. These acrosomal filaments are among the best studied examples of the actin-containing microfilament system of cells; unlike other microfilament systems (Wessels *et al.*, 1971; Spudich, 1973), the assembly of the acrosomal filament is not inhibited by cytochalasin B (Sanger and Sanger, 1975).

The mammalian acrosome reaction found by Austin and Bishop (1958) in rodent sperm has a substantially different morphological progression from that of marine invertebrates. Whereas fusion between acrosomal and plasma membranes with attendant release of acrosomal contents (as shown in Fig. 2 for mouse sperm) is a feature held in common, there is no actin filament assembly in mammalian sperm. Another common feature of the acrosome reaction in diverse phyla is the elaboration of a site for fusion with the egg. In invertebrate sperm, this site is located at the tip of the acrosomal filament, and thus its relationship to an actin-mediated membrane rearrangement is clear. In mammalian sperm, the relationship of membrane rearrangement, as detected by an altered distribution of lectin binding sites and intramembranal particles (e.g., Friend *et al.*, 1977; Kinsey and Koehler, 1978; Koehler, 1978), to the cytoskeletal elements is not defined. However, cytoskeletal elements are associated with mammalian sperm as with those of invertebrates: both actin (Talbot and Kleve, 1978) and tubulin (Stambaugh and Smith, 1978) are found in the sperm head. Perhaps the cytoskeleton of both invertebrates and mammals has a common role in rearranging the sperm membrane for its ultimate fusion with the egg.

The clearest association between cytoskeletal elements and membrane rearrangement in sperm is seen in *Limulus;* here actin filaments may be involved in preparing the region of fusion between the acrosomal vesicle and the plasma membrane (Tilney *et al.*, 1979), for actin filament distribution parallels that of adjacent intramembranous particles and both move with membrane fusion in a manner reminiscent of the opening of a camera lens.

We will consider only certain biochemical features of the acrosome reaction, since several reviews have dealt with the phenomenon (Dan, 1956, 1967; Austin, 1975; Green, 1978a; Meizel, 1978; Metz, 1978). After briefly considering the

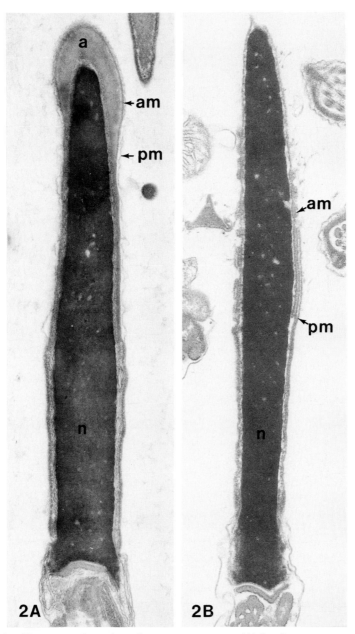

Fig. 2. The acrosomal reaction of mouse spermatozoa. (A) is of a mouse spermatozoon (Swiss–Webster) fixed immediately upon release from the cauda epididymis. The acrosome, acrosomal membrane, and plasma membrane (slightly ruffled) are intact. (B) is of a mouse spermatozoon incubated after release from the cauda for 2 hours in Toyoda's medium to achieve capacitation. The ''inner'' acrosomal membrane now covers the anterior portion of the sperm head. Some attached membrane vesicles persist as well as the equatorial segment (right side). a, Acrosome; am, acrosomal membrane; pm, plasma membrane; n, nucleus. (Pictures courtesy of Drs. James Koehler and Kodo Sato.)

nature of the components released by the acrosome reaction, we will concentrate on the mechanism of triggering, as inferred from data with different species.

2. *Components Released from the Sperm*

Much attention has been paid to the lytic enzymes that are released by sperm during the acrosome reaction, and the subject has been extensively reviewed (McRorie and Williams, 1974; Stambaugh, 1978); we will briefly discuss their role in fertilization and direct the interested reader elsewhere for details. Although enzymes are associated with acrosomal fractions of disrupted mammalian sperm, neither their location within the sperm nor their role in fertilization is clear. The best studied enzymes are a hyaluronidase and a trypsin-like protease (acrosin), and both seem to be involved in fertilization. From a study of the effect of antibodies directed against hyaluronidase or inhibitors of acrosin on fertilization, the enzymes have been inferred to effect passage of the sperm through the cumulus oophorus and zona pellucida, respectively. Although hyaluronidase is relatively loosely associated with the sperm and, in fact, can be detected even before the acrosome reaction occurs, acrosin is quite tightly bound to the acrosomal membrane. This bound form, present at greater than 1 mM, is thought to participate in fertilization by dissolving a channel through the zona pellucida (e.g., Brown and Hartree, 1976). In fact, free acrosin, like other proteases, inhibits fertilization (Gwatkin *et al.*, 1977), so tight association with the sperm head may be required to allow penetration without cleaving off sperm receptor sites nonspecifically. Thus, the enzyme may act as a localized catalytic drill.

Acrosin exists in the unreacted mammalian sperm as a zymogen, proacrosin (Meizel and Mukerji, 1975), whose mode of activation is not clear. A thermolysin-like enzyme ("acrolysin") has been postulated to be the activator, since it is present in sperm; but there is no compelling evidence that it is located in the acrosome, and many proteolytic enzymes can activate proacrosin *in vitro*. A uterine fluid glycosaminoglycan can also activate proacrosin *in vitro*, perhaps by inducing a change in structure that allows a few molecules to be activated; the process then proceeds autocatalytically (Wincek *et al.*, 1979). Thus, activation of only a few molecules of proacrosin need take place, for the acrosin thus formed could activate the remainder. A related effect of sperm acrosin on its own physiology is in allowing the enzyme to be released from the sperm head; acrosin slowly diffuses away from the acrosome-reacted sperm, and this release is prevented by acrosin inhibitors (Green, 1978b). Although most of our knowledge of the role of acrosin in fertilization has come from studies with mammalian sperm, it may play similar roles in invertebrates. Acrosin exists in sea urchin sperm and its activity is seen only upon triggering the acrosome reaction (Levine *et al.*, 1978). As with the mammal, inhibitors of acrosin inhibit fertilization, supporting the idea that localized proteolysis may be a general mechanism for penetration of the coats of the egg.

3. *Triggering Mechanisms*

Both the responsible agent and triggering mechanism are better defined for the invertebrate acrosome reaction, but many results with mammalian systems suggest strong parallels. There is some limited species specificity in triggering the invertebrate acrosome reaction, but the specificity is less for the acrosome reaction than for the binding of gametes to each other (Summers and Hylander, 1975; reviewed in Metz, 1978). Thus, the binding reaction seems to be more the agent of specificity in fertilization than acrosome reaction triggering, although both play a role.

Since the experiments of Dan (1952), which demonstrated that "egg water" triggers the acrosome reaction, it was assumed that some component of the loosely adhering jelly coat of the egg was the active agent. However, although the triggering material may be washed free from eggs, the location of this substance in the intact egg is still unclear. In the case of *Limulus*, the acrosome reaction only occurs after sperm bind to the egg surface; this is followed by tighter sperm–egg interaction and filament extension (Brown, 1976; Tilney *et al.*, 1979). In the sea urchin *Pseudocentrotus*, sperm inside isolated jelly coats do not undergo the acrosome reaction; this is triggered only upon contact with the vitelline coat (Aketa and Ohta, 1977). Also, contact of sperm with extensively dejellied eggs (where both acid and dithiothrietol were used to remove jelly components) allows the acrosome reaction to occur (Kimura-Furakawa *et al.*, 1978). These studies suggest that a tightly adhering component of the egg surface is the triggering agent and that it might be part of the vitelline layer that slowly diffuses into the surrounding sea water; it may be extracted by treating eggs at acid pH, the normal method of obtaining egg jelly.

Echinoderm jelly is heterogeneous in composition. For example, that from the sea urchin *S. purpuratus* was shown by a combination of immunological, electrophoretic, and hydrodynamic techniques (Lorenzi and Hedrick, 1973) to have at least five macromolecular components. Treatment of jelly with trypsin and Pronase to remove 80% of the amino acids, or boiling, or extended dialysis, did not affect the ability of the material to trigger the acrosome reaction (Ishihara and Dan, 1970). However, after periodate oxidation, the activity was lost. Hydroxyapatite fractionation of jelly led to the isolation of a sialoprotein component, which comprised about 0.3% of the jelly and which could agglutinate sperm and trigger the acrosome reaction (Isaka *et al.*, 1970; Hotta *et al.*, 1970). However, a more recent purification of jelly coat components by a combination of gel filtration and ion-exchange chromatography showed that a sialoglycoprotein accounting for 20% of the jelly was ineffective as a trigger, whereas a fucose–sulfate polymer had all the acrosome reaction triggering activity, as well as exhibiting the expected species specificity (SeGall and Lennarz, 1978). The major difference between the properties of the crude and purified jelly fractions was that the latter required more calcium to effect triggering. Since Ca^{2+} by itself under certain conditions can trigger the acrosome reaction, the higher Ca^{2+} requirement for triggering with the fucose–sulfate polymer suggests that another component may be required for normal triggering. Starfish egg jelly was also fractionated into two components: a macromolecular one, which was excluded from Sephadex G-100 and which induced the acrosome reaction; and a low-molecular-weight, nondialyzable component, which was identical to asterosaponin A (a sulfated steroid glycoside) and was responsible for sperm agglutination (Uno and Hoshi, 1978). Such attempts to fractionate jelly are of great interest, for unfractionated jelly leads to a number of changes in sperm physiology and biochemistry: by using purified preparations to elicit either the acrosome reaction or isoagglutination, we will be better able to identify the molecular processes associated with the different phenomena.

In early experiments on the kinetics of the acrosome reaction, Dan (1952, 1954, 1956) suggested that an increased permeability to Ca^{2+} is the triggering event and that triggering involves removal of the plasma membrane from the tip of the sperm, followed by filament extension. Recent experiments on the mechanism support and extend these ideas. In *A. punctulata,* triggering is dependent upon both Ca^{2+} and pH: whereas normal triggering occurs at pH 8 (that of sea water) and few reactions occur below 7.5, spontaneous acrosome reactions (in the absence of jelly) occur above pH 8.5 (Gregg and Metz, 1976). A similar pH dependence was found in several other studies (Decker *et al.,* 1976; Collins and Epel, 1977), and it was found additionally that the divalent ionophore A23187 could induce the reaction in *Arbacia* (Decker *et al.,* 1976) and in *S. purpuratus* (Collins and Epel, 1977). In all cases, triggering by any agent requires Ca^{2+}, although with the ionophore the required concentration is lower. Ba^{2+} and Sr^{2+} can substitute for Ca^{2+}, and La^{3+} and local anaesthetics (agents that block Ca^{2+} movement in other systems) block the acrosome reaction. The Ca^{2+} dependence of the acrosome reaction probably reflects a requirement for this cation for exocytosis, in agreement with suggestions from other cell systems (reviewed in Baker, 1974; Rubin, 1970).

The sea urchin acrosome reaction is a particularly attractive form of exocytosis to study, for it occurs synchronously in the entire population within seconds. For this reason, attempts are being made to identify discrete steps in the overall pathway of exocytosis and filament extension of invertebrate sperm. Tilney and co-workers (1978), in several types of echinoderm sperm, found that both the divalent ionophores A23187 and X537A and the monovalent ionophore, nigericin, could elicit the reaction. The ionophores lead to actin polymerization with concomitant release of acid when the reaction is triggered in isotonic NaCl or

KCl; Ca^{2+} is not required for actin polymerization. This was seen at pH 8; at pH 6.5, there was neither actin polymerization nor acid release (Tilney *et al.*, 1978). Because of the correlation between acid release and actin polymerization, the increased intracellular pH may permit actin to be released from a bound form, and thereby assemble into the acrosomal filament. The nucleation site for this assembly appears to be a structure called the actomere, that nestles in the center of the cup of assembling actin (Tilney, 1978). However, in the absence of Ca^{2+}, exocytosis did not occur, but actin polymerized in a lump around the intact acrosomal vesicle. These experiments both supported the idea that Ca^{2+} plays a role in exocytosis and additionally suggested that the acid release was related to polymerization of the actin filament.

Schackmann *et al.* (1978) also found acid release when the acrosome reaction of the sea urchin *S. purpuratus* was triggered with either egg jelly or nigericin. The Ca^{2+} requirement of the acrosome reaction was reflected in Ca^{2+} uptake. With nigericin, filament extension, acid release, and Ca^{2+} uptake were delayed, in contrast to triggering with egg jelly, suggesting that the processes are linked. By the time the acrosomal filaments extended, the sperm accumulated about 1 nmole of Ca^{2+} per 10^8 sperm, giving the sperm a Ca^{2+} concentration several orders of magnitude higher than the free Ca^{2+} concentration of the egg (Schackmann *et al.*, 1978). Some of this Ca^{2+} is likely to be associated with calmodulin, the Ca^{2+} regulatory protein that is found in high levels in sperm (Jones *et al.*, 1978) and localized to the sperm head. We postulate that the Ca^{2+} that is taken up may serve to activate the cortical reaction (see following) as a sort of "calcium bomb." Ca^{2+} continues to accumulate in sea urchin sperm even after the acrosome reaction is completed and sperm lose the ability to fertilize. This long-term Ca^{2+} uptake is probably mitochondrial; this hypothesis is supported by the following observations. Inhibitors of mitochondrial function block the long-term Ca^{2+} uptake (Schackmann *et al.*, 1978) under conditions where the acrosome reaction is not inhibited. Additionally, Ca^{2+} uptake is found by X-ray microanalysis to be localized over the mitochondrial region, both in sea urchin (Cantino and Schackmann, unpublished observations) and mammalian sperm (Babcock *et al.*, 1978).

In keeping with the evidence for a Ca^{2+} requirement, inhibitors of Ca^{2+} channels in excitable cells (e.g., verapamil and D600) inhibit the acrosome reaction (Schackmann *et al.*, 1978). Likewise, an inhibitor of potassium movement (tetraethylammonium ion) also inhibits the acrosome reaction, as do elevated levels of potassium in the sea water. The acrosome reaction is accompanied by a net efflux of potassium, as determined with either $^{42}K^+$ or by potassium-sensitive electrode measurements; this is a late event, like the massive Ca^{2+} uptake. External Na^+ is an absolute requirement for eliciting the acrosome reaction with jelly, and Na^+ uptake occurs with a fixed temporal relationship to H^+ efflux and after triggering with jelly or nigericin. This observation is one of a series used to arrange the diverse ion movements in the acrosome reaction (Schackmann and Shapiro, 1980). In these studies, we have employed different inhibitory conditions and several means of triggering the reaction to attempt to place the steps in a sequence, much like the sequence of a biochemical reaction pathway. For example, when egg jelly is added in the presence of inhibitors so that the reaction does not occur, and then the jelly and inhibitors are washed away, the sperm become resistant to subsequent triggering by readdition of jelly. This is the case with inhibitors such as verapamil, tetraethylammonium, or potassium, as well as with low pH (pH 6), but not for Ca^{2+}-free medium. When the sperm are mixed with jelly in a Ca^{2+}-free medium, the reaction is not triggered, but after washing away the jelly and resuspending in normal seawater (containing Ca^{2+}) sperm can be triggered upon readdition of jelly (Schackmann, unpublished data). We interpret these data to mean that, in the absence of Ca^{2+}, sperm can not even enter the pathway of the acrosome reaction when they contact jelly, whereas with the other inhibitors, they enter the pathway and proceed to a state, indicated by Q in Scheme I, from which they can not be rescued by subsequent addition of jelly. However, they can be rescued from this state by addition of the ionophore, nigericin. With experiments like this, and others where ion movements were correlated with the kinetics of the acrosome reaction, the sequence of Scheme I was obtained (Schackmann and Shapiro, 1980):

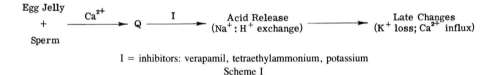

I = inhibitors: verapamil, tetraethylammonium, potassium
Scheme I

In this scheme, a $Na^+:H^+$ exchange correlates temporally with filament extension, and all of the inhibitors of the reaction block this and subsequent steps. A small amount of Ca^{2+} is taken up by the sperm under all conditions of triggering in the presence of Ca^{2+}, so some uptake may be associated with the triggering event per se; this is overshadowed by a massive Ca^{2+} uptake into the mitochondria, which follows the reaction and which is not verapamil sensitive. Even at this preliminary stage of our understanding, it is clear that the sperm behave like other excitable cells and respond to specific stimuli by altering the membrane permeability to ions. We are presently interested in whether we can correlate specific morphological concomitants of the acrosome reaction with specific ion movements, in the fashion begun by Tilney et al. (1978).

In addition to a role for ion movements in the acrosome reaction mechanism, proteolytic enzymes may be involved. Meizel and Lui (1976) showed that inhibitors of acrosin inhibited the acrosome reaction of hamster sperm, and Levine and Walsh (1978) found similar data for the sea urchin. In the latter studies, acrosin could be exposed by treatment of sperm under conditions that did not trigger the acrosome reaction (with low jelly concentrations or with 20 mM Ca^{2+}); if the exposed protease was inactivated with specific inhibitors, the reaction could not be subsequently triggered with jelly. Thus, exposure of acrosin may be a preliminary step in the reaction sequence and may play some role in the overall process.

In addition to the potential involvement of proteolytic enzymes, several other aspects of the acrosome reaction of mammals parallel that of sea urchins. However, because of the difficulties associated with triggering the reaction synchronously and rapidly, as well as in separating capacitation from the acrosome reaction, the process is less well defined. Additionally, the rodent systems in which techniques for eliciting the reaction have been best defined do not provide enough gametes for extensive biochemical analyses.

Much effort has been spent in defining the triggering agents in the mammal, which are located in the female reproductive tract rather than on the egg surface (as with sea urchins). Both dialyzable and nondialyzable components of the female are involved in eliciting the acrosome reaction. Yanagimachi (1970) showed that albumin could replace the nondialyzable factor, and Bavister *et al.* (1976) showed that the adrenal gland had high concentrations of a dialyzable factor of low molecular weight that increased sperm motility. Subsequent studies showed that different preparations of albumin varied in their efficacy as the nondialyzable factor (Lui and Meizel, 1977) and that, by removing the fatty acids bound to bovine serum albumin, the protein became a much more effective triggering agent. If palmitic or oleic acids were readded to defatted albumin, its triggering activity decreased. Even peptic hydrolysates of bovine serum albumin can elicit the reaction, but the nature of the active fragment is unclear. Cornett and Meizel (1978) showed that the dialyzable factor could be replaced by adrenergic amines like epinephrine and that such compounds, when added with bovine serum albumin and an undefined adrenal cortex component, were sufficient to elicit the arosome reaction. Both α- and β-adrenergic antagonists block the acrosome reaction, suggesting that both types of receptors are involved. Phenylephrine, an α-adrenergic agonist, stimulated the acrosome reaction but did not activate sperm motility, whereas isoproterenol, a β-adrenergic agonist, stimulated motility but not the acrosome reaction. Thus, the sperm seems to have specific receptors, which are akin to those in the autonomic nervous system and which may mediate some of its properties. Not only adrenergic amines, but also cholinergic ligands may alter sperm behavior (Nelson, 1978), at least at pharmacologic concentrations. Recently, similar receptors have been found in eggs (see later).

The study of the mammalian acrosome reaction was facilitated by Yanagimachi and Usui (1974), who found that guinea pig sperm could be capacitated in Ca^{2+}-free medium at the same rate as in the presence of Ca^{2+}; upon subsequent addition of 0.2 mM Ca^{2+}, the acrosome reaction occurred relatively synchronously and rapidly. In keeping with this Ca^{2+} requirement was the observation that A23187, the Ca^{2+} ionophore, elicits the acrosome reaction in mammalian sperm (Talbot et al., 1976; Green, 1978a) as long as Ca^{2+} is present. The ionophore also shortens the time required for capacitation, and so it is difficult to separate the two processes, except in the system of Yanagimachi and Usui (1974).

Since Ca^{2+} is intimately involved with the mammalian acrosome reaction, a Ca^{2+}-stimulated ATPase on the acrosomal membrane and adjacent plasma membrane was thought to mediate Ca^{2+} movements related to triggering (Gordon, 1973; Gordon et al., 1978). However, from studies of similar enzymes in other cells, the more likely role is to remove Ca^{2+} from the cell and thus to keep intracellular $[Ca^{2+}]$ low. Perhaps the triggering mechanism involves inhibition of such a Ca^{2+} efflux mechanism, with subsequent Ca^{2+} flooding of the sperm cytoplasm playing the decisive role in triggering. An analogous situation exists in the sea urchin (Schackmann, unpublished data) where a sodium-dependent Ca^{2+} efflux system was found; with low sodium concentrations, the sperm accumulates Ca^{2+}, and the reaction is triggered spontaneously (without egg jelly). Not only is Ca^{2+} required for the mammalian acrosome reaction, but during capacitation, guinea pig sperm take up Ca^{2+} (Singh et al., 1978) in a reaction that is stimulated by inducers of the acrosome reaction and blocked by its inhibitors. There is both $^{45}Ca^{2+}$ uptake and net Ca^{2+} accumulation; Mg^{2+} inhibits both the Ca^{2+} uptake and the acrosome reaction.

With a pH indicator dye, the pH of the hamster acrosome was estimated to be below 5 (Meizel and Deamer, 1978), and the suggestion was made that this low pH inhibited activation of the zymogen, proacrosin. The relationship between the low pH of the acrosomal contents in hamster sperm and the acid release seen in the echinoderm acrosome reaction is of some interest for it suggests that some of the H^+ release could be due to exocytosis of a weak acid rather than a Na^+:H^+ exchange. In contrast to the sea urchin system, metabolic inhibitors such as oligomycin inhibit the hamster acrosome reaction and fertilization but do not affect motility (Rogers et al., 1977). In these studies, sperm were incubated a long time with the inhibitors, whereas incubation was brief with the sea urchin experiment (Schackmann et al., 1978); thus in the experiments with the hamster, extensive changes in ATP levels and other metabolic alterations may have occurred and some of these may reflect additional requirements for the overall reaction. For, if sea urchin sperm are similarly treated for long periods of time, they fail to react (Schackmann, unpublished data). Another point of agreement between echinoderm and mammalian systems is in the effect of verapamil and tetraethylammonium ion. These inhibitors of the sea urchin acrosome reaction (Schackmann et al., 1978) also block mouse fertilization in vitro (Shellenbarger and Shapiro, 1980) at similar concentrations, although it is not yet clear that the latter effect is due to inhibition of the acrosome reaction. Taken together, the results suggest that some common mechanisms may be employed for triggering the acrosome reaction in both mammals and sea urchins.

4. *Motility, Respiration, and Isoagglutination*

Jelly (and perhaps other factors) released from eggs lead to alterations in sperm behavior that were only partially dealt with above. These events are difficult to define because different factors may be responsible for the behavioral changes than for the acrosome reaction (or for capacitation) and because some behavioral changes may be part of an overall process initiated by a single egg component. Additionally, following the acrosome reaction, there is a dramatic decrease in the fertilizing ability of sperm (e.g., Dan, 1954; Kinsey *et al.,* 1978), the relationship of which to the other events is unresolved.

Sea urchin sperm have increased motility when the acrosome reaction is triggered (Hathaway, 1963); the motility triggering factor was separated from that responsible for the acrosome reaction and found to be dialyzable, heat stable, and alcohol soluble. Sperm suspensions that have decreased respiration and motility due to acidification of the sea water as they swim can be reactivated by addition of jelly (Ohtake, 1976a). The factor responsible for this has a molecular weight around 630 and a pI = 5.3 and is nonvolatile, alcohol soluble, labile to acid- and alkaline-hydrolysis, as well as to Pronase digestion, and is ninhydrin positive; it contains no fucose, hexose, or hexosamine (Ohtake, 1976b). In a related series of studies, Garbers and Kopf (1978) found that certain factors released from eggs lead to a 20-fold increase in sperm cyclic AMP (cAMP) concentrations within 1 minute. The effect is most striking in the presence of theophylline (a cyclic AMP phosphodiesterase inhibitor) and is not seen with several hormones, nucleosides, nucleotides, anesthetics, ionophores, metals, neurotransmitters, or mammalian tissue extracts. This factor has a molecular weight of about 1800, is inactivated by Pronase, elevates cyclic AMP and cyclic GMP levels in sperm, increases sperm respiration, and increases the oxidation of long- and medium-chain fatty acids (for example, arachidonic acid oxidation was increased 14-fold and linolenic acid 15-fold) (Hansbrough *et al.,* 1979). Taken together, these data suggest that a low-molecular-weight component is responsible for activation of motility and metabolism in the sea urchin sperm as it approaches the egg.

Certain related events are seen in mammalian sperm. For example, increased cAMP levels increase sperm motility (Garbers *et al.,* 1971; Hoskins, 1973; Garbers and Hardman, 1975); this phenomenon is best defined in bovine sperm, the motile behavior of which changes from a random twitching to a directed, forward motion as sperm move through the epididymis. The factors responsible for this change in behavior are a seminal plasma glycoprotein composed of 37,500 MW subunits and elevated cAMP levels, working synergistically (Acutt and Hoskins, 1978). Ca^{2+} may act along with cyclic nucleotide in stimulating the motility of mammalian sperm (Morton *et al.*, 1974; Babcock *et al.*, 1975). When Ca^{2+} is displaced from the mitochondria into the extramitochondrial space by incubating bovine epididymal sperm with A23187, the respiration and motility increase (Babcock *et al.*, 1975), suggesting that Ca^{2+} stimulates the flagellar apparatus. Thus both intracellular messengers, Ca^{2+} and cAMP, may regulate sperm motility.

Sperm motility plays an important role in isoagglutination, a phenomenon found over 70 years ago (Lillie, 1919) and extensively discussed (e.g., Metz, 1967, 1978). This phenomenon, initially thought to reflect a specific sperm–egg interaction like antigen–antibody complex formation, seems rather to reflect the fact that sea urchin sperm swim into a clump in the presence of certain factors released from eggs in what appears to be a chemotactic phenomenon that is independent of the acrosome reaction (Collins, 1976). Whether chemotaxis plays a role in sea urchin and mammalian fertilization is not known, although it clearly does in plants (e.g., Rothschild, 1956) and other animals (e.g., Miller, 1977).

III. Sperm–Egg Fusion

The moment of union of sperm and egg is the most dramatic of the events surrounding fertilization. How does the sperm interact with the egg of its own species? How does this unique membrane fusion, which is necessary for making a new organism, take place? What role does the sperm have in subsequent development? Questions like this have occupied the thoughts of embryologists since Aristotle, even before the responsible cells were identified. The remarkable species specificity of fertilization suggests that there must be a positive recognition system of high resolution that allows self to recognize self. In addition, in some cases (e.g., with hermaphrodites like ascidians) the system must operate more stringently, allowing discrimination in a negative sense. Here self-sterility is the rule, for eggs can not be fertilized by even high concentrations of autogenous sperm (Morgan, 1904, 1940). This resolution transcends that of species-dependent recognition systems, and works at the level of the individual, although there may be classes of exclusion systems within any species. The exclusion of autogenous sperm in ascidian fertilization seems to be at the level of the chorion for, with its removal, self-sterility disappears (Rosati and de Santis, 1978).

It has been known for a long time that species specificity is disturbed by fertilizing echinoderm eggs under conditions of high pH or Ca^{2+} concentration (Loeb, 1916), and the relationship between specific sperm–egg interaction and specific components of the gamete surfaces has been a matter of continued debate. The species-specific reaction probably occurs between the acrosome-reacted sperm and a component of the egg glycocalyx and precedes the fusion of gamete membranes. This is the predominant view, aspects of which have been reviewed elsewhere (Summers *et al.*, 1975; Metz, 1978; Yanagimachi, 1977, 1978). Although some interaction between sperm and egg may take place prior to the acrosome reaction, as for example in the Ca^{2+}-dependent binding of mouse epididymal sperm to the zona pellucida (Saling *et al.*, 1978), such interactions probably play a preliminary role in the overall process. After the specific binding has been accomplished, membrane fusion occurs (reviewed in Colwin and Colwin, 1967), but the fate of the sperm surface after fertilization is still unknown. Mg^{2+} is needed for fertilization by acrosome-reacted sperm but not for parthenogenic activation of the sea urchin egg with butyric acid (Sano and Mohri, 1976). Thus, this cation may play a specific role in the binding or fusion reactions.

A related aspect of gamete fusion is that the sperm can activate the egg without actually entering it, i.e., the fertilization membrane or the changes associated with fertilization can be seen under conditions where sperm penetration is inhibited. An early example of this phenomenon was provided by Lillie (1912) using *Nereis*, where penetration of the sperm into the egg takes 40–60 minutes. Centrifugation of inseminated eggs strips away the partially penetrated sperm but allows the cortical reaction to take place without entry of the sperm nucleus. Likewise, Loeb (1916) found that starfish sperm could activate *S. purpuratus* eggs, but the nucleus of the sperm did not penetrate or remain with the egg. A modern series of experiments on the same problem has used the properties of cytochalasin B, which in fertilization of sea urchin *Spisula* and *Urechis* eggs, does not interfere with activation or the cortical reaction but does block pronuclear incorporation (Gould-Somero *et al.*, 1977; Longo, 1977, 1978; Byrd *et al.*, 1977). Although the data are not as explicit for mouse fertilization, a similar phenomenon may be occurring (Niemerko and Komer, 1976). Thus, cytochalasin is not affecting the membrane fusion of the cortical reaction, and may not block sperm–egg fusion per se, although the latter point has not been resolved; it does block pronuclear incorporation, which then may be inferred to result from a microfilament-dependent system. This raises the question of whether a transient sperm–egg membrane fusion occurs in cytochalasin-treated eggs, followed by removal of sperm components in the wave of exocytosis of the cortical reaction. In such a case, activation of the egg may be mediated by release of Ca^{2+} from the sperm into the egg upon fusion. Such an hypothesis is compatible with the cortical reaction being Ca^{2+}-mediated (see below) and with the fact that the sperm accumulates Ca^{2+} during the acrosome reaction (Schackmann *et al.*, 1978). Thus, we postulate that the sperm may act as a "Ca^{2+} bomb" at membrane fusion, dumping Ca^{2+} into the egg to trigger the cortical reaction. Of course, this proposal is still untested, and alternative possibilities exist.

The following sections will consider the types of sites that have been suggested to account for sperm–egg binding and will use the paradigm that specific sperm–egg binding substances exist. Then, after a brief consideration of other factors involved in sperm–egg interaction, we will consider the fate of the sperm surface in the fertilized egg.

A. The Binding Site of Sperm

Acrosomal components of invertebrate sperm were suggested in earlier studies to be essential for binding of sperm to egg (reviewed in Summers et al., 1975; Hylander and Summers, 1977). Vacquier and collaborators (Vacquier and Moy, 1977) have initiated a series of experiments to define the structure and function of one such molecule, which they call bindin. Bindin was isolated as a particulate component of a Triton X-100 extract of sperm. This preparation is composed of a single protein of 30,500 MW that contains no carbohydrate; it agglutinates eggs in a reaction that is blocked by isolated vitelline layers and vitelline layer peptides. Either trypsin or periodate treatment of eggs, under conditions that block sperm binding and fertilization, also block bindin-dependent egg agglutination. Agglutination is species-specific (Glabe and Vacquier, 1977b), at least between S. purpuratus and S. fransicanus, even at high bindin concentrations. Bindin is an insoluble preparation; when solubilized, it still causes agglutination, but without species specificity. Also, upon aging, eggs no longer can be agglutinated in a species-specific fashion. When the physical properties of bindins from S. purpuratus and S. fransicanus were compared (Bellet et al., 1977), they were found to be similar, but not identical; both bindins had the same molecular weight and had many of the same tryptic peptides, but each had a distinct subset of tryptic peptides. These experiments have been extended to the oyster Crassostrea (Brandriff et al., 1978), where the isolated bindin is not homogeneous but consists principally of a glycoprotein of 65,000 MW that agglutinates oyster eggs. Whether similar binding components exist in mammalian fertilization is still not clear, because no studies at this level of biochemical sophistication have been reported. However, a factor has been isolated from a mixture of capacitated hamster sperm and the zona pellucida (egg coat) that induces premature binding between gametes (Hartmann and Hutchison, 1977). No clear relationship between this factor and bindin has been shown, except that capacitated (acrosome-reacted) sperm were required to release the factor.

Although the data from Vacquier's group suggest that a single protein component (most likely from the acrosome) participates in sperm–egg binding, other experiments by Aketa and co-workers are contradictory. This group has been looking principally at the sperm receptor on eggs (see later); they have isolated a component from a supernatant fraction obtained after trichloroacetic acid precipitation of sperm that inhibits fertilization of homologous eggs without causing their agglutination (Aketa et al., 1978). This component works at substantially higher concentrations than does bindin and is principally carbohydrate, with only

5% of the mass as protein; its effect is neutralized by the egg factor that Aketa's group studied. Since Aketa found (1975) that concanavalin A blocked fertilization by affecting the sperm of *Anthocidaris* and *Hemicentrotus,* this also suggested that the recognition substance was a carbohydrate component of sperm. Thus, Aketa's carbohydrate factor bears little relationship to Vacquier's bindin. The resolution of the problem is not obvious, although concanavalin A inhibition could be explained by several mechanisms; and the sperm factor of Aketa was not shown to be homogeneous. This factor may mediate a different aspect of sperm–egg binding: i.e., there may be several types of interactions that occur between sperm and egg, and both factors may be important.

B. The Binding Site of Eggs

The locus on the egg for interaction with sperm varies from species to species. For example, in amphibian eggs, most fertilization occurs in the animal half (Elinson, 1975). The fertilization site is recognized as a small, microvillus-free area soon after insemination and begins to contain microvilli after a few hours. However, the entry point remains as a scar that is susceptible to hypotonic lysis (Elinson and Manes, 1978). In the sea urchin, fertilization can occur anywhere with respect to the animal-vegetal axis (cf. Czihak, 1975; Giudice, 1973; Gabel *et al.,* 1979b). In some cases (Schatten and Mazia, 1976), but not all (Eddy and Shapiro, 1976), microvillous projections of the plasma membrane appear to surround the fertilizing sea urchin sperm; similar protrusions of the egg membrane at the site of fertilization have been seen in teleost fish (e.g., Iwamatsu and Ohta, 1978).

The nature of the receptive substance on the egg surface is beginning to be explored. Early results of Hagstrom (1956) suggested that the receptor was altered by digestion with trypsin-like proteases. This was confirmed by Vacquier *et al.* (1972a) and Aketa *et al.* (1972), who showed that digestion of the sea urchin egg surface with trypsin decreased fertilizability. A putative receptor protein was isolated from the surface of the sea urchin, *Hemicentrotus,* which, when incorporated into air bubbles, led to the binding of sperm to the bubbles (Aketa and Tsuzuki, 1968); this material was not characterized as to its homogeneity but appeared to be a 2.3S glycoprotein (Aketa *et al.,* 1968). Antiserum directed against this component blocked fertilization (Aketa and Onitake, 1969), and sperm treated with the component lost their fertilizing ability, but not their motility (Aketa, 1973). In fact, sea urchin sperm treated with the component did not agglutinate or undergo the acrosome reaction, and thus the factor seems to bind sperm *before* the acrosome reaction has occurred. This makes it less likely to be the component involved in the species-specific binding, although the antisera directed against it do block fertilization in a species-specific fashion (Aketa *et al.,* 1972) and do not block fertilization of trypsin-treated eggs. Thus, this material has some, but not all, the binding characteristics expected from the physiological and morphological studies of sperm–egg interaction. A similar approach was used to detect a component from the hamster zona pellucida that was suggested to be involved in the binding reaction (Gwatkin and Williams, 1977), but the component showed some cross reactivity in the mouse system, and has not been extensively studied (see Gwatkin, 1977, for a general discussion of the work on properties of the mammalian receptors).

Another factor suspected of being the sperm receptor of the egg has been obtained from a membrane protein preparation of *Arbacia* eggs that inhibited fertilization of eggs of *Arbacia* but not of *S. purpuratus*. Trypsin treatment of this component blocked the effect, and a soluble form of the component bound to sperm and prevented fertilization (Schmell *et al.*, 1977); it is bound to con-canavalin A, which itself blocked fertilization. Concanavalin A binding sites exist on both the plasma membrane and vitelline layer of *S. purpuratus* eggs (Veron and Shapiro, 1977). The high affinity sites on the vitelline layer appear to be involved in fertilization, since with partial occupancy of these sites, fertilization was completely inhibited (Veron and Shapiro, 1977). Yet another approach to obtaining the sperm receptor was to isolate the vitelline layer by Triton X-100 and EDTA treatment of eggs (Glabe and Vacquier, 1977a). These vitelline layer preparations contained microvillous projections similar to those seen on the egg, and sperm adhered to them only on the external surface; similar binding of sperm to only the external surface of a vitelline membrane–egg cortex preparation has recently been described (Decker and Lennarz, 1979). When the vitelline layer preparations were iodinated and examined by polyacrylamide gel elec-trophoresis, two high-molecular-weight components were seen (Glabe and Vac-quier, 1977a). A high-molecular-weight, trypsin-sensitive glycoprotein obtained from this preparation (Glabe and Vacquier, 1978) has species-specific affinity for bindin. Glycoprotein preparations are difficult to fractionate and work with, so that at this stage of analysis it is hard to make comparisons between the work of the several groups. However, hopefully, with solubilized preparations of bindin, it will be possible to characterize completely the interaction of these components and to test whether they are indeed the mediators of the species-specific events of fertilization.

C. Other Possibilities for Sperm–Egg Interaction

Several enzymes have been implicated in the process of sperm-egg interaction. For example, Monroy (1953) found phospholipase activity in sea urchin sperm and Conway and Metz (1976) found that the activity was expressed 15 seconds after the acrosome reaction was triggered, along with extension of the acrosomal filament. The enzyme was suggested to play a role in sperm–egg membrane fusion, for it could lead to a transient state of membrane instability. In mouse

fertilization, a stimulation of sialyl transferase activity was detected by histochemical methods when sperm and eggs were mixed, with a resultant patch of
silver grains seen on the egg when an appropriate substrate (CMP-[^{3}H]N-acetyl
neuraminic acid) was added (Durr *et al.*, 1977). This could have reflected an
interaction of the enzyme of the sperm with an acceptor on the egg surface, but
the data implicating either phospholipase or sialyl transferase in fertilization are
not definitive enough to allow unambiguous interpretation.

Fluorescein dyes block fertilization (Carroll and Levitan, 1978a,b) at μM
concentrations (assayed by induction of the cortical reaction or insertion of the
sperm pronucleus). This effect is reversible if the fluorescent dyes are washed
away. Parthenogenesis is not inhibited, nor is there any effect on sperm motility,
agglutination, or respiration. Although there is a slightly reduced binding of
sperm to egg, the event that seems to be most affected is membrane fusion that
follows binding. Fertilization is inhibited not only in the sea urchin, but also in
chordates, molluscs, annelids, and other echinoids. Such studies, along with the
complexity seen in the "receptor" preparations, support the idea that there are
multiple levels for the interaction of sperm and egg that are interposed between
the initial contact and the initiation of the cortical reaction. This is an old idea
(reviewed in Giudice, 1973; Czihak, 1975) that has received other experimental
support (Baker and Presley, 1969). However, the sequence of reactions has not
yet been unraveled.

D. EFFECT OF MEMBRANE FUSION ON THE EGG SURFACE AND THE FATE OF THE SPERM SURFACE

The most dramatic effect of fertilization on the egg surface is the induction of
the cortical reaction (to be discussed later). The properties of the egg plasma
membrane change enormously after cortical granule exocytosis. For example, a
topographic mosaicism that seems to be related to the insertion of new cortical
vesicle membrane (Eddy and Shapiro, 1976) persists for at least the first hour
after fertilization, even in the presence of colchicine or cytochalasin B. These
results suggest that a rigidification of the egg cortex occurs with fertilization, as
suggested by previous observations of a cortical stiffening (Mitchison and
Swann, 1955; reviewed in Rothschild, 1956; Hiramoto, 1974). These experiments were done with sea urchins, but a similar phenomenon seems to occur in
mouse fertilization (Johnson and Edidin, 1978) where lateral diffusion in the
plasma membrane of the egg decreases enormously at fertilization. The latter
studies, which utilized laser-induced photobleaching of fluorescence, showed
that the mobility of a surface protein and lipid in unfertilized eggs was equivalent
to that of many other cells, whereas upon fertilization, this mobility was greatly
reduced. This result appears to contradict others obtained when the mobility of a
spin-labeled fatty acid was estimated before and after fertilization of two sea

urchin species (Campisi and Scandella, 1978); however, this label (5-doxyl stearate) probably equilibrated with all the lipid in the cells, and thus was likely not to be assaying for fluidity of the plasma membrane per se. In general, the results suggest that fertilization leads to rigidification of the plasma membrane (and associated components). In amphibian fertilization, sperm entry leads to a cortical contraction that occurs upon activation of the egg (Elinson, 1977), and the sperm entry phase may be dissociated from the cortical contraction by fertilizing immature oocytes and artifically activating them later.

To trace the fate of the sperm surface after fertilization, we have designed a reagent that permits labeling of the sperm membrane for subsequent fluorescent, radiochemical, and immunological analysis. The reagent diiodofluorescein isothiocyanate may be synthesized with ^{125}I to high specific radioactivity, is fluorescent and has similar properties to its congener fluorescein isothiocyanate (Gabel and Shapiro, 1978). This former reagent is excluded from the erythrocyte where it labels membrane proteins, and thus appears to be a vectorial reagent. In sea urchins and mammals, labeling with either fluorescein isothiocyanate or the radioactive derivative is most concentrated over the sperm midpiece region (Gabel et al., 1979a,b). Several proteins of the sea urchin sperm surface are labeled with the radioactive reagent, and about one-quarter of the radioactivity is found in lipid species; in the hamster, all of the radioactivity is in several polypeptide components. Sperm are fully viable after labeling, as tested by their ability to fertilize eggs (Gabel et al., 1979a,b). Upon fertilization, the label persists in the embryo throughout early development: in the case of the sea urchin, to the pluteus larval stage (Gabel et al., 1979a,b), and in the mouse, at least until the eight cell stage, which is as far as it has been followed. A remarkable feature of the transfer of label is that it persists as a patch in the developing embryo of both the sea urchin and the mouse (Gabel et al., 1979a,b), i.e., after fertilization and upon first cleavage, a patch is present in one of the two blastomeres, showing that at this point the two cells are differentiated with respect to the presence of sperm surface components (Fig. 3). In the case of the sea urchin, there is no unusual distribution of the patch with respect to the animal-vegetal axis, as assessed by examining 16-cell embryos (Gabel et al., 1979b), but the relationship to the dorsal-ventral axis or to other aspects of development has not yet been determined.

The presence of sperm surface components in early mouse embryos has been seen in other systems. For example, sperm antigens that are detected with an antiserum to teratocarcinoma cells (the "F9 antigen") (Gachelin et al., 1976) are present uniquely on sperm and on early cleavage embryos. Rabbit antiserum that is directed against mouse sperm and extensively absorbed with other mouse cells detects an antigen present both on sperm and on mouse embryos (Menge and Fleming, 1978). In addition, certain rabbit isoantigens are found on sperm and on

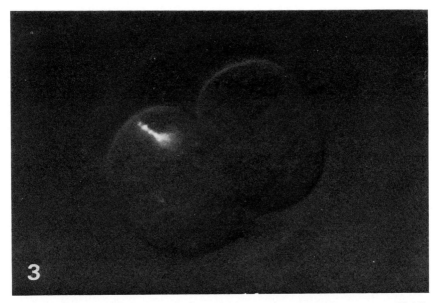

FIG. 3.   Fluorescence micrograph of a two-cell mouse embryo recovered from a mouse artificially inseminated 24 hours earlier with tetramethylrhodamine-labeled sperm. The fluorescent patch is due to the presence of labeled sperm surface components in the early embryo. (From Gabel *et al.*, 1979b; *Cell,* copyright MIT Press.)

fertilized rabbit eggs, but not on unfertilized eggs (assayed with cytotoxic antibodies in a complement dependent lysis reaction; O'Rand, 1977). None of these experiments rigorously excluded the possibility that latent components, which were not observed in the unfertilized egg, were subsequently exposed upon fertilization and thus reacted with the antibody, i.e., they were not sperm-derived. However, in our experiments using the artificial reagent diiodofluorescein isothiocyanate, the sperm were labeled prior to fertilization; so the presence of the label in the embryo is unequivocal evidence of the persistance of sperm components following fertilization.

The significance of a patch of sperm surface that is localized to one of the blastomeres of the developing embryo is not yet clear, but poses some interesting questions. The determination of the dorsal–ventral axis in amphibians is related to the locus of sperm penetration (reviewed in Brachet, 1977). A similar phenomenon can be artificially induced by injection of sperm components into the egg, which triggers an axis in relation to the injection site (Manes and Barbieri, 1977). Several other events in development also occur in relation to the site of sperm penetration (reviewed in Gabel *et al.*, 1979b). The sperm is not obligatory for determination of developmental axes, as a host of experiments with parthenogenesis have shown. Nonetheless, parthenogenesis is a relatively ineffective phenomenon. For example, in the sea urchin (Brandriff *et al.*, 1975), there is very low survival rate of parthenogenotes and a successive loss of embryos at each developmental stage. It is therefore intriguing to speculate that the sperm surface may play an organizing role in development, perhaps a focus of discontinuity that could generate a gradient of positional information (see Wolpert, 1971).

IV. The Response of the Egg to the Sperm

A. THE CORTICAL REACTION

1. *The Nature of the Cortical Reaction*

In most types of eggs, sperm–egg fusion is followed by secretion: vesicles that underlie the plasma membrane (the cortical granules) undergo an exocytosis that spreads from the point of the fertilizing sperm in a wave on the surface of the egg (Fig. 4). The released material alters the egg surface and decreases its receptivity to sperm; the egg surface coat is modified to act as a protective barrier and a block to further sperm entry. This reaction, which was first noticed by Derbes (1847) in his discovery of the elevation of the fertilization membrane in the sea urchin, has been most extensively studied in that animal (reviewed in Runnstrom, 1966): probably because elevation of the fertilization membrane is a dramatic event. However, the reaction is a general one; an analogous secretion occurs in mammals (Braden *et al.*, 1954; Austin and Braden, 1956) where it results in an alteration of the properties of the zona pellucida and causes it to become less receptive to sperm penetration. However, many exceptions to this general rule exist. For example, in the mouse, a substantial amount of cortical granule release occurs upon exposure of eggs to sperm, but before sperm–egg fusion (Nicosia *et al.*, 1977), suggesting that fusion is not the triggering event. A similar situation is seen in the annelid *Sabellaria* (Pasteels, 1965) where the cortical reaction occurs when the eggs are shed into sea water, and upon fertilization, the only additional change is an alteration in the microvilli. In *Urechis*, a subset of the cortical granules are released at fertilization; the majority are released substantially later, with elevation of the vitelline membrane (e.g., Paul, 1975a), yet a block to polyspermy occurs immediately after insemination of *Urechis*. At the extreme end of the spectrum are the ascidians, which do not have cortical granules. In the case of *Ciona intestinalis*, there is a release of cytoplasmic material at the time of fertilization (some of which may come from subcortical vesicles) that seems to be related to an increased osmotic pressure in the egg (Rosati *et al.*, 1977) but is not a typical cortical reaction. However, these exceptions serve to underlie the general mechanism that pervades animal

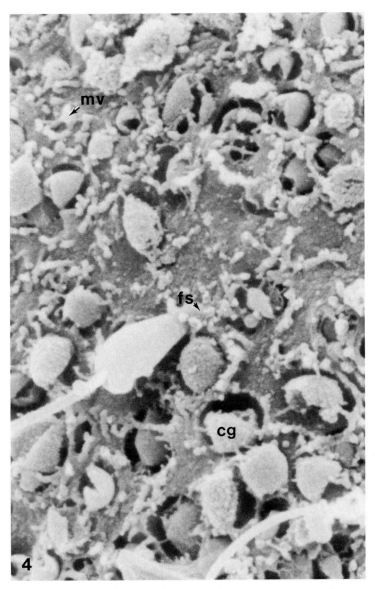

FIG. 4. Cortical reaction of an egg of the sea urchin *S. purpuratus* fixed 25 seconds after insemination. Cortical granule exocytosis spreads as a wave from the site of fertilization and results in release of cortical granule contents and insertion of the perigranular membrane into the egg surface to produce a mosaic membrane. The vitelline layer was altered by dithiothreitol treatment to expose surface membrane events. cg, Cortical granule components being released; mv, microvilli of egg surface; fs, site of sperm-egg fusion. (Reprinted by permission, Eddy and Shapiro, 1976.)

fertilization—the release of vesicles in response to the sperm—and that serves to decrease the probability of other sperm penetrating the egg.

2. Mechanism of Triggering

Although the mechanism by which cortical vesicles fuse with the plasma membrane and the nature of intermediates in the reaction are no more clear than with any other exocytosis, Ca^{2+} certainly plays a role. The current paradigm that places free intracellular Ca^{2+} ion as an effector of the reaction has its roots in experiments of Mazia (1937), who found that the free Ca^{2+} concentration of eggs increased at fertilization, and of Moser(1939), who found that a number of parthenogenic agents that led to an elevated fertilization membrane did not work if the eggs were pretreated with oxalate or citrate. Recent experiments with radioactive isotopes have confirmed that the exchangeable Ca^{2+} pool increases upon fertilization (Nigon and Do, 1965) and that the exchangeable Ca^{2+} is found in the microsomal fraction (Nakamura and Yasumasu, 1974), whereas a nonexchangeable fraction is in the mitochondria. In the latter study, a Ca^{2+}-binding protein of 35,000 MW could be extracted from the microsomal fraction with 0.6 M KCl. Other experiments using $^{45}Ca^{2+}$ show that an increase in the exchange of internal Ca^{2+} with sea water follows fertilization, both in sea urchin (Nakazawa et al., 1970; Azarnia and Chambers, 1976; Paul and Johnston, 1978) and in Urechis (Johnston and Paul, 1977). This exchange reflects an increase both in uptake and efflux; the latter dominates, so that there is a net loss of Ca^{2+} from the egg at fertilization. This released Ca^{2+} then binds to the egg coats (Azarnia and Chambers, 1976). In contrast to what is seen with fertilized eggs, unfertilized eggs have a remarkably low exchange rate for Ca^{2+}.

One of the stronger pieces of evidence indicating that alteration in Ca^{2+} metabolism is related to triggering of the cortical reaction was found by using divalent ionophores such as A23187, which lead to a (delayed) cortical reaction in sea urchins (Chambers et al., 1974; Steinhardt and Epel, 1974) as well as in starfish, amphibians, and mammals (Steinhardt et al., 1974). The finding that triggering is not dependent upon extracellular Ca^{2+} suggested that Ca^{2+} was being released from an intracellular store. Triggering with A23187 leads to many of the changes seen with fertilization, including increased protein and DNA synthesis and nuclear membrane breakdown. An interesting fact was uncovered in these studies: the level of protein synthesis attained was lower and the egg nucleus did not break down if Na^+ was not present in the sea water. The signficance of this result, which provided a clue to another activation mechanism, is discussed later. Further support for the role of Ca^{2+} as a mediator of the reaction was obtained with cortical granules isolated in the absence of Ca^{2+}, which lysed upon addition of Ca^{2+} (Vacquier, 1975).

The most striking correlation between elevated free intracellular Ca^{2+} and triggering of the cortical reaction, however, came from experiments with the

Ca^{2+}-sensitive photoprotein aequorin, the luminescence of which is dependent upon free Ca^{2+} concentration (Ridgway et al., 1977; Steinhardt et al., 1977). The most graphic of these studies used the teleost fish medaka, which had been well characterized as a model for embryological study (e.g., Yamamoto, 1961) because it has a large egg. The cortical granules of the egg fire off quite slowly and progressively from the micropyle, which is the site where the sperm penetrates and fertilizes the egg. If one examines the luminescence of aequorin previously injected into the medaka egg, a wave of Ca^{2+} release precedes the cortical reaction and occurs from the micropyle over the surface of the sphere (Ridgway et al., 1977). When A23187 is used to activate the medaka egg, there is no monotonic progression, but rather, several foci of increased Ca^{2+} concentration occur at different sites on the egg surface and serve as generating points for waves that ultimately coalesce. A similar experiment in the sea urchin (Steinhardt et al., 1977), which has a much smaller egg, showed that the level of intracellular free Ca^{2+} increases with both fertilization and ionophore activation and independently of the presence of extracellular Ca^{2+}. These experiments suggest that Ca^{2+} released from one site leads to Ca^{2+} release from an adjacent one, i.e., that Ca^{2+}-mediated Ca^{2+} release is an important aspect of the overall reaction. The data are supported by observations that EGTA, when microinjected into eggs, prevents the cortical reaction (Zucker and Steinhardt, 1978), as well as by the finding that several intracellular stores of Ca^{2+} exist in the sea urchin egg (Zucker et al., 1978), one of which may be accessed by both fertilization and parthenogenic agents that trigger the cortical reaction. The nature of the Ca^{2+} store is not clear, but energy may be required to fill it, since depletion of ATP blocks the cortical reaction (Baker and Whitaker, 1978), and an ATP dependent binding of $^{45}Ca^{2+}$ was seen in isolated cortices. After its depletion, the store of Ca^{2+} requires some 40 minutes to fill (estimated by aequorin luminescence; Zucker et al., 1978). The finding of Nakamura and Yasumasu (1974) of Ca^{2+} accumulation in a microsomal fraction suggests that a structure analogous to the sarcoplasmic reticulum of muscle may exist in the egg cortex and may use an ATP-requiring pump to accumulate Ca^{2+}. Although no such "ooplasmic reticulum" has been seen, either with definitive biochemical or electron microscopic experiments, the possibility is enticing. In addition to ATP-dependent Ca^{2+} binding, we might expect to find a Ca^{2+}-dependent ATPase if the structure is analogous to the sarcoplasmic reticulum of muscle. One Ca^{2+} binding protein has been found in the sea urchin egg: a protein with the properties of the Ca^{2+} regulator protein or calmodulin (Head et al., 1979). The sea urchin protein was able to activate a calmodulin-dependent enzyme and bound four Ca^{2+} per 17,000 MW; this is comparable to the binding of calmodulin found in other tissues.

We suggested previously (Schackmann et al., 1978) that one reason sperm accumulate Ca^{2+} during the acrosome reaction might be to deposit it in the egg and thus initiate triggering of the cortical reaction. As mentioned earlier, the

sperm had substantial quantities of calmodulin so that, in addition to mitochondrial accumulation of Ca^{2+}, a fraction of the Ca^{2+} should remain in the extramitochondrial space. This calmodulin–Ca^{2+} complex might be involved in triggering the cortical reaction, since the Ca^{2+} is already bound to an intracellular
effector. This is, of course, speculative, but we suggest the following hypothesis:
that a Ca^{2+}-calmodulin complex of sperm is the active component that initiates
the cortical reaction in the egg. An extension of this hypothesis would be that a
Ca^{2+}-calmodulin complex is the mediator of the Ca^{2+}-dependent Ca^{2+} release
seen to precede exocytosis of the cortical reaction; as Ca^{2+} is released from the
ooplasmic reticulum, it binds to egg calmodulin and then acts as the adjacent site
for release. Thus, a network of "ooplasmic reticulum" could release Ca^{2+},
which then might bind to egg calmodulin and lead to the release of adjoining
ooplasmic reticulum sites or to other Ca^{2+}-activated steps in development. This
hypothesis, although fanciful and as yet unsubstantiated, is compatible with the
autocatalytic nature of the cortical reaction. It also could explain the observation
that eggs can only be activated after sperm–egg membrane fusion (for microin-
jection of sperm into eggs of the sea urchin does not lead to the cortical reaction;
Hiramoto, 1962), as well as the finding that cortical granules distant from the
oolemma and presumably not in contact with the putative ooplasmic reticulum
network do not participate in the cortical reaction (Longo and Anderson, 1970).

Although this hypothesis is attractive to us, any one of a number of other
mechanisms could equally obtain. A number of observations on the cortical
reaction need to be integrated into any final hypothesis, and there are some that
are not easily accommodated in the Ca^{2+} paradigm discussed earlier. These include
observations suggesting that it is possible to obtain a localized cortical reaction
without propagation of the cortical wave. For example, Allen (1953), fertilized
eggs in tight fitting glass capillaries, and the cortical wave did not pass over the
whole egg surface. Areas compacted against the wall of the capillary remained
unactivated, and the cortical reaction occurred only on the side where the sperm
were injected. Sugiyama (1956) showed that detergents could lead to localized
cortical granule release, with no propagation of the cortical wave, and eggs
treated with narcotic agents did not propagate the cortical wave. A similar finding is that the fertilization wave can pass through cortical areas that have been
depleted of cortical granules by local application of detergent, i.e., transmission
did not require the presence of the granules (Nehara and Katou, 1972). These
experiments might be explained by different mechanisms of Ca^{2+} sequestration
that do not allow release after local surface manipulation or, perhaps, by the need
for another event (e.g., membrane depolarization) that is transmitted along the
plasma membrane before the cortical wave. They certainly indicate that
additional components should be added to any hypothesis that involves the
generalized elevation of intracellular Ca^{2+} as the mediator of the response. As in
the case of exocytosis of the acrosomal granule, there is some evidence that a

trypsin-like protease is involved in the exocytosis of the cortical granules (Longo and Schuel, 1973; Schuel *et al.*, 1976), since proteolytic enzyme inhibitors seem to decrease the rate of cortical granule dehiscence. In general, although Ca^{2+} ion is clearly a mediator of the reaction, the source and fate of the Ca^{2+}, as well as the participation of other mechanisms in the cortical reaction, remain unknown.

3. *Effects on the Plasma Membrane*

The dramatic change in plasma membrane structure that occurs at the time of the cortical reaction is reflected in alterations in psysiological and biochemical properties. One striking change occurs in the membrane potential. The membrane potential of the unfertilized sea urchin egg is negative. Upon fertilization, there is first a transient positive potential (the "fertilization action potential"), and then the potential reaches a stable negative value again (Steinhardt *et al.*, 1971; Ito and Yoshioka, 1972, 1973). The potential of the unfertilized egg is sensitive to the extracellular anion composition (Steinhardt *et al.*, 1971; Ito and Yoshioka, 1973), whereas the stable negative potential of the fertilized egg is a K^+-sensitive potential. The potential of the unfertilized egg was estimated to be -8 to -10 mV (Steinhardt *et al.*, 1971; Ito and Yoshioka, 1973) by microelectrode measurements or as low as -70 mV (Jaffe and Robinson, 1978) by a combination of tracer efflux and microelectrode measurements. Fertilization leads to dramatic changes in membrane resistance (Jaffe and Robinson, 1978) and capacitance (Cole and Spencer, 1938). The final resting potential of the fertilized egg takes about 6 to 8 minutes to be achieved.

The fertilization action potential has been suggested to cause the early block to polyspermy (Jaffe, 1976). As summarized by Rothschild (1954), there are several mechanisms for blocking fertilization by excess sperm. One class of effects, called the early block to polyspermy, is incomplete; this partial block is then followed by a delayed, absolute block. Jaffe (1976) found that when the potential of the egg was positive, sperm could not fertilize the sea urchin egg; when the membrane potential was lowered to a negative value, the eggs were fertilized. This phenomenon suggests how eggs have a decreased receptivity to sperm in the early moments after fertilization before the definitive changes in the egg coats occur (see following) and was proposed to mediate the rapid block to polyspermy. Although changes in the membrane potential of mouse eggs have been shown to occur during oocyte maturation and after fertilization (Cross *et al.*, 1973; Okamoto *et al.*, 1977; Powers and Tupper, 1977), no direct correlation with a block to polyspermy has yet been made.

In addition to the properties of electrical excitability, a striking recent finding is that the egg, as an excitable cell, also has receptors for mediation of neurotransmitter effects. For example, *Xenopus* oocytes are sensitive to acetylcholine (Kusano *et al.*, 1977), which leads to depolarization and an increase in membrane permeability to Cl^- (with some variability in the response). The effect of

acetylcholine was blocked by atropine but not by other inhibitors of the cholinergic recptor, nor was it evoked by many other neurotransmitters. Similar results
have been seen with mouse oocytes (Eusebi *et al.*, 1979), but in this case, the
acetylcholine sensitivity disappears after fertilization. The latter is of interest
with regard to the potential effect of sperm cytoplasmic components in development: spontaneously or parthenogenically activated mouse eggs retained their
sensitivity to acetylcholine, unlike fertilized eggs.

The new plasma membrane that is inserted due to the cortical reaction would
be expected to alter the total surface area of the egg. A rough estimate of the
change in surface area is about 2-fold (Eddy and Shapiro, 1976); this estimate is
supported by observations by scanning electron microscopy, as well as by a
doubling of plasma membrane concanavalin A binding sites seen after fertilization (Veron and Shapiro, 1977). However, the increase in total surface area may
be less than that predicted from the sum of the preexisting membrane area and the
cortical granule membrane area (Schroeder, 1978). Some of the increased surface area is taken up by elongation of microvilli (Eddy and Shapiro, 1976;
Schroeder, 1978). Microvilli contain actin filaments (Burgess and Schroeder,
1977) that elongate normally in the presence of colchicine, but not of cytochalasin B (Eddy and Shapiro, 1976). It was suggested that the elongation of microvilli that can be seen with some physical means of parthenogenic activation of
eggs may be an important event in the activation sequence (Mazia *et al.*, 1975);
while this is an interesting speculation, the correlation between activation and
microvillar elongation is not absolute. Similar alterations have been seen in the
plasma membrane of other eggs after fertilization; for example, in *Rana pipiens,*
the discharged cortical granules leave craters in the surface for up to 30 minutes
(Kemp and Istock, 1967).

Some changes at the plasma membrane level might be related to the release of
a proteolytic enzyme from the cortical granules. Such an enzyme was inferred
from work of Hagstrom (1956) who showed that sea urchin eggs fertilized in the
presence of soybean trypsin inhibitor elevated their fertilization membranes ab-
normally and became polyspermic. Several groups (Vacquier *et al.*, 1972a, 1973;
Grossman *et al.*, 1973; Schuel *et al.*, 1973, 1976) found that a protease was released from the egg at fertilization and that by inhibiting this protease both an
abnormal elevation of the fertilization membrane occurred, as well as an alteration
in the block to polyspermy. The enzyme was localized in the cortical granule fraction (Schuel *et al.*, 1973) and was released with the kinetics appropriate for the
cortical reaction (Vacquier *et al.*, 1973). The protease was suggested to have two
activities: one involved in cleaving the attachment of the vitelline layer to the
plasma membrane and the other in proteolysis of sperm attachment sites. In fact,
a purification has been described (Carroll and Epel, 1975a,b; Carroll and Baginski,
1978b) that suggests that the two activities belong to distinct molecular species.
However, in the published experiments, the enzymes were not purified to homo-

geneity, and other explanations could explain the apparent separation of the two activities. Nonetheless, by biological assay, both peaks that had esterase activity had different effects on eggs: as a "vitelline delaminase" activity and a "sperm receptor hydrolase activity." In fact, the vitelline delaminase has been used to remove vitelline layers from eggs, for it is better than heterologous proteases in this regard (Carroll *et al.*, 1977). However, an independent purification of the egg protease (Fodor *et al.*, 1975) resulted in a homogeneous trypsin-like enzyme of different properties. The enzyme is released in a particulate form (Grossman *et al.*, 1973; Fodor *et al.*, 1975), but when purified to homogeneity, it is freed from the particle and behaves like a typical soluble enzyme of 22,500 MW. A similar enzyme was found in unfertilized eggs in an inactive form; it could be activated by dialysis at pH 4.6. The activated form of the enzyme was identical to the soluble form of the released protease. Thus, the data suggest that the protease exists in the cortical granules in some type of aggregate that can be disrupted by appropriate chemical treatment. One component of the aggregate may include modifying proteins that give specific sperm receptor hydrolase or vitelline delaminase activities to a trypsin-like subunit, but this has yet to be rigorously demonstrated. One function of the protease is to perform limited proteolytic cleavages of certain plasma membrane proteins (Shapiro, 1975). In these experiments where surface proteins were labeled with lactoperoxidase-catalyzed iodination, some proteins were released at fertilization and others were cleaved to lower-molecular-weight forms that persisted throughout development to the pluteus larval stage. The persistance of these proteins is striking, but the significance of this observation is not clear. It was suggested (Johnson and Epel, 1975) that some of the released proteins were involved in the metabolic activation, i.e., that their role in the unfertilized egg was to suppress metabolic activity. However, the experiments leading to this hypothesis have not been reproduced (Epel, 1978), and although it was an attractive idea, it now seems unlikely.

One component of the mammalian block to polyspermy occurs at the egg plasma membrane (Barros and Yanagimachi, 1972; Wolf and Hamada, 1977; Wolfe, 1978). This block exists independently of the alteration in the zona pellucida (discussed next) by components released from the cortical granules at fertilization. It has recently been suggested, based on the inability of the divalent ionophore A23187 (which triggers cortical granule release) to cause the plasma membrane block to polyspermy, that the plasma membrane block may not be related to cortical granule discharge, but rather to some other aspect of sperm–egg interaction (Wolf and Nicosia, 1978a,b).

4. *Effects on the Surface Coat (Vitelline Layer or Zona Pellucida)*

The change in the surface of the sea urchin egg, first noted by Derbes (1847), is an example of a general phenomenon that occurs in the extracellular coats of many egg types and in which structural change is induced by the influence of components released from the cortical granules. This alteration serves two functions: to prevent access of additional sperm to the plasma membrane of the egg and to protect the developing embryo in its early stages. Another characteristic of this event is its annulment later in development when the embryo hatches out of its coat to become a free-living larval form in the sea urchin or to implant in the uterine wall in the mammal. The alteration in the cell coat has been exhaustively analyzed in the sea urchin and the amphibian where it has some interesting similarities that parallel events in mammalian fertilization.

For example, in mammals, fertilization leads to a biochemical change in the properties of the zona pellucida (Barros and Yanagimachi, 1971; Gould *et al.*, 1971) that alters its solubility. An alteration in the zona can be caused by acrosomal extracts (Gould *et al.*, 1971) or by the product released from fertilized eggs, which, when added to unfertilized eggs, changes the zona so that it cannot be penetrated by sperm. The alteration in the properties of the zona pellucida appear to be due to a trypsin-like protease released from the cortical granules (Gwatkin *et al.*, 1973); this enzyme has properties similar to the enzyme released from the sea urchin egg at fertilization. The activity of this protease, which shows no species specificity, probably accounts for the activity of acrosomal extracts to alter the zona pellucida; the alteration may be due in part to a limited proteolytic cleavage of zona pellucida proteins detected by microelectrophoretic techniques (Repin and Akimova, 1976). Thus, at least one role of the cortical granule contents in the mammal is to alter the properties of the zona pellucida by proteolysis; whether other components are inserted stoichiometrically into the structure, as is the case with amphibians and sea urchins (see following), is still not clear.

In an elegant series of experiments by Hedrick and co-workers, the surface alteration in the coat of the amphibian egg (*Xenopus*) was shown to be a lectin-mediated process in which a lectin released from the egg at fertilization reacts with a component of the egg coat and alters its receptivity to sperm. The cortical granule lectin interacts with the J_1 layer of the extracellular coat in a Ca^{2+}-dependent reaction that can be detected by precipitin lines in double diffusion assays (Wyrick, *et al.*, 1974). If the galactose-specific lectin is added to the outside of the egg coats, the altered layer is formed in an appropriate place, by reaction with the ligand, which is a sulfated macromolecule of the jelly coat (Birr and Hedrick, 1979). The reaction, which is completed some 9 minutes after fertilization (Gray *et al.*, 1974), leads to formation of a fertilization envelope that cannot be dissolved by trypsin and that has a decreased solubility in mercaptan solutions (Wolf, 1974). One polypeptide from the cortical granule exudate was identified in the final fertilization envelope (Wolf, 1974; Yurewicz *et al.*, 1975), and some components of the vitelline layer are cleaved by limited proteolysis during the cortical reaction and in concert with the altered solubility properties of the fertilization envelope (Wolf *et al.*, 1976). The altered fertilization envelope is the structure that constitutes the block to polyspermy; isolated vitelline envelopes can be penetrated by sperm from both sides, whereas isolated fertilization envelopes are impenetrable and act as a barrier to the escape of macromolecules from the perivitelline space (Grieve and Hedrick, 1978). Thus, the amphibian surface alteration of the cortical reaction has one feature that is shared with the mammal—limited proteolysis of surface components; it has an additional aspect, as yet undiscovered in mammals, in which a specific interaction of components from the cortical granule with those from the surface coat make a new structure with novel properties.

In the case of the sea urchin, both of the processes seen in amphibians seem to occur. For example, the proteolytic enzyme released from sea urchin cortical vesicles cleaves surface coat material (Shapiro, 1975); so limited proteolysis is shared by the mammal, amphibian, and sea urchin. The altered sea urchin egg coat exhibits one special property: this structure (the fertilization membrane) becomes extremely hard and resists solubilization by reagents that do not break peptide bonds (reviewed in Giudice, 1973; Runnstrom, 1966). This hardening process is striking, occurs within minutes after fertilization, is caused by a peroxidatic system, and involves cross-linking of tyrosyl residues in adjacent polypeptide chains by a reaction catalyzed by an ovoperoxidase released from the cortical granules (Foerder and Shapiro, 1977; Hall, 1978; Fig. 5) (see following).

Although earlier observations had suggested that the sea urchin egg became less fragile after fertilization, the first evidence for hardening of the fertilization membrane appears to be that of Harvey (1910), who found that the solubility of the egg surface in sulfuric acid decreased tremendously at fertilization. Chase (1935) found that the vitelline layer was "soft" in sea urchin and sand dollar eggs, but became "hard" (insoluble in urea) upon fertilization. Motomura (1941) added several insights to the problem of hardening: (a) components of the cortical granule participate stoichiometrically in fertilization membrane assembly; (b) hardening of the fertilization membrane could be inhibited by sulfite in a reaction that was reversed by permanganate (1954); and (c) a "third factor" (1957) was required for the hardening reaction. In these experiments, Motomura suggested that an oxidative mechanism was at work; this has been recently substantiated (Foerder and Shapiro, 1977; Foerder *et al.*, 1978). Ca^{2+} was shown to be required for hardening of the fertilization membrane (Markman, 1958), and this may relate to the release from cortical granules of a paracrystalline material that aggregates in the presence of Ca^{2+} (Bryan, 1970a,b). Hardening of the fertilization membrane can be blocked by glycine ethyl ester (Lallier, 1970), as well as by penicillamine, isoniazid, and other aldehyde reactants such as benzhydrazide and semicarbazide (Lallier, 1971). A convenient assay for the harden-

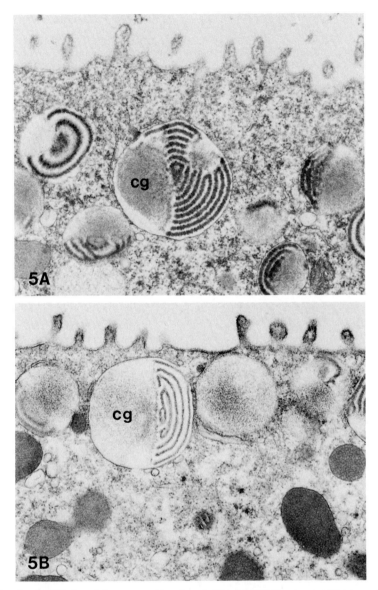

FIG. 5.   Localization of the ovoperoxidase in unfertilized and fertilized eggs of the sea urchin S. *purpuratus*. Eggs were treated with diaminobenzidine and hydrogen peroxide and then fixed with osmium tetroxide and processed for electron microscopy. The reaction product indicative of an endogenous peroxidase is present in the cortical granules of unfertilized eggs (A) and in the fertiliza-

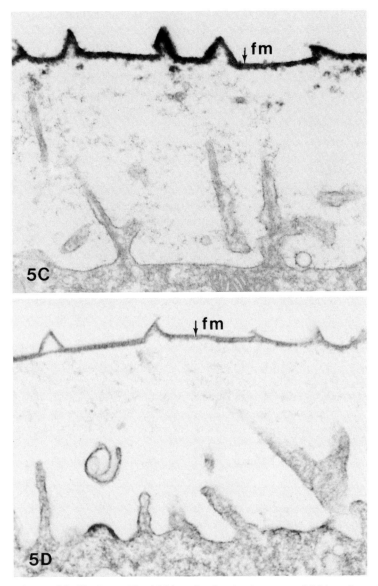

tion membrane of fertilized eggs (C). Addition of aminotriazole, a potent inhibitor of peroxidase, results in cortical granules (B) and fertilization membranes (D) similar in electron density to those of eggs unreacted for peroxidase. cg, Cortical granules; fm, fertilization membrane. (Reprinted by permission, Klebanoff *et al.*, 1979.)

ing of the fertilization membrane is its insolubility in dithiothreitol; whereas nascent, soft fertilization membranes are readily solubilized, the hard structures are resistant to this treatment (Paul and Epel, 1971; Veron *et al.*, 1977).

A distinct morphological change occurs in the fertilization membrane of the sea urchin *S. purpuratus* as it hardens. Finger-like projections, which are casts of the microvillar tips and are igloo-shaped ("I forms"), are converted to tent-like projections ("T forms") concomitant with hardening (Tegner and Epel, 1973; Veron *et al.*, 1977). However, this morphological change precedes the actual hardening event, as determined by using sulfite and glycine ethylester, which are two inhibitors of hardening that work differentially on mercaptan solubility and on the morphologic transition (Veron *et al.*, 1977). In fact, by using the characteristics of solubility in dithiothreitol, the morphologic transition, and permeability of the fertilization membrane to macromolecules like concanavalin A, the hardening reaction was found to occur in a defined sequence upon triggering of the cortical reaction (Veron *et al.*, 1977). By further analyzing the properties of the inhibitors of hardening and by analysis of the composition of the hardened fertilization membrane, a dityrosine-cross-linking mechanism was postulated (Foerder and Shapiro, 1977; Hall, 1978). Thus, dityrosine was found in the fertilization membrane, with one cross-link occurring per 50,000–100,000 MW of fertilization membrane protein. The ovoperoxidase is released from *S. purpuratus* eggs concomitant with the hardening reaction; all of the inhibitors of the hardening reaction inhibit the peroxidase activity. When known inhibitors of peroxidases were tested in the hardening assay, they were likewise found to inhibit hardening. Further support for the role of the ovoperoxidase in hardening was the finding that the enzyme is found in cortical granules before fertilization (Katsura and Tominaga, 1974; Klebanoff *et al.*, 1979) and in the fertilization membrane after fertilization (Hall, 1978; Klebanoff *et al.*, 1979) (Fig.5). In fact, the enzyme is localized so that virtually all of the activity associated with the egg (that which is not washed away in the sea water) after fertilization is in the fertilization membrane where it can catalyze iodination and other reactions charactersitic of the peroxidase. The ovoperoxidase uses hydrogen peroxide as a substrate; hydrogen peroxide is synthesized in a burst after fertilization (Foerder *et al.*, 1978). This burst of hydrogen peroxide synthesis is of interest in relation to the stimulation of respiration at fertilization (first studied by Warburg, 1908) and is the classic hallmark of metabolic activation at fertilization. This oxygen uptake occurs in a burst after fertilization or with parthenogenic activation (Ohnishi and Sugiyama, 1963) and has been called the "respiratory burst" of fertilization; along with the burst, there is an increased steady state level of oxygen consumption by the fertilized egg. When careful measurements of hydrogen peroxide production are made and compared with the amount of oxygen taken up in the first 15 minutes after fertilization, two-thirds of the oxygen uptake can be accounted for by hydrogen peroxide production (Foerder *et al.*, 1978).

Thus, much of this "respiratory burst" is not respiration at all, but rather, is indicative of increased oxidase activity, with hydrogen peroxide as the product.

Another feature of the peroxidative system activated at fertilization is the production of light; this chemiluminescence is inhibited by inhibitors of the ovoperoxidase. Thus, two of the hallmarks of fertilization in the sea urchin *S. purpuratus* are that the egg gasps and flashes as it is activated; this reflects the triggering of a complex peroxidative system. The peroxidative system of the egg is similar in many respects to that triggered by the phagocytic white blood cell upon encountering foreign invaders such as bacteria (Klebanoff *et al.*, 1979). For this reason, we have suggested that the peroxidative system of the sea urchin egg may play a spermicidal role in addition to hardening the fertilization membrane. The peroxidatic system of the white blood cell is spermicidal (Smith and Klebanoff, 1970) toward mammalian sperm, and we have found (Shapiro, unpublished observations) that lactoperoxidase-catalyzed iodination kills sea urchin sperm, although under some conditions, the sperm may resist such treatment (Lopo and Vacquier, 1978). Thus, the peroxidatic system of the egg may play a double role: in hardening the fertilization membrane, as well as in killing sperm (discussed further in Foerder *et al.*, 1978).

The peroxidatic system of the sea urchin egg has not yet been seen in other types of eggs, but it may be more general. This system is only one aspect of the assembly of the sea urchin fertilization membrane. As mentioned earlier, other components secreted from the cortical granule after fertilization doubtlessly play a role in the assembly process. The fertilization membrane is substantially thicker than the vitelline layer (see Giudice, 1973; Runnstrom, 1966), and the assembly process is markedly specific. For example, the ovoperoxidase is inserted with great fidelity into the fertilization membrane; perhaps a lectin-mediated mechanism, like that used in the assembly of the fertilization envelope of *Xenopus,* may also play some role in the sea urchin. The substrate for cross-linking may be released from the cortical granules, since the fertilization product does not catalyze hardening of the vitelline layer (Carroll and Epel, 1975a). Likewise, other noncovalently bound components are present in the fertilization envelope: in 6 M urea and 1.5 M β-mercaptoethanol at alkaline pH, about 70% of the protein of the fertilization membrane can be extracted as five major polypeptide species, from 18,000–92,000 MW (Carroll and Baginski, 1978a). Although the extraction conditions were quite severe, the regular pattern of polypeptides suggested that random destruction of peptide bonds was not occurring. Obviously much more work needs to be done on the assembly reaction. Since Ca^{2+} is required for assembly, the cortical granule components that aggregate in the presence of Ca^{2+} (Bryan, 1970a,b) are interesting candidates for reactants in the assembly. Cortical granule components of the sand dollar do not disperse if eggs are fertilized in the presence of concanavalin A, and the fertilization membrane does not harden under these circumstances (Vanquier and O'Dell,

1975), pointing to the involvement of glycoproteins in the assembly process. Sulfated acidic polysaccharides of large size are released during the cortical reaction (Immers, 1961; Ishihara, 1964). Additionally, an enzyme that catalyzes polysaccharide hydrolysis, a β-glucanase, is released from eggs at fertilization and may be located in the cortical granules (Epel et al., 1969; Schuel et al., 1972).

Methods to isolate cortical granules have been developed, and these may be useful in studying fertilization membrane assembly (Schuel et al., 1972; Detering et al., 1977). For example, a method to prepare cortical granules reported by Detering et al. (1977) is quite interesting, since it gave a 40-fold purification of the ovoperoxidase and of iodinated surface components and a greater than 30-fold purification of the cortical granule protease. In this preparation, there was lysis of cortical granules upon addition of Ca^{2+} and also assembly of a fertilization membrane but no generation of dityrosine residues (Decker and Lennarz, 1979; Foerder, unpublished data).

B. Metabolic Activation

This fascinating area of the biochemistry of the fertilized egg will not be discussed at length here, partly because it has recently been extensively reviewed (Epel, 1978) and also because it will be the focus for a article soon to appear in this series (Steinhardt and Epel, in preparation). The modern analysis of this process has its roots in two discoveries made almost a half century ago: the findings that at fertilization eggs release acid (Runnstrom, 1933) and that the process of fertilization can be reversed if eggs are placed in acid sea water immediately after fertilization (Tyler and Schultz, 1932). These observations, along with the findings that sodium influx increases after fertilization of the sea urchin egg (Chambers and Chambers, 1949) and that metabolic activation can be reversed by lowering the pH (Allen, 1953), set the stage for the current hypothesis that an increase in intracellular pH is necessary and sufficient for metabolic activation of many of the fertilization-related events. This idea came about from observations that by treating eggs with ammonia, many aspects of metabolic activation could be induced (Steinhardt and Mazia, 1973; Epel et al., 1974) and that this activation is due to ammonia behaving as a permeant weak base (Winkler and Grainger, 1978) that serves to increase the intracellular pH. The Na^+ requirement for activation of the egg (Chambers, 1975, 1976) and the discovery that a sodium–proton exchange reaction occurs at the time of fertilization (Johnson et al., 1976) led to the suggestion that acid efflux is coupled to sodium uptake and thereby increases the intracellular pH. The finding of increased intracellular pH was made in egg homogenates and by direct microelectrode recording (Shen and Steinhardt, 1978). However, the mechanism by which the sodium–proton exchange is activated is not clear, nor is the means by which the increase in intracellular pH leads to an alteration of metabolic activities such as stimulation of protein synthesis. One possibility is that increased pH is reflected in a change in the covalent structure of certain enzymes and thus affords them differential activity. For example, a change in protein kinase activity toward specific egg substrates is seen after fertilization or metabolic activation of the egg (Keller and Shapiro, 1977; Keller *et al.*, 1979), suggesting that altered protein phosphorylation may be one means by which the increased intracellular pH may be read out in metabolism. However, the evidence for the increased phosphorylation being a physiological event is indirect, and more work needs to be done to clarify the mechanisms attendant upon this interesting process.

V. Afterword

By studying mechanisms of gamete interaction, we encounter problems in biochemistry and cell biology that are common to all cells. Sperm and eggs are often ideal systems to study such general mechanisms and occasionally provide surprising insights. Examples include the assembly of actin filaments, as found in the acrosome reaction of invertebrate sperm; the patch of sperm surface that remains after fusion with the egg, in contrast to other cell fusion systems; the activation of a peroxidative system in the fertilized egg that is similar to that seen in the phagocytic white blood cell; and the activation of egg metabolism by increasing intracellular pH. A great pleasure of working on problems of fertilization is the legacy from the past, so that novel approaches with sophisticated instruments allow us to take another look at important ideas that go back well over a century. Such is the excitement of working in a field with a strong intellectual tradition, where major ideas extend throughout time and bind today's workers with the founders of experimental embryology.

ACKNOWLEDGMENTS

We are grateful to C. Gabel, G. Gundersen, D. Shellenbarger, and R. Schackmann for their criticisms of the paper and to the National Institutes of Health (GM23950) and National Science Foundation (PCM 7720472) for supporting the work from our laboratories described herein.

REFERENCES

Acutt, T. S., and Hoskins, D. D. (1978). *J. Biol. Chem.* **253**, 6744–6750.
Aketa, K. (1973). *Exp. Cell Res.* **80**, 435–411.
Aketa, K. (1975). *Exp. Cell Res.* **90**, 56–62.

Aketa, K., and Ohta, T. (1977). *Dev. Biol.* **61**, 366–372.
Aketa, K., and Onitake, K. (1969). *Exp. Cell Res.* **56**, 84.
Aketa, K., and Tsuzuki, H. (1968). *Exp. Cell Res.* **50**, 675–676.
Aketa, K., Tsuzuki, H., and Onitake, K. (1968). *Exp. Cell Res.* **50**, 676–679.
Aketa, K., Onitake, K., and Tsuzuki, H. (1972). *Exp. Cell Res.* **71**, 27–32.
Aketa, K., Miyazoki, S., Yoshida, M., and Tsuzuki, H. (1978). *Biochem. Biophys. Res. Commun.* **80**, 917–922.
Allen, R. P. (1953). *Biol. Bull.* **105**, 213–239.
Austin, C. R. (1951). *Aust. J. Sci. Res.* **B 4**, 581–596.
Austin, C. R. (1975). *J. Reprod. Fertil.* **44**, 155–166.
Austin, C. R., and Bishop, M. W. H. (1958). *Proc. R. Soc.* **B 149**, 241–248.
Austin, C. R., and Braden, A. W. H. (1956). *J. Exp. Biol.* **33**, 358–365.
Azarnia, R., and Chambers, E. L. (1976). *J. Exp. Zool.* **198**, 65–78.
Babcock, D. F., First, N. L., and Lardy, H. A. (1975). *J. Biol. Chem.* **251**, 3881–3886.
Babcock, D. F., Stammerjohn, D. M., and Hutchinson, T. (1978). *J. Exp. Zool.* **204**, 391–400.
Babcock, D. F., Singh, J. P., and Lardy, H. A. (1979). *Dev. Biol.* **69**, 85–93.
Baccetti, B., and Afzelius, B. A. (1976). The Biology of the Sperm Cell. *In* "Monographs in Devel. Biol." (A. Wolsky, ed.), Vol. 10. S. Karger, Basel, New York.
Baker, P. F. (1974) *Adv. Physiol.* **9**, 51–86.
Baker, P. F., and Presley, R. (1969). *Nature (London)* **221**, 488–489.
Baker, P. F., and Whitaker, M. J. (1978). *Nature (London)* **276**, 513–515.
Barros, C., and Yanagimachi, R. (1971). *Nature (London)* **233**, 268–269.
Barros, C., and Yanagimachi, R. (1972). *J. Exp. Zool.* **180**, 251–256.
Bavister, B. D., Yanagimachi, R., and Teichman, R. J. (1976). *Biol. Reprod.* **14**, 219–224.
Bedford, J. M. (1970). *Biol. Reprod. Spol.* **2**, 128–158.
Bedford, J. M., and Cooper, G. W. (1978). *In* "Membrane Fusion" (G. Poste and G. L. Nicolson, eds.). *Cell Surf. Rev.* **5**, 66–127. North Holland, Amsterdam.
Bellett, N. F., Vacquier, J. P., Vacquier, V. P. (1977). *Biochem. Biophys. Res. Commun.* **79**, 159–165.
Birr, C., and Hedrick, J. C. (1979). *Fed. Prod.* **38**, 1252.
Brachet, J. (1977). *Curr. Top. Dev. Biol.* **11**, 133–186.
Braden, A. W. H., Austin, C. B., and David, H. A. (1954). *Aust. J. Biol. Sci.* **7**, 391.
Brandriff, B., Hinegardner, R. T., and Steinhardt, R. (1975). *J. Exp. Zool.* **192**, 13–24.
Brandriff, B., May, G. W., and Vacquier, V. D. (1978). *Gamete Res.* **1**, 89–99.
Brown, C. R., and Hartree, E. F. (1976). *Hoppe-Seylers Z. Physiol. Chem.* **357**, 57–65.
Brown, G. G. (1976). *J. Cell Sci.* **22**, 547–562.
Bryan, J. (1970a). *J. Cell Biol.* **45**, 606–614.
Bryan, J. (1970b). *J. Cell Biol.* **44**, 635–644.
Burgess, D. R., and Schroeder, T. E. (1977). *J. Cell Biol.* **74**, 1032–1037.
Byrd, E. Q., Jr., and Collins, F. D. (1975). *Nature (London)* **257**, 675–677.
Byrd, W., Perry, G., and Weidener, E. (1977). *J. Cell Biol.* **75**, 267a.
Campisi, J., and Scandella, C. J. (1978). *Science* **199**, 1336–1337.
Carroll, E. J., and Baginski, R. M. (1978a). *J. Cell Biol.* **79**, 162a.
Carroll, E. J., Jr., and Baginski, R. M. (1978b). *Biochemistry* **17**, 2605–2612.
Carroll, E. J., Jr., and Epel, D. (1975a). *Dev. Biol.* **44**, 22–32.
Carroll, E. J., Jr., and Epel, D. (1975b) *Dev. Biol.* **79**, 162.
Carroll, E. J., Jr., and Levitan, H. (1978a). *Dev. Biol.* **3**, 432–440.
Carroll, E. J., Jr., and Levitan, H. (1978b). *Dev. Biol.* **64**, 329–331.
Carroll, E. J., Jr., Byrd, E. W., and Epel, D. (1977). *Exp. Cell Res* **108**, 365–374.
Chambers, E. L. (1975). *J. Cell Biol.* **67**, 609.

Chambers, E. L. (1976). *J. Exp. Zool.* **197,** 149.

Chambers, E. L., and Chambers, R. (1949). *Am. Natur.* **83,** 269.

Chambers, E. L., Pressman, B. C., and Rosen, B. (1974). *Biochem. Biophys. Res. Commun.* **60,** 126–132.

Chang, M. C. (1951). *Nature (London)* **168,** 697–698.

Chang, M. C. (1957). *Nature (London)* **175,** 258–259.

Chase, H. Y. (1935). *Biol. Bull.* **69,** 159–184.

Cole, K. S., and Spencer, J. M. (1938). *J. Gen. Physiol.* **21,** 538–590.

Collins, F. (1976). *Dev. Biol.* **49,** 381–394.

Collins, F., and Epel, D. (1977). *Exp. Cell Res.* **106,** 211–222.

Colwin, L. H., and Colwin, A. L. (1967). *In* "Fertilization: Comparative Morphology, Biochemistry and Immunology" (C. Metz and A. Monroy, eds.), Vol. 1, pp. 295–367. Academic Press, New York.

Conway, A. F., and Netz, C. B. (1976). *J. Exp. Zool.* **198,** 39–48.

Cornett, L. E., and Meizel, S. (1978). *Proc. Natl. Acad. Sci. U.S.A.* **75,** 4954–4958.

Cross, M. H., Cross, P. C., and Brinster, R. C. (1973). *Dev. Biol.* **33,** 411–416.

Czihak, G., ed. (1975). "The Sea Urchin Embryo: Biochemistry and Morphogenesis." Springer-Verlag, Berlin and New York.

Dalcq, A. M. (1938). "Form and Causalty in Early Development." Cambridge.

Dan, J. C. (1952). *Biol. Bull.* **103.** 54–66.

Dan, J. C. (1954). *Biol. Bull.* **107,** 335–349.

Dan, J. C. (1956). *Int. Rev. Cytol.* **5,** 365–393.

Dan, J. C. (1967). In "Fertilization: Comparative Morphology, Biochemistry and Immunology" (C. B. Metz and A. Monroy, eds.), Vol. I, pp. 237–281. Academic Press, New York.

Dan, J. C., and Wada, S. K. (1955). *Biol. Bull.* **109,** 40–55.

Dan, J., Ohri, Y., and Kushida, H. (1964). *J. Ultrastruct. Res.* **11,** 508–524.

Davis, B. K. (1971). *Proc. Natl. Acad. Sci. U.S.A.* **68,** 951–955.

Decker, G. L., and Lennarz, W. J. (1979). *J. Cell Biol.* **81,** 92–103.

Decker, G. L., Joseph, D. B., and Lennarz, W. J. (1976). *Dev. Biol.* **53,** 115–125.

Derbes, M. (1847). *Ann. Sci. Nat. III Ser. Zool.* **8,** 80–98.

DeRosier, D., Mandelkow, E., Silliman, A., Tilney, G., and Kane, R. (1977). *J. Mol. Biol.* **113,** 675–695.

Detering, N. K., Decker, G. L., Schmell, E. D., and Lennarz, W. J. (1977). *J. Cell Biol.* **75,** 899–914.

Durr, R., Shur, B., and Ruth, S. (1977). *Nature (London)* **265,** 547–548.

Eddy, E. M., and Shapiro, B. M. (1976). *J. Cell Biol.* **71,** 35–48.

Elinson, R. P. (1975). *Dev. Biol.* **47,** 257–268.

Elinson, R. P. (1977). *J. Embryol. Exp. Morphol.* **37,** 187–201.

Elinson, R. P., and Manes, M. E. (1978). *Dev. Biol.* **63,** 67–75.

Epel, D. (1977). *Sci. Am.* **237,** 128–138.

Epel, D. (1978). *Curr. Top. Biol.* **12,** 185–246.

Epel, D., and Vacquier, V. D. (1978). *In* "Membrane Fusion" (G. Poste and G. L. Nicolson, eds.). *Cell Surf. Rev.* **5,** 2–65. North Holland, Amsterdam.

Epel, D., Weaver, A. M., Muchmore, A. V., and Schimke, R. J. (1969). *Science* **163,** 294–296.

Epel, D., Steinhardt, R., Humphreys, T., and Mazia, D. (1974). *Dev. Biol.* **40,** 245–255.

Eusebi, F., Mangia, F., and Alfei, L. (1979). *Nature (London)* **277,** 651–653.

Fodor, E. J. B., Ako, H., and Walsh, K. A. (1975). *Biochemistry* **14,** 4923–4927.

Foerder, C. A., and Shapiro, B. M. (1977). *Proc. Natl. Acad. Sci. U.S.A.* **74,** 4214–4218.

Foerder, C. A., Klebanoff, S. J., and Shapiro, B. M. (1978). *Proc. Natl. Acad. Sci. U.S.A.* **75,** 183–187.

Friend, D. S. (1977). *In* "Immunobiology of Gametes" (M. Edidin and M. V. Johnson, eds.), pp. 5–30. Cambridge Univ. Press, Cambridge.

Friend, D. S., Orci, L., Perrelet, A., and Yanagimachi, R. (1977). *J. Cell Biol.* **74,** 561–577.

Gabel, C. A., and Shapiro, B. M. (1978). *Anal. Biochem.* **86,** 396.

Gabel, C. A., Eddy, E. M., and Shapiro, B. M. (1977). *J. Cell Biol.* **75,** 416a.

Gabel, C. A., Eddy, E. M., and Shapiro, B. M. (1978). *J. Cell Biol.* **79,** 162a.

Gabel, C. A., Eddy, E. M., and Shapiro, B. M. (1979a). *J. Cell Biol.* **82,** 742–754.

Gabel, C. A., Eddy, E. M., and Shapiro, B. M. (1979b). *Cell* **18,** 207–215.

Gachelin, G., Fellons, M., Gueret, J. L., and Jacob, F. (1976). *Dev. Biol.* **50,** 310–320.

Garbers, D. L., and Hardman, J. G. (1975). *Nature (London)* **257,** 677–678.

Garbers, D. L., and Kopf, G. S. (1978). *J. Reprod. Fertil.* **52,** 135–140.

Garbers, D. L., Lust, W. D., First, N. L., and Lardy, H. A. (1971). *Biochemistry* **10,** 1825–1831.

Giudice, G. (1973). "Developmental Biology of the Sea Urchin Embryo." Academic Press, New York.

Glabe, C. G., and Vacquier, V. D. (1977a). *J. Cell Biol.* **75,** 410–421.

Glabe, C. G., and Vacquier, V. D. (1977b). *Nature (London)* **267,** 836–838.

Glabe, C. G., and Vacquier, V. D. (1978). *Proc. Natl. Acad. Sci. U.S.A.* **75,** 881–885.

Gordon, M. (1973). *J. Exp. Zool.* **185,** 111–120.

Gordon, M., Dandeker, P. V., and Eager, P. R. (1978). *Anat. Rec.* **191,** 123–133.

Gould, K., Zanerveld, L. J. D., Seivastore, P. N., and Williams, W. L. (1971). *Proc. Soc. Exp. Biol. Res.* **136,** 6–10.

Gould-Somero, M., Holland, L., and Paul, M. (1977). *Dev. Biol.* **58,** 11–22.

Gray, R. D., Wolf, D. P., and Hedrick. J. L. (1974). *Dev. Biol.* **36,** 44–61.

Green, D. P. L. (1978a). *In* "The Mechanism of the Acrosome Reaction in the Development in Mammals" (M. H. Johnson, ed.), pp. 65–81. North Holland, Amsterdam.

Green, D. P. L. (1978b). *J. Cell Sci.* **32,** 137–151.

Green, D. P. L. (1978c). *J. Cell Sci.* **32,** 152–168.

Gregg, K. W., and Metz, C. B. (1976). *Biol. Reprod.* **14,** 405–411.

Grey, R. D., Working, P. K., and Hedrick, J. L. (1976). *Dev. Biol.* **54,** 52–60.

Grieve, L. C., and Hedrick, J. L. (1978). *Gamete Res.* **1,** 13–18.

Grossman, A., Levy, M., Troll, W., and Weissman, G. (1973). *Nature (London) New Biol.* **243,** 277–278.

Gwatkin, R. B. L. (1977). "Fertilization Mechanisms in Man and Mammals." Plenum, New York.

Gwatkin, R. B. L., and Williams, D. T. (1977). *J. Reprod. Fertil.* **49,** 55–59.

Gwatkin, R. B. L., Williams, D. J., Hartmann, J. F., and Kniazuk, M. (1973). *J. Reprod. Fertil.* **32,** 259–265.

Gwatkin, R. B., Wudl, L., Hartree, E. F., and Fink, E. (1977). *J. Reprod. Fertil.* **50,** 359–361.

Hagstrom, B. F. (1956). *Ark. Zool.* **10,** 307–315.

Hall, H. G. (1978). *Cell* **15,** 343–355.

Hansbrough, J. R., Kopf, G. S., and Garbers, D. L., (1979). *Fed. Proc.* **38,** 465.

Hartmann, J. F., and Hutchison, C. F. (1977). *J. Cell Physiol.* **93,** 41–48.

Harvey, E. N. (1910). *J. Exp. Zool.* **8,** 355–376.

Hathaway, R. R. (1963). *Biol. Bull.* **125,** 486–498.

Hayashi, T. (1946). *Biol. Bull.* **90,** 177–187.

Head, J. F., Mader, S., and Kaminer, B. (1979). *J. Cell Biol.* **80,** 211–218.

Hiramoto, Y. (1962). *Exp. Cell Res.* **27,** 416–426.

Hiramoto, Y. (1974). *Exp. Cell. Res.* **89,** 320.

Hoskins, D. D. (1973). *J. Biol. Chem.* **248,** 1135–1140a.

Hotta, K., Hamazaki, H., and Kurukawa, M. (1970). *J. Biol. Chem.* **245,** 5434.

Humphreys, W. J. (1967). *J. Ultrastruct. Res.* **17,** 314.

Hylander, B. L., and Summers, R. G. (1977). *Cell Tissue Res.* **182**, 469–489.

Immers, J. (1961). *Exp. Cell Res.* **24**, 356–378.

Isaka, S., Akino, M., Hotta, K., and Kurakawa, M. (1970). *Exp. Cell Res.* **59**, 37–42.

Ishihara, K. (1964). *Exp. Cell Res.* **36**, 354–367.

Ishihara, K., and Dan, J. C. (1970). *Dev. Growth Differ.* **12**, 179–188.

Ito, S., and Yoshioka, K. (1972). *Exp. Cell Res.* **72**, 547–551.

Ito, S., and Yoshioka, K. (1973). *Exp. Cell Res.* **78**, 191–200.

Iwamatsu, T., and Ohta, T. (1978). *J. Exp. Zool.* **205**, 157–180.

Jaffe, L. A. (1976) *Nature (London)* **261**, 68–71.

Jaffe, L. A., and Robinson, K. R. (1978). *Dev. Biol.* **62**, 215–228.

Johnson, M., and Edidin, M. (1978). *Nature (London)* **272**, 448–450.

Johnson, J. D., and Epel, D. (1975). *Proc. Natl. Acad. Sci. U.S.A.* **72**, 4474–4478.

Johnson, J. D., Epel, D., and Paul, M. (1976). *Nature (London)* **262**, 661–664.

Johnston, R. N., and Paul, M. (1977). *Dev. Biol.* **57**, 364–374.

Jones, H. P., Bradford, M. M., McRorie, R. A., and Cormier, M. J. (1978). *Biochem. Biophys. Res. Commun.* **82**, 1264–1271.

Katsura, S., and Tominaga, A. (1974). *Dev. Biol.* **40**, 292–297.

Keller, C. H., and Shapiro, B. M. (1977). *J. Cell Biol.* **75**, 408a.

Keller, C. H., Gunderson, G. G., and Shapiro, B. M. (1980). *Dev. Biol.* **74**, 86–101.

Kemp, N. E., and Istock, N. L. (1967). *J. Cell Biol.* **34**, 111–122.

Kimura-Furakawa, J., Sugemitsu, T., and Ishihara, K. (1978). *Exp. Cell Res.* **114**, 143–151.

Kinsey, W. H., and Koehler, J. K. (1978). *J. Ultrastruct. Res.* **64**, 1–13.

Kinsey, W. H., Se Gall, G. K., and Lennarz, W. J. (1978). *J. Cell Biol.* **79**, 167a.

Klebanoff, S. J., Foerder, C. A., Eddy, E. M., and Shapiro, B. M. (1979). *J. Exp. Med.* **149**, April.

Koehler, J. K. (1978). *Int. Rev. Cytol.* **54**, 73–108.

Kusano, K., Miledi, R., and Stinnakre, J. (1977). *Nature (London)* **270**, 739–741.

Lallier, R. (1970). *Exp. Cell Res.* **63**, 460–462.

Lallier, R. (1971). *Experientia* **27**, 1323–1324.

Lallier, R. (1977). *Experientia* **33**, 1263–1267.

Levine, A. E., and Walsh, K. A. (1978). *J. Cell Biol.* **79**, 1690.

Levine, A. E., and Walsh, K. A. (1979). *Dev. Biol.* **72**, 126–137.

Levine, A. E., Walsh, K. A., and Fodor, E. J. B. (1978). *Dev. Biol.* **63**, 299–306.

Lillie, F. R. (1912). *J. Exp. Zool.* **12**, 413–477.

Lillie, F. R. (1919). "Problems of Fertilization." Univ. of Chicago Press, Chicago, Illinois.

Loeb, J. (1916). "The Organism as a Whole." Putnam, New York.

Longo, F. J. (1973). *Biol. Reprod.* **9**, 149–215.

Longo, F. (1977). *J. Cell Biol.* **75**, 44a.

Longo, F. (1978). *Dev. Biol.* **67**, 249–265.

Longo, F. J., and Anderson, E. (1970). *J. Cell Biol.* **47**, 646–665.

Longo, F. J., and Schuel, D. (1973). *Dev. Biol.* **34**, 187–199.

Lorenzi, M., and Hedrick, J. L. (1973). *Exp. Cell Res.* **79**, 417.

Lui, C. W., and Meizel, S. (1977). *Differentiation* **9**, 59–66.

Lui, C. W., and Meizel, S. (1979). *J. Exp. Zool.* **207**, 173–186.

Lopo, A., and Vacquier, V. D. (1978). *J. Cell Biol.* **79**, 1660.

McRorie, R. A., and Williams, W. L. (1974). *Annu. Rev. Biochem.* **43**, 777–803.

McRorie, R. A., Turner, T. B., Bradford, M. M., and Williams, W. L. (1976). *Biochem. Biophys. Res. Commun.* **71**, 492–498.

Manes, M., and Barbieri, F. (1977). *J. Embryol. Exp. Morphol.* **40**, 187–197.

Markman, B. (1958). *Acta Zool.* **39**, 103–115.

Mazia, D. (1937). *J. Cell. Comp. Physiol.* **10**, 291–304.

Mazia, D., Schatten, G., and Steinhardt, R. A. (1975). *Proc. Natl. Acad. Sci. U.S.A.* **72,** 4465-4473.

Meizel, S. (1978). *In* "Development in Mammals" (M. H. Johnson, ed.), Vol. 3, pp. 1-64. North Holland, Amsterdam.

Meizel, S., and Deamer, D. W. (1978). *J. Histochem. Cytochem.* **26,** 98-105.

Meizel, S., and Lui, C. W. (1976). *J. Exp. Zool.* **195,** 137-144.

Meizel, S., and Mukerji, S. K. (1975). *Biol. Reprod.* **13,** 83-93.

Menge, A. C., and Fleming, C. H. (1978). *Dev. Biol.* **63,** 111-117.

Metz, C. B. (1967). *In* "Fertilization" (C. B. Metz and A. Monroy, eds.), Vol. 1, pp. 163-236. Academic Press, New York.

Metz, C. B. (1978). *Curr. Top. Dev. Biol.* **1.**

Miller, R. L. (1977). *In* "Advances in Invertebrate Reproduction" (K. G. Adiyodi and R. G. Adiyodi, eds.), Vol. 1, pp. 99-119. Peralam, Kenoth, India.

Millette, C. F. (1977). *In* "Immunobiology of gametes" (M. Edidin and M. V. Johnson, eds.), pp. 51-57. Cambridge Univ. Press, Cambridge.

Mitchison, J. M., and Swann, M. M. (1955). *J. Exp. Biol.* **32,** 734-750.

Monroy, A. (1953). *Experientia* **9,** 424-425.

Montaigne, M. (1576). Apology for Raymond Sebond. *In* "The Complete Essays of Montaigne" (D. M. Frame, trans.) Stanford Univ. Press, 1958, Stanford, California.

Morgan, T. H. (1904). *J. Exp. Zool.* **1,** 135.

Morgan, T. H. (1940). *J. Exp. Zool.* **85,** 1-32.

Morton, B., Herrigan-Lum, J., Albagli, L., and Jooss, T. (1974). *Biochem. Biophys. Res. Commun.* **56,** 372-375.

Moscona, A. A., and Monroy, A., eds. (1978). "Current Topics in Developmental Biology," Vol. 12. Academic Press, New York.

Moser, F. (1939). *J. Exp. Zool.* **80,** 448-471.

Motomura, I. (1941). *Sci. Rep. Tohoku Imp. Univ.* **16,** 245-363.

Motomura, I. (1954). *Sci. Rep. Tohuku Univ.* **20,** 158-162.

Motomura, I. (1957). *Sci. Rep. Tohuku Univ.* **23,** 167-181.

Nakamura, M., and Yasumasu, I. (1974). *J. Gen. Physiol.* **63,** 374-388.

Nakazawa, J., Asami, K., Shuger, R., Fujiwara, A., and Yasumasu, I. (1970). *Exp. Cell. Res.* **63,** 143-146.

Nehara, T., and Katou, K. (1972). *Dev. Growth Diff.* **14,** 175-184.

Nelson, L. (1978). *Fed. Proc.* **37,** 2543-2547.

Nicosia, S. V., Wolf, I. P., and Inoue, M. (1977). *Dev. Biol.* **57,** 56-74.

Niemerko, A., and Komer, A. (1976). *J. Reprod. Fertil.* **48,** 279-284.

Nigon, V., and Do, F. (1965). *C.R. Acad. Sci.* **257,** 2178-2180.

Ohnishi, T., and Sugiyama, M. (1963). *Embryologia* **8,** 79-88.

Ohtake, H. (1976a). *J. Exp. Zool.* **198,** 313-322.

Ohtake, H. (1976b). *J. Exp. Zool.* **198,** 303-312.

Okamoti, H., Takahashi, K., and Yamashita, N. (1977). *J. Physiol.* **267,** 465-495.

O'Rand, M. (1977). *J. Exp. Zool.* **202,** 267-273.

Pasteels, J. J. (1965). *J. Embryol. Exp. Morphol.* **13,** 327-340.

Paul, M. (1975a). *Exp. Cell Res.* **90,** 137-142.

Paul, M. (1975b). *Dev. Biol.* **43,** 259-312.

Paul, M., and Epel, D. (1971). *Exp. Cell Res.* **65,** 281-289.

Paul, M., and Johnston, R. N. (1978). *J. Exp. Zool.* **203,** 143-149.

Popa, G. T. (1927). *Biol. Bull.* **52,** 238-257.

Powers, R. D., and Tupper, J. T. (1977). *Dev. Biol.* **56,** 306-315.

Repin, U. S., and Akimova, I. M. (1976). *Biokhimiya* **41,** 50-58.

Reyes, A., Oliphant, G., and Brackett, B. G. (1975). *Fertil. Steril.* **26**, 148–157.
Ridgway, E. B., Gilkey, J. C., and Jaffe, L. F. (1977). *Proc. Natl. Acad. Sci. U.S.A.* **74**, 623–627.
Rogers, B. J. (1978). *Gamete Res.* **1**, 165–223.
Rogers, B. J., Ueno, M., and Yanagimachi, R. (1977). *J. Exp. Zool.* **199**, 129–136.
Rosati, F., and de Santis, R. (1978). *Exp. Cell Res.* **112**, 111–115.
Rosati, F., Monroy, A., and de Prisco, P. (1977). *J. Ultrastruct. Res.* **58**, 261–270.
Rothschild, L. (1954). *Q. Rev. Biol.* **29**, 332.
Rothschild, L. (1956). "Fertilization." Methuen, London.
Rubin, R. P. (1970). *Pharmacol. Dev.* **22**, 389–428.
Runnstrom, J. (1933). *Biochem. Z.* **258**, 257–259.
Runnstrom, J. (1966). *Adv. Morphogen.* **5**, 221–325.
Saling, P. M., Storey, B. T., and Wolf, D. P. (1978). *Dev. Biol.* **65**, 515–525.
Sanger, J. W., and Sanger, J. M. (1975). *J. Exp. Zool.* **193**, 441–442.
Sano, K., and Mohri, H. (1976). *Science* **192**, 1339–1340.
Schackmann, R. W., and Shapiro, B. M. (1980). *Dev. Biol.,* in press.
Schackmann, R. W., Eddy, E. M., and Shapiro, B. M. (1978). *Dev. Biol.* **65**, 483–495.
Schatten, G., and Mazia, D. (976). *Exp. Cell Res.* **98**, 325–337.
Schmell, E., Earles, B. J., Breaux, C., and Lennarz, W. J. (1977). *J. Cell Biol.* **72**, 35–46.
Schroeder, T. J. (1978a). *J. Cell Biol.* **79**, 1710.
Schroeder, T. J. (1978b). *Dev. Biol.* **64**, 342–346.
Schuel, H., Wilson, W. C., Bressler, R. S., Kelly, J. W., and Wilson, J. R. (1972). *Dev. Biol.* **29**, 307–320.
Schuel, H., Wilson, W. L., Chen, K., and Lorand, L. (1973). *Dev. Biol.* **34**, 175–180.
Schuel, H., Troll, W., and Lorand, L. (1976). *Exp. Cell Res.* **103**, 442–447.
SeGall, G. K., and Lennarz, W. J. (1978). *J. Cell Biol.* **79**, 170a.
Shapiro, B. M. (1975). *Dev. Biol.* **46**, 88–102.
Shapiro, B. M. (1977). *Horizons Biochem. Biophys.* **4**, 201–243.
Shen, S. S., and Steinhardt, R. A. (1978). *Nature (London)* **272**, 153–154.
Shellenbarger, D. L., and Shapiro, B. M. (1980). *Gamete Res.,* in press.
Singh, J. P., Babcock, D. F., and Lardy, H. A. (1978). *Biochem. J.* **172**, 549–556.
Smith, D. C., and Klebanoff, S. J. (1970). *Biol. Reprod.* **3**, 229–235.
Spudich, J. A. (1973). *Cold Spring Harbor Symp. Quant. Biol.* **37**, 585–593.
Stambaugh, R. (1978). *Gamete Res.* **1**, 65–85.
Stambaugh, R., and Smith, M. (1978). *J. Exp. Zool.* **203**, 135–141.
Steinhardt, R. H., and Epel, D. (1974). *Proc. Natl. Acad. Sci. U.S.A.* **71**, 1915–1919.
Steinhardt, R. A., and Mazia, D. (1973). *Nature (London)* **241**, 400–401.
Steinhardt, R. A., Lundin, L., and Mazia, D. (1971). *Proc. Natl. Acad. Sci. U.S.A.* **68**, 2426–2430.
Steinhardt, R. A., Epel, D., Carroll, E. J., and Yanagimachi, R. (1974). *Nature (London)* **252**, 41–43.
Steinhardt, R., Zucker, R., and Schatten, G. (1977). *Dev. Biol.* **58**, 185–196.
Sugiyama, M. (1956). *Exp. Cell Res.* **10**, 364–376.
Summers, R. G., and Hylander, B. L. (1975). *Exp. Cell Res.* **96**, 63–68.
Summers, R. G., Hylander, B. L., Colwin, L. H., and Colwin, A. L. (1975). *Am. Zool.* **15**, 523–557.
Takahashi, Y. M., and Sugiyama, M. (1973). *Dev. Growth Differ.* **15**, 261–267.
Talbot, P., and Franklin, L. E. (1978). *J. Exp. Zool.* **203**, 1–14.
Talbot, P., and Kleve, M. G. (1978). *J. Exp. Zool.* **204**, 131–136.
Talbot, P., Summers, R. G., Hylander, B. L., Keough, F. M., and Franklin, L. E. (1976). *J. Exp. Zool.* **198**, 383–392.
Tegner, M. J., and Epel, D. (1973). *Science* **179**, 685–688.

Thomas, D. (1957). If I was Tickled by the Rub of Love. *In* "Collected Poems of Dylan Thomas." New Directions, New York.

Tilney, L. (1978). *J. Cell Biol.* **77**, 851–864.

Tilney, L. G., Hatano, S., Ishikawa, H., and Mooseker, M. S. (1973). *J. Cell Biol.* **59**, 109–126.

Tilney, L. G., Kiehart, D. P., Sardet, C., and Tilney, M. (1978). *J. Cell Biol.* **77**, 536–550.

Tilney, L. G., Clain, J. G., and Tilney, M. S. (1979). *J. Cell Biol.* **81**, 229–253.

Tyler, A., and Schultz, J. (1932). *J. Exp. Zool.* **63**, 509–532.

Uno, Y., and Hoshi, M. (1978). *Science* **200**, 50–59.

Vacquier, V. D. (1975). *Dev. Biol.* **43**, 62–74.

Vacquier, V. D., and O'Dell, D. S. (1975). *Exp. Cell Res.* **90**, 465–468.

Vacquier, V. D., and Moy, G. W. (1977). *Proc. Natl. Acad. Sci. U.S.A.* **74**, 2456–2460.

Vacquier, V. D., Epel, D., and Douglas, L. A. (1972a). *Nature (London)* **237**, 34–36.

Vacquier, V. C., Tegner, M. J., and Epel, D. (1972b). *Nature (London)* **240**, 352–353.

Vacquier, V. C., Tegner, M., and Epel, D. (1973). *Exp. Cell Res.* 80, 111.

Veron, M., and Shapiro, B. M. (1977). *J. Biol. Chem.* **252**, 1286–1292.

Veron, M., Foerder, C., Eddy, E. M., and Shapiro, B. M. (1977). *Cell* **10**, 321–328.

Wada, S. K., Collier, J. R., and Dan, J. C. (1956). *Exp. Cell. Res.* **10**, 168–180.

Warburg, O. (1908). *Z. Physiol. Chem.* **57**, 1–16.

Wessels, N. K., Spooner, B. S., Ash, J. F., Bradley, M. D., Ludwena, M. A., Taylor, E., Wren, J. T., and Yamada, K. M. (1971). *Science* **171**, 135–143.

Wincek, T. J., Parrish, R. F., and Polokoski, K. L. (1979). *Science* **203**, 553–554.

Winkler, M. M., and Grainger, J. L. (1978). *Nature (London)* **273**, 536–538.

Wolf, D. P. (1974). *Dev. Biol.* **38**, 14–29.

Wolf, D. P. (1978). *Dev. Biol.* **64**, 1–10.

Wolf, D. P., and Hamada, M. (1977). *Biol. Reprod.* **17**, 350–354.

Wolf, D. P., and Nicosia, S. V. (1978). *J. Cell Biol.* **79**, 160a.

Wolf, D. P., Nishihara, T., West, D. M., Wyrick, R. E., and Hedrick, J. L. (1976). *Biochemistry* **15**, 3671–3677.

Wolpert, L. (1971). *Curr. Top. Dev. Biol.* **6**, 183–224.

Wyrick, R. E., Nishihara, T., and Hedrick, J. L. (1974). *Proc. Natl. Acad. Sci. U.S.A.* **71**, 2067–2071.

Yamamoto, T. (1961). *Int. Rev. Cytol.* **12**, 361–405.

Yanagimachi, R. (1970). *Biol. Reprod.* **3**, 147–153.

Yanagimachi, R. (1977). *In* "Immunobiology of Gametes" (M. Edidin and M. H. Johnson, eds.), pp. 255–295. Cambridge Univ. Press, Cambridge.

Yanagimachi, R. (1978). *Curr. Top. Dev. Biol.* **12**, 83–105.

Yanagimachi, R., and Usui, N. (1974). *Exp. Cell Res* **89**, 161–174.

Yurewicz, E. C., Oliphant, G., and Hedrick, J. L. (1975). *Biochemistry* **14**, 3707.

Zucker, R. S., and Steinhardt, R. A. (1978). *Biochim. Biophys. Acta* **541**, 455–466. *Dev. Biol.* **65**, 285–295.

Zucker, R. S., Steinhardt, R. A., and Winkler, M. M. (1978). *Dev. Biol.* **65**, 285–295.

INTERNATIONAL REVIEW OF CYTOLOGY, VOL. 66

Perisinusoidal Stellate Cells (Fat-Storing Cells, Interstitial Cells, Lipocytes), Their Related Structure in and around the Liver Sinusoids, and Vitamin A-Storing Cells in Extrahepatic Organs

KENJIRO WAKE

Department of Anatomy, Tokyo Medical and Dental University, Tokyo, Japan

I. Introduction

The stellate cells ("Sternzellen") in the liver were first described by Kupffer in 1876, using the gold chloride method. However, 22 years after the discovery, a misconception by himself (Kupffer, 1898, 1899a,b) generated new misconceptions and led to a great confusion in the liver histology. The true nature of these cells remained enigmatic for over several decades until they were rediscovered by Wake (1971). It was observed that the stellate cells are quite different from the phagocytic Kupffer cells being located in the space of Disse. They are identical to the perisinusoidal cells such as "fat-storing cells" (Ito, 1951) and "interstitial cells'' (Suzuki, 1958). It was further found that vitamin A is stored in their cytoplasm.

In this article, emphasis will be laid on the complicated history of the stellate cells. In order to understand the systematic history of the stellate cells, it appears essential to obtain a deep insight into various aspects of their nature. Various methods that have been adopted for the demonstration of the stellate cells by several authors will be considered. Morphological criteria of the stellate cells will be described at light and electron microscopic levels. Further, the present state of our knowledge on the origin and functions of these cells will be critically evaluated. Finally, different aspects of the vitamin A-storing cells in extrahepatic organs will be dealt with in this article. However, no attempt will be made to provide an exhaustive review of phagocytic Kupffer cells and the endothelial cells in the liver sinusoids, but only the relevant data on these cells related to the perisinusoidal stellate cells will be incorporated.

II. History of the Stellate Cells

A. THE FIRST PERIOD (1876–1897)

First description of the stellate cells by Kupffer (1876) is available in the form of a letter to Waldeyer, which begins with the following: ''Im Verlaufe von andauernden, aber leider noch immer vergeblichen Bemühungen, die Nerven der Leberläppchen nachzuweisen, bin ich auf ein bisher nicht bekanntes, oder jedenfalls nicht genügend beachtetes Strukturverhältniss an der gesunden Leber von Säugethieren und des Menschen gestossen, von dem ich Ihnen Kenntnisse geben möchte.'' His method included cutting of fresh liver sections with a ''Doppelmesser'' and their treatment with the Gerlach's gold chloride solution (gold chloride 1, HCl 1, water 10,000), after brief rinse in 0.05% chromic acid. Kupffer's efforts for the demonstration of nerve fibers in the liver lobules bore no fruit, but he found black *stars* scattered in the liver lobules. He indicated that these cells were located perisinusoidally, always attached to the sinusoidal capillaries and also to the parenchymal cells. He thought that these cells probably belonged to the Waldeyer's ''perivasculäre Bindegewebszellen'' or adventitia cells. These cells, designated as ''Sternzellen (stellate cells)'' were observed in liver of various mammals such as, rat, rabbit, pig, bull, and dog (Fig. 1) and also in the human liver.

Six years after the discovery of the stellate cells, the gold chloride method was employed for demonstration of these cells in other mammals and a bird by Rothe (1882), who was a pupil of Kupffer. He confirmed Kupffer's description with regard to their shape, distribution, and localization (Figs. 2 and 3). He depicts clear inclusions in their cytoplasm as ''small nuclei'' (Fig. 4). Because of their

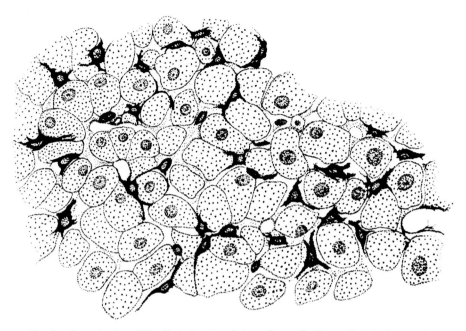

Fɪɢ. 1.   Reproduction of Kupffer's drawing of the stellate cells (Sternzellen) in the dog liver
(from V. von Ebner, 1902), showing dark stellate-shaped cells intercalated between the parenchymal
cells. Gold chloride method.

affinity for gold chloride and their regular distribution in the liver lobules, Rothe
assumed that the stellate cells were nervous elements, although he could not be
certain about them.

During the two subsequent decades, the perisinusoidal stellate cells in the liver
attracted the attention of pathologists. Platen (1878) showed that these cells
contained lipid droplets during fatty degeneration and also after oil perfusion.
These findings were supported by Asch (1884), who reported the existence of
lipid droplets within the cells even in normal liver. From their localization and
profiles these liver droplets were actually Kupffer's stellate cells. Disse (1890)
also described stellate cells located in the perisinusoidal lymph space. These
authors were of the same opinion that the stellate cells are located in the
perisinusoidal site in the liver lobules.

B. Tʜᴇ Sᴇᴄᴏɴᴅ Pᴇʀɪᴏᴅ (1898–1970)

1. *Confusion about the Stellate Cells*

In 1898, Kupffer read another paper on the so-called stellate cells at the
Twelfth Congress of Anatomists held in Kiel. In this paper, he changed his

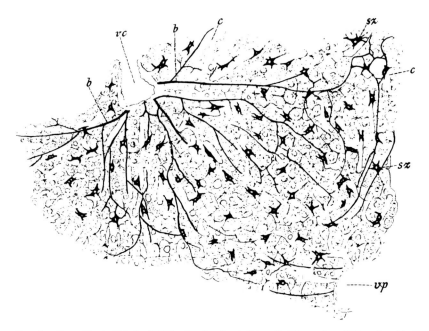

FIG. 2.   From Fig. 1 of Rothe (1882), showing the stellate cells (Sternzellen, SZ) scattered in the liver lobule of a rat. Gold chloride method. b, Connective tissue fibers; c, capillary; vc, central vein; vp, portal vein.

earlier opinion and concluded that the stellate cells are the special endothelial cells of the sinusoids (''integrierende Bestandtheile der Capillarwand''). In his detailed paper on this subject (Kupffer, 1899a), he presented several criteria given below in order to explain why the gold reactive stellate cells were identical with the endothelial cells: (a) The distribution pattern of gold-reactive stellate cells in the human liver (Fig. 5a) was strikingly similar to that of phagocytic cells of rabbit liver (Fig. 5b) as revealed by studying sections of liver after India ink perfusion. (b) Profiles of both types of cells were star-shaped with cytoplasmic processes, although the cell bodies of phagocytic cells were somewhat elongated. (c) Perikarya of the gold-reactive cells were protruded into the lumen of the sinusoids (Fig. 6a), and their location is identical with that of the phagocytic endothelial cells (Fig. 6b). (d) Having compared the vacuolar inclusions of the gold-reactive stellate cells (Fig. 6c), which Rothe (1882) noticed and called small nuclei, with the structure of erythrophagy in dye-stained preparations of the liver (Fig. 6d), Kupffer concluded that the inclusions in the stellate cells might be fragments of erythrocytes.

Kupffer's renewed conclusions can be summarized as follows. The stellate cells are phagocytic in nature, are integral elements of the wall of liver sinusoids, and are endothelial cells. In later years, this new concept led to deep-rooted

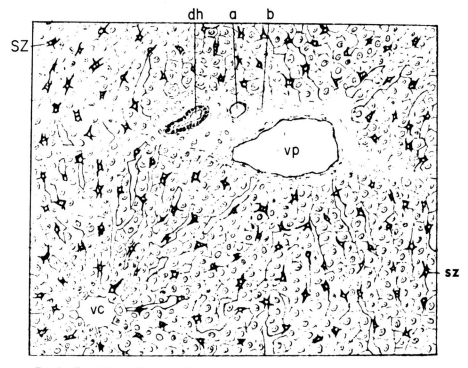

FIG. 3.   From Fig. 2 of Rothe (1882), depicting the stellate cells (Sternzellen, SZ) in the sheep liver. Gold chloride method. a, Hepatic artery; b, connective tissue; dh, bile duct; vp, portal vein; vc, central vein.

confusions in liver histology. In the present article, this stage is referred to as the second period in the history of the stellate cells.

Phagocytic cells in the liver were studied by Ponfick (1869) and Hoffmann and Langerhans (1869). These authors proposed, from their observations using perfusion of cinnaber, that cinnaber-uptake cells (''Zinnoberzellen'') were located in the perisinusoidal site. The extravascular localization of Zinnoberzellen prompted Heidenhein (1880) to suggest that they might be identical with the Sternzellen of Kupffer. From his studies using perfusion of cinnaber or carmine into the jugular vein, Asch (1884) pointed out that extravascular ''stellate cells'' took up cinnaber or carmine. Löwit (1888) mentioned that ''stellate cells'' played an important role not only in the destruction of erythrocytes, but in the transportation of bile pigment produced in the parenchymal cells. Thus, it is reasonable to assume that the development of Kupffer's later concepts was probably introduced by the controversies on the localization of the phagocytic cells in the liver lobules.

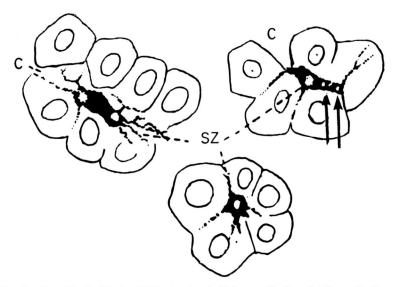

FIG. 4.   From Fig. 3 of Rothe (1882), showing the high magnification of stellate cells (Sternzellen, SZ) in the sheep liver. Cytoplasmic processes extend along the wall of capillaries and between the parenchymal cells. (Further explanation appears in the text.) c, Capillary. Arrows, introduced by the present author, indicate the "small nuclei" (Rothe, 1882) of the stellate cells.

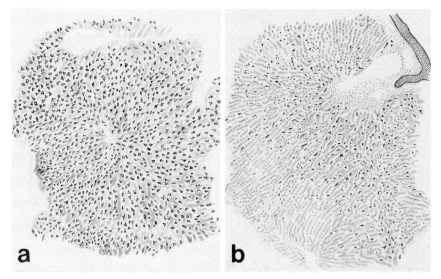

FIG. 5.   (a) Figure 1 of Kupffer (1899), showing a lobule of the human liver. Gold chloride method. (b) Figure 16 of Kupffer (1899) depicting the rabbit liver lobules injected with india ink. Kupffer attempted to show similar distribution patterns of gold-reactive cells and the phagocytic cells in the liver lobule. (Further explanation appears in the text.)

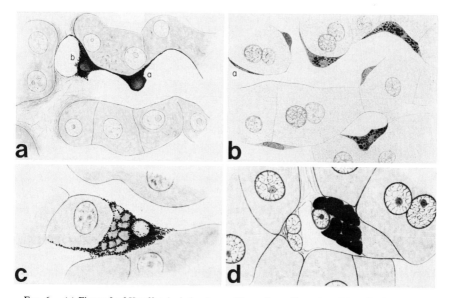

Fig. 6. (a) Figure 2 of Kupffer depicting two gold reactive cells projected into the lumen of the sinusoid in the human liver. Gold chloride method. (b) Figure 17 of Kupffer (1899), showing endothelial cells that have incorporated india ink. (c) Figure 10 of Kupffer (1899), showing the vacuolated structures other than the larger round nucleus in the gold preparation of the human liver. (d) Figure 11 of Kupffer (1899) depicting the erythrophagy of the phagocytic cells. Comparing 5a and b, 6a and b, and 6c and d, he concluded that the "Sternzellen" is the phagocytic endothelial cells of the sinusoidal capillaries. (Further explanation appears in the text.)

Kupffer's new concept was accepted by Mayer (1899), Schilling (1909), and others. However, it is noteworthy that as early as in 1898, Browicz in Cracow reported evidence against the Kupffer's new concept. He stated that the phagocytic cells are not the integral element of the sinusoidal wall, but are pear-shaped cells hanging in the lumen of the sinusoid with their processes attached on the inner surface of the sinusoidal walls. These cells later were called "endocytes" by Zimmermann (1928). According to Pfuhl (1932), the above disagreement of Browicz was resolved with the interpretation of the endocytes as active forms of phagocytic endothelial cells of Kupffer.

Subsequently, phagocytic cells in the liver sinusoids attracted attention and were studied with perfusion of colloidal silver (Cohn, 1904) or colloidal dyes (Ribbert, 1904; Kiyono, 1914). These studies stimulated the development of concepts of the reticuloendothelial cell system by Aschoff (1924). Subsequent investigations revealed other characteristics of the reticuloendothelial Kupffer cells. Three of these features (cell shape, "Verfettung," and vitamin A storage of the phagocytic Kupffer cells) will be reviewed briefly.

a. *Cell Shape.* After Kupffer (1898) changed his opinion, reticuloendothelial phagocytic cells in the liver lobules have been called ''Sternzellen.'' However, because histological characteristics of phagocytic cells could not accurately reveal stellate shape, authors' opinions on their cell shape have been contradictory. Higgins and Murphy (1928) reported that in certain animals the cells were more numerous, considerably larger, and possessed a more distinct stellate form; in others, the cells were more stocky and without the finer protoplasmic processes that were usually seen. Wolf-Heidegger (1941) stated that flat phagocytic cells without cell processes were unusual. Aterman (1958, 1963) stated that although stellate-shaped cells were found in freeze-dry preparations, they were rather exceptional. He interpreted the stellate shape of the ''stellate cells'' seen in Fig. 4 of the report of Kupffer (1899b) as an artifact resulting from the chromic acid treatment (Aterman, 1963, 1977).

   b. *''Verfettung.''*   Pathologists often found that in the human liver the phagocytic Kupffer cells are laden with cytoplasmic fat droplets even when no fatty degeneration of the parenchymal cells is evident. This phenomenon was termed ''Verfettung'' of the reticuloendothelial Kupffer cells. The Verfettung was reported in 40% (Helly, 1911) or 70% (Fischer, 1912) of the autopsies of whites, and in contrast, only in 13% of Japanese (Yoshida, 1940). Levine (1932) reported occurrence of fat droplets in the Kupffer cells of all but two of the 43 persons suffering a sudden, violent death, and concluded that neutral fat was normally found in the Kupffer cells. Jaffé (1938) also stated that fat in the Kupffer cells is not pathologic, but indicates their greater participation in fat metabolism.

   c. *Vitamin A-Storage.*   Histochemical studies on the intrahepatic localization of vitamin A were first carried out by Hirt and his co-workers in 1929. They described bright yellow fluorescent droplets in the walls of the sinusoids of the unstained liver. Querner (1935) described a luminescent substance (Leuchtstoff X) in the isotropic fat droplets of the hepatic parenchyma cells. This substance was soluble in the lipid solvents (Querner and Sturm, 1934) and was thought to be related to fat-soluble carotenoids (Querner, 1935). Querner observed the fluorescence in parenchymal cells, whereas Kudo (1938) stated that fluorescence is emitted not from the parenchymal cells but from reticuloendothelial Kupffer cells, and that vitamin A storage in the Kupffer cells was inhibited by the reticuloendothelial blockade. Uotila and Simola (1938) described that massive sudanophilic substance occurred in the reticuloendothelial Kupffer cells in the hypervitaminosis A liver. Popper and his co-workers (Popper, 1941; Meyer *et al.*, 1941) examined the liver in 240 autopsy cases and found green fading fluorescence in the Kupffer and parenchymal cells. Intense fluorescence has been observed in the large Kupffer cells filled with lipid droplets in the liver of experimental animals following the administration of excess vitamin A (Domagk and Dobeneck, 1933; Querner, 1935; Popper, 1944; Rodahl, 1950; Popper and Schaffner, 1957; Lane, 1968).

It was independently reported by Nakane (1963) and Wake (1964) that vitamin A is not stored in the phagocytic Kupffer cells but in the perisinusoidal cells of the liver.

2. *Other Perisinusoidal Cells*

Ponfick's Zinnoberzellen and Kupffer's Sternzellen, both of which were described first as perisinusoidal cells, were treated as the integral elements of the sinusoidal capillaries by Kupffer in 1899. In addition, other perisinusoidal cells have been reported since the end of the last century, and different aspects of various types of these cells are reviewed in this section.

a. *Granular Cells* (Berkley, 1893).   Using Golgi and picric acid silver impregnation methods, Berkley (1893) showed scattered, yellowish, granular cell bodies lying between the hepatocytes and the adjacent vascular walls (Fig. 7). The extensions from the granular cell bodies are ordinarily quite short and rapidly attenuate along the side of the vessel or between the hepatocytes, but rarely were long processes that were as granular as the body found (Fig. 7). Berkley pointed out the similarities of his granular cells with the stellate cells; the majority of the granular cells are undeniably the stellate cells described by Kupffer (1876). Their position corresponds precisely and is always constant, (i.e., remaining in direct contact with the portal capillaries and even encircling the vessels in a ringlike fashion as Kupffer (1876) described). Based on the nature of these granules, he

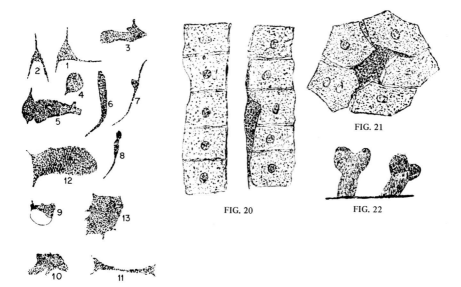

Fig. 7.   Drawing of Berkley (1893), showing the "granular cells," located in the perisinusoidal space. (See text.) Golgi and picric acid–silver impregnation method.

described that these cells do not take up a dark staining even after exposure to osmic acid solution, and hence were probably not fatty in nature.

b. *Pericytes* (Zimmermann, 1923). Zimmermann (1923) described a sort of dendritic perisinusoidal cell in the liver lobules with the Golgi–Kopsch chrome-silver method. Their branching processes surround the wall of the sinusoids. Some stems of these processes, as shown in Fig. 8, reach the neighboring sinusoids and surround them. He concluded that these cells are identical with Rouget's pericyte in the wall of capillaries.

c. *Fat-Storing Cells* (Ito, 1951). In 1951, Ito observed cells containing lipid droplets in the human liver lobules. These cells were located in the space of Disse and were surrounded by reticular fibers. He presumed that lipid droplets in these cells were derived from the blood stream and designated these cells "Fettspeicherungszellen" (fat-storing cells, or in Japanese, *shibô-sesshu saibô: shibô,* fat; *sesshu,* intake; *saibô,* cell). However, an experiment by Ito's pupils (Satsuki *et al.,* 1956), which involved intravenous injections of various kinds of fat emulsions and cod-liver oil, revealed that these cells failed to show an increase in the lipid droplets in their cytoplasm, contrary to Ito's expectations. Surprisingly, a decrease was observed between 40 minutes and 17 hours after lipid administration. On the other hand, these lipid droplets increased remarkably after administering glucose and insulin (Kano, 1952; Sunaga, 1954a), so Ito changed his earlier opinion and concluded that they contained fat (triglyceride) converted from glycogen and proposed the name "*shibô-chozô saibô*" (*chozô* means store) for these cells (Ito, 1956).

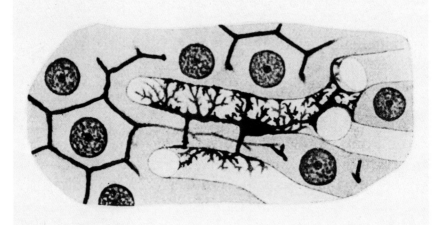

FIG. 8. Drawing of Zimmermann (1923), demonstrating the "pericyte" in the rat liver. Note the dendritic processes that surround the wall of the sinusoid. Silver impregnation method.

The existence of the fat-storing cells was reported in fishes (Ito *et al.*, 1952, 1962; Schmidt, 1956), amphibia (Ito *et al.*, 1952), reptiles (Ito *et al.*, 1952; Watari, 1959), birds (Ito *et al.*, 1952, 1960b; Tanaka, 1960; Umahara, 1963; Kitagewa, 1960), mammals (Ito *et al.*, 1952, 1953, 1960a; Ito and Nemoto, 1956; Kitamura *et al.*, 1956; Yamagishi, 1958, 1959; Takahashi, 1959; Satsuki, 1954; Sunaga, 1954a; Tanikawa *et al.*, 1965), and man (Ito, 1951; Ito and Nemoto, 1952; Tahira, 1958; Ito and Shibasaki, 1968). Variations in the number of lipid droplets have been reported in different seasons (Takahashi, 1959; Tanaka, 1960; Ito *et al.*, 1960a) and under some nutritional conditions (Kitagawa, 1960; Satsuki, 1954, 1955; Sunaga, 1954a; Ito *et al.*, 1960a). These extensive experiments and discussions on the fat-storing cells seem to have been carried out in order to understand their relationships with glycogen metabolism.

d. *Interstitial Cells* (Suzuki, 1958).   Suzuki (1958) found stellate-shaped cells located in the space of Disse, using his own silver impregnation method. According to Suzuki, within the liver lobules, branches of the autonomic nerve fibers were attached to these cells, coursing through the space of Disse. The cytoplasm of these cells showed argyrophil mesh-like structure and their branching processes covered the surface of the parenchymal cells (Suzuki, 1958, 1963; Watanabe, 1960) (Fig. 9). Since these cells are intercalated between nerve fibers and the effector cells, Suzuki refers to them as "interstitial cells (*neurones sympathiques interstitiels*)" (Cajal, 1909) in the liver. He put forward his ideas in the form of a hypothesis that these cells transmit the impulses from the autonomic nerve fibers to the parenchymal cells in the liver. Riegele (1928, 1932) showed a close relationship between the phagocytic cells and the autonomic fibers in the liver lobules. But Watanabe (1960) and Suzuki (1963) pointed out that Riegele might be confusing phagocytic cells with the interstitial cells.

Suzuki's contribution, which has been of great value, however, suggests that the interstitial cells are probably the original Sternzellen (Kupffer, 1876). This suggestion was based upon the following facts. First, the interstitial cells lie in the perisinusoidal space of Disse. Second, their well-developed cytoplasmic processes make the cells look like stars. Third, these cells are impregnated more intensely with the gold chloride treatment used routinely in his silver impregnation method.

Though both Ito's fat-storing cells and Suzuki's interstitial cells lie in the space of Disse, homology of the two cells had remained undiscussed till the report of Wake in 1971.

C. THE THIRD PERIOD (1971 ONWARD)

In 1971, concomitant application of the classic gold impregnation method (Kupffer, 1876; Rothe, 1882), the silver impregnation method (Suzuki, 1958),

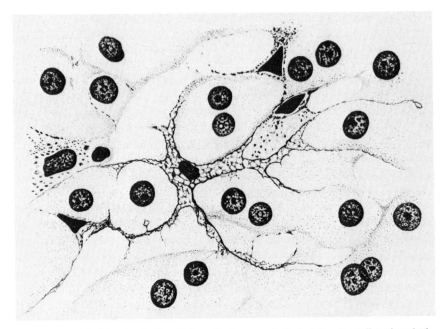

FIG. 9. Original drawing by Suzuki (in author's possession), depicting the stellate-shaped "interstitial cells" in the rabbit liver. Note the vacuolated structure of the cytoplasm. Suzuki's silver impregnation method.

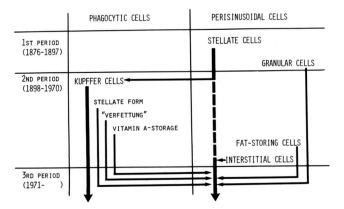

FIG. 10. A summarized history of the cell in and around the liver sinusoids. (See the text.)

fluorescence, and electron microscopy by Wake revealed that the Sternzellen first described by Kupffer are quite different from the phagocytic Kupffer cells in the sinusoids and are identical to the "fat-storing cells" and also to the "interstitial cells." The lipid droplets in the stellate cells of normal and hypervitaminosis A animals imparted an intense vitamin A-fluorescence. These droplets also reacted intensely with gold chloride, whereas the stellate cells in vitamin A-deficient animals failed to react with gold chloride. The gold reaction has been shown to demonstrate vitamin A histochemically at light and electron microscopical levels, as well as *in vitro* (Wake, 1973). Attention has also been paid to the mechanism underlying the storage of vitamin A in the stellate cells (Wake, 1974, 1975b). Detailed description of the stellate cells in the liver of various mammals (Wake, 1976b) and in human liver (Wake, 1978) by the gold chloride method has been given. The possible role of these cells in the fibrogenesis of the liver has also been discussed (see Section VI, B), and the vitamin A-storing cells in the extrahepatic organs have been reported (see Section VIII). The history of the stellate cells is summarized in Fig. 10.

III. Terminology

The cells, which are stained with gold chloride and were originally designated as Sternzellen, have been given various names. The designations given by various authors to the phagocytic cells that were confused with the stellate cells are excluded from this list.

Stellate cells (Kupffer, 1876; Wake, 1971)
Granular cells (Berkley, 1893)
Pericytes (Zimmermann, 1928)
Fat-storing cells (Ito, 1951)
Metalophil cells (Marshall, 1956)
Interstitial cells (Suzuki, 1958)
Lipophagic cells (Horiuchi, 1960)
Perisinusoidal cells (Wood, 1963)
Transmittal cells (Suzuki, 1963)
Fat-storage cells (Ladman, 1964)
Mesenchymal cells (Rhodin, 1964a)
Lipocytes (Bronfenmajer *et al.,* 1966)
Adventitious connective tissue cells (Schnack *et al.,* 1967; Stockinger, 1967)
Disse-space cells (Takeuchi *et al.,* 1967)
Sinusoidal mesenchymal cells (Rubin *et al.,* 1970)

Ito cells (Hruban *et al.*, 1974)
Lipid-storing cells (Hruban *et al.*, 1974)
Lipocytus perisinusoideus (terminology in Histology, 1975)
Vitamin A-storing cells (Yamada and Hirosawa, 1976)
Vitamin A-uptake cells (Kusumoto and Fujita, 1977)

The variety of these names suggests that not only the classification but also the nature of these cells has long been controversial. The various designations cited above can be classified by the morphological and functional properties of these cells: stellate in their form (stellate cells); their location in the liver lobules (pericytes, perisinusoidal cells, Disse-space cells, subendothelial cells); their cytoplasmic appearance (granular cells); their affinity to metals (metalophil cells); their intimate relations to the autonomic nerve fibers (interstitial cells, transmittal cells); lipid droplets in their cytoplasm (fat-storing cells, lipocytes, lipid-storing cells); their relations to the fibrogenesis in the liver (adventitious connective tissue cells, sinusoidal mesenchymal cells).

The term "fat-storing cells" is frequently used, but this term does not appropriately reflect the function of these cells (Wake, 1974), because these cells store vitamin A (Nakane, 1963; Wake, 1964, 1971, 1974; Hirosawa and Yamada, 1973) and not fat (glycerides). The morphological characteristics of the lipid droplets in their cytoplasm are different from the fat droplets located in the glycogen areas of the hepatic parenchymal cells (Wake, 1974, 1975b; Blouin *et al.*, 1977). McGee and Patrick (1972) remarked that it would appear unjustifiable to regard these cells as specialized simply for the storage of lipid because it is likely that the collagen fibers in the space of Disse are formed by these cells. The present knowledge of the function of these cells is too limited to enable us to label these cells based on their function alone.

In this article, the author will designate these cells as "stellate cells" or "perisinusoidal stellate cells," as did Kupffer (1876) in his earlier report. Indeed, star-like profiles of these cells are well demonstrated by the gold and silver impregnation methods (see Section IV, A and D). The author believes that this *morphological* designation might be most appropriate, at least until the origin and functions of these cells are further clarified.

It has been suggested (Aterman, 1963) that it might not be appropriate to designate the phagocytic Kupffer cells as "stellate cells," in spite of the proposal of Rüttner *et al.* (1956) to designate only the stellate-shaped phagocytic cells because the number of phagocytic cells that are classified in this category is very low (Aterman, 1963). Above all, the designation of phagocytic cells as "the stellate cells of Kupffer" was introduced by the misconception of Kupffer (1898). In this article, the phagocytic cells in the sinusoids will be called "Kupffer cells."

IV. Light Microscopy of the Stellate Cells

A. The Gold Chloride Method

Since the first report by Kupffer (1876), no attempt to demonstrate the stellate cells with his gold chloride method has been made except by of Rothe (1882) and Wake (1971, 1976b, 1978). This method is very unstable and it is difficult to obtain satisfactory results constantly, as was pointed out by earlier investigators.

With the help of well-impregnated gold preparations, almost all classic descriptions by Kupffer (1876) and Rothe (1882) were confirmed by Wake (1971, 1976b, 1978). Hepatic parenchymal cells are stained red, whereas interlobular connective tissue takes a pale red color. Numerous scattered star-shaped cells in the liver lobules appeared black against a red background (Figs. 11-15). They are generally less dense in the central zone of the lobule than in the peripheral

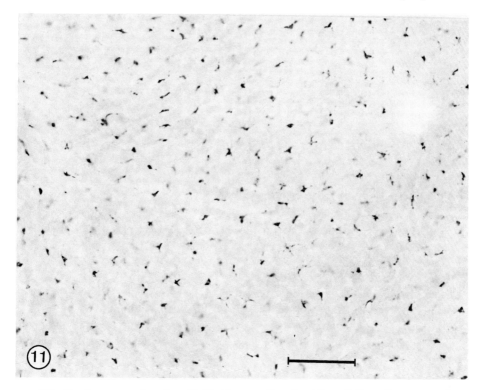

Fig. 11. Low-power view of a gold chloride preparation of a rat liver. Stellate cells are scattered regularly in the liver lobule. Kupffer's gold chloride method. No counterstaining. Bar, 100 μm. ×180.

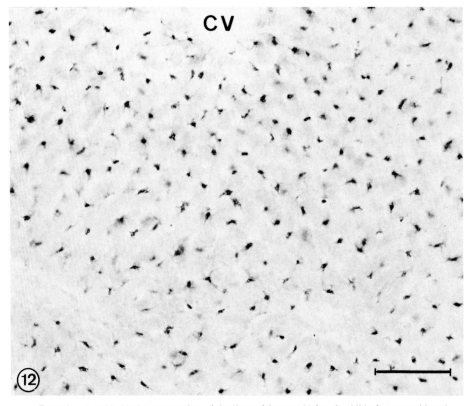

FIG. 12. A gold chloride preparation of the liver of 2-year-old female child after an accidental death. Kupffer's gold chloride method. No counterstaining. CV, Central vein. Bar, 100 μm. ×200.

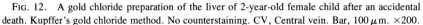

area, and the cells in the peripheral zone possess more cytoplasm than those in the central zone (Fig. 13). The cytoplasm of stellate cells contains numerous fine grains of reduced gold, whereas the nuclei are stained reddish brown. The cells reveal characteristically vacuolated cytoplasm with a mesh-like appearance (Figs. 16-20). The internal structures were described by Rothe (1882) as "small nuclei" (Fig. 4) and by Kupffer (1899a) as "fragments of erythrocytes" (Fig. 6c). However, these vacuoles in the cytoplasm are in fact vitamin A-containing lipid droplets as has been clearly demonstrated by the concomitant application of gold chloride method, Sudan III staining, fluorescence, and electron microscopy (Wake, 1971).

The size and number of vitamin A-containing lipid droplets vary from species to species. In cat (Figs. 19 and 20), rat (Figs. 21 and 22), rabbit, and man (Figs. 23 and 24), the lipid droplets are numerous and rather small and are also distrib-

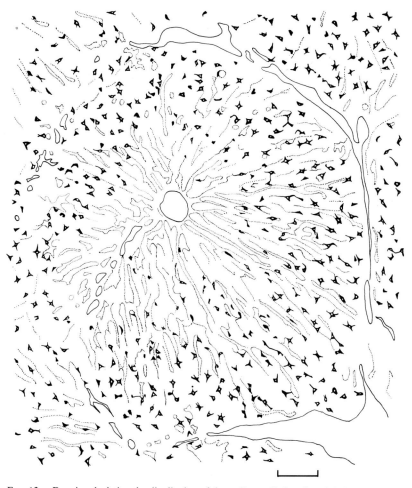

FIG. 13. Drawing depicting the distribution of the stellate cells in a liver lobule of a rabbit and showing their star-shaped profile. Kupffer's gold chloride method. Bar, 100 μm. (From Wake, 1971, reproduced by permission of the Editor of *American Journal of Anatomy*.)

uted in the cytoplasmic processes, which are well developed and branched out (Wake, unpublished data). In other mammals investigated such as pig (Figs. 16, 17, and 18), hamster, and bull, the stellate cells have one or two large lipid droplets adjacent to the nuclei, which show deep indentations on the surface apposed to the lipid droplets. In these species the cytoplasmic processes are not so conspicuous as in the rat and rabbit. In man, the size and number of the vitamin A-containing lipid droplets show marked individual variations.

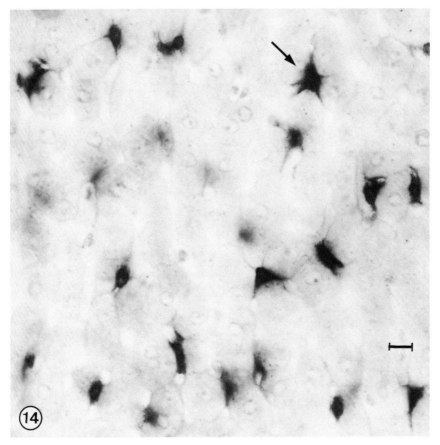

Fig. 14. The stellate cells from the rabbit liver. An arrow shows a stellate-pentapolar cell surrounded with parenchymal cells. Kupffer's gold chloride method. Bar, 10 μm. ×600.

B. Histochemistry of the Gold Chloride Reaction

In preparations of various mammalian livers well impregnated by the Kupffer's original gold method, only the scattered perisinusoidal stellate cells were stained black with metallic gold deposits; otherwise, no black-stained elements can be seen either in the liver lobules or in the interlobular connective tissues (Kupffer, 1876; Rothe, 1882; Wake, 1971, 1976b, 1978). Thus, this gold reaction is highly specific for the stellate cells. Kupffer (1898) believed that the gold-reactive substance in the stellate cells might represent glucose.

Since it has become possible to stain the stellate cells with gold chloride after formalin fixation instead of chromic acid treatment (Wake, 1964, 1971), the histochemical nature of the gold-reactive substance has been established to some

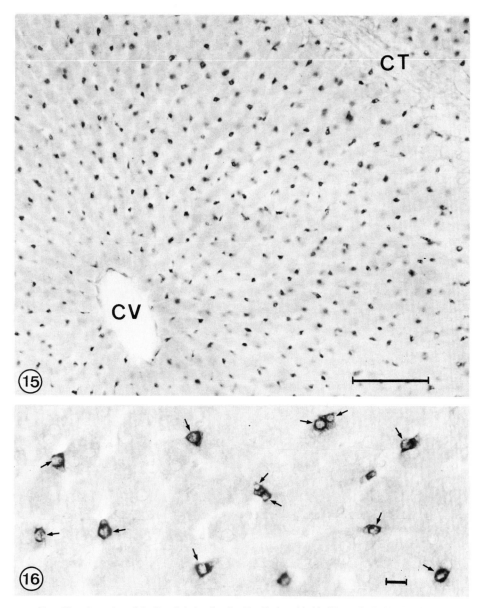

FIG. 15. A portion of the liver lobule of a pig. Kupffer's gold chloride method. CT, Interlobular connective tissue; CV, central vein. Bar, 100 μm. ×210.

FIG. 16. Micrograph of stellate cells of the pig liver. Small arrows indicate large lipid droplets in their cytoplasm. Kupffer's gold chloride method. Bar, 10 μm. ×500.

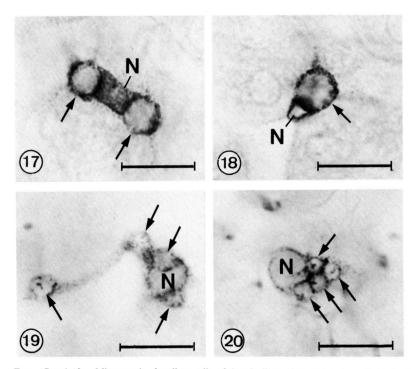

FIGS. 17 and 18.   Micrograph of stellate cells of the pig liver. One or two large lipid droplets (arrows) located at poles of the nucleus (N). Kupffer's gold chloride method. Bar, 10 μm. ×2000.

FIGS. 19 and 20.   Micrograph of stellate cells of the cat liver. Arrows show the lipid droplets in their cytoplasm. Kupffer's gold chloride method. Bar, 10 μm. ×2000.

extent. Double staining with gold chloride and Sudan III (Wake, 1964, 1971) and application of electron microscopy after gold reaction (Wake, 1971; 1974) demonstrated the fact that the deposition of the metal occurred on the surface of the lipid droplets in the cytoplasm of the stellate cells (Fig. 25). Under a fluorescence microscope, these lipid droplets displayed quickly fading vitamin A fluorescence (Nakane, 1963; Wake, 1964, 1971) (Figs. 26–29).

The gold chloride reaction only occurs below pH 4.2, increases with the reduction of pH value, and is strong at pH 2.6 (Wake, 1973). On the other hand, if the sections are irradiated with ultraviolet light, no reaction occurs (Wake, 1973). The reaction also fails to occur in the lipid droplets of parenchymal cells in normal condition (Wake, 1973). Vitamin A acetate, palmitate, and cod liver oil react with gold chloride *in vitro* in the presence of detergents to produce a violet color with black precipitation (Wake, 1973).

From the above results, it is apparent that the gold chloride reaction is a reduction reaction by vitamin A, and probably double bonds of vitamin A play a

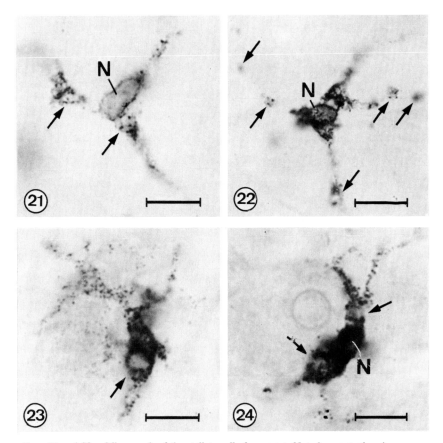

FIGS. 21 and 22.   Micrograph of the stellate cells from a rat. Note long cytoplasmic processes containing small lipid droplets (arrows). Kupffer's gold chloride method. Bar, 10 μm. ×1400.

FIGS. 23 and 24.   Micrograph of the stellate cells from the liver of 2-year-old female infant after an accidental death. Reduced gold precipitation occurs in the branching cytoplasmic processes and the surface of the lipid droplets (arrows). Kupffer's gold chloride method. Bar, 10 μm. ×1400.

role in this reduction. Thus, it is interesting to note that Kupffer demonstrated vitamin A in the liver histochemically over one hundred years ago, while using the gold chloride method.

C. FLUORESCENCE MICROSCOPY

Histologists and pathologists have traditionally believed that the fluorescence of vitamin A is emitted from the phagocytic Kupffer cells (see Section II). Indeed, scattered cells in the liver lobules impart intense vitamin A fluorescence.

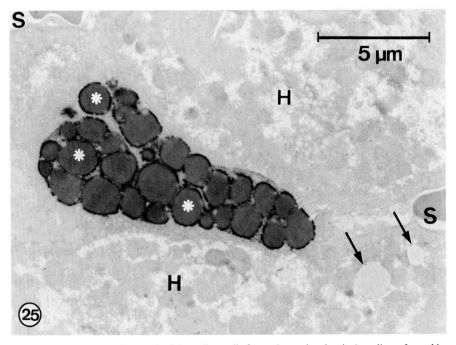

FIG. 25.   Electron micrograph of the stellate cells from a hypervitaminosis A rat liver after gold reaction. Note the well preserved lipid droplets containing vitamin A (asterisks) with a reduced gold precipitation on their surface. The reaction fails to occur in the lipid droplets (arrows) in the parenchymal cells (H). Glutaraldehyde in phosphate buffer (pH 7.4) for 2 hours, 0.02% gold chloride solution (pH 2.8) for 16 hours in darkness. Postfixed with osmium tetroxide. S, Sinusoid. ×6000. (From Wake, 1973, reproduced by permission of the publisher, North-Holland Publishing Co., Amsterdam, The Netherlands.)

However, these cells are not the Kupffer cells but the perisinusoidal stellate cells (Nakane, 1963; Wake, 1964, 1971, 1974). The statement on the form of fluorescent cells by Popper (1941), who extensively investigated liver vitamin A under the fluorescence microscope, seems to be noteworthy: "In the Kupffer cells vitamin A fluorescence was noted as droplets which filled the cytoplasm, thus *imparting an outline of the characteristic stellate form of Kupffer.*" This description implies that fluorescent cells are literally the perisinusoidal stellate cells.

Fluorescent lipid droplets are disposed in the form of a rosette around the nucleus or as elongated beads (Fig. 27). Developing small lipid droplets also impart vitamin A fluorescence. These lipid droplets appear to correspond to the membrane-bound (Type I) lipid droplets (Figs. 40 and 42) in the stellate cells (Wake, 1974). After the administration of excess vitamin A, increased lipid droplets in the stellate cells impart an intense fluorescence. Soon after the administration of vitamin A, fine granules in the cytoplasm of phagocytic Kupffer

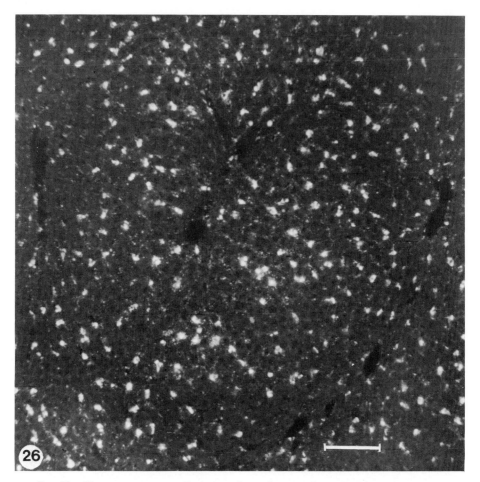

Fɪɢ. 26. Fluorescence micrograph from a liver of a rat after subcutaneous injections of 1,320,000 IU/kg vitamin A acetate. Intense vitamin A fluorescence is observed in the cytoplasm of the stellate cells distributed regularly throughout the lobules. A frozen section after calcium–formalin fixation. Bar, 100 μm. ×1500.

cells also emit a weak fluorescence (Wake, 1975b) (Fig. 29). The fluorescence is emitted from vitamin A-chylomicrons in the phagosomes and/or phagolysosomes (Fig. 35) in the Kupffer cells (see Wake, 1975b).

D. Sɪʟᴠᴇʀ Iᴍᴘʀᴇɢɴᴀᴛɪᴏɴ Mᴇᴛʜᴏᴅ

In the silver-impregnated preparations, the stellate-shaped cytoplasmic processes are obvious (Figs. 30 and 31). Dendritic processes extend either over the

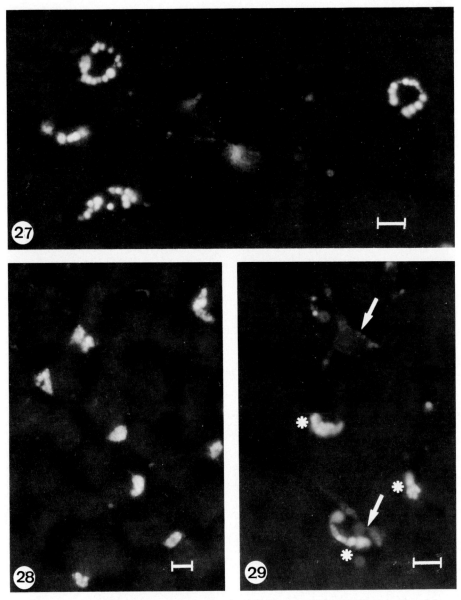

FIG. 27.   Fluorescence micrograph showing vitamin A fluorescence from the lipid droplets of the stellate cells 2 weeks after subcutaneous injections of excess vitamin A acetate. Bar, 10 μm. ×800.

FIG. 28.   Fluorescence micrograph showing vitamin A in the stellate cells of a 48-year-old man who died from carcinoma of the tongue. Bar, 10 μm. ×500.

FIG. 29.   Fluorescence micrograph showing vitamin A fluorescence from the stellate cells (asterisks) and from Kupffer cells (arrows) 1 week after subcutaneous injections of excess vitamin A palmitate. Note weak fluorescence emitted from the vacuolar structures in the cytoplasm of the Kupffer cells. Bar, 10 μm. ×700.

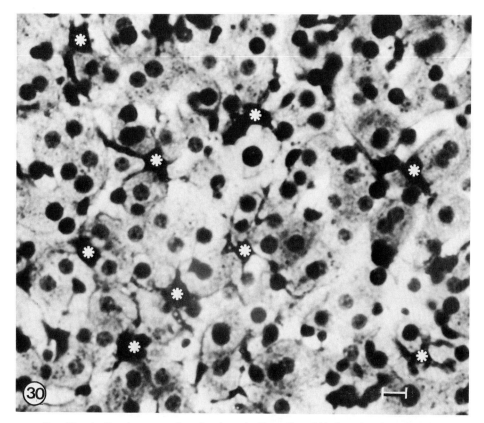

FIG. 30. A silver-impregnated section from the liver of a rabbit 2 weeks after subcutaneous injections of 200,000 IU/kg vitamin A acetate. Star-shaped stellate cells (asterisks) are scattered. Suzuki's silver impregnation method. Bar, 10 μm. ×600.

parenchymal cells or around the sinusoidal capillaries (Zimmermann, 1928; Suzuki, 1958; Wake, 1971). Their cytoplasm shows conspicuous vacuolar structures (Fig. 31), which correspond with the vitamin A-containing lipid droplets in the cytoplasm (Wake, 1971). After the administration of excess vitamin A, the stellate cells acquire an enhanced affinity for silver nitrate, and their cytoplasmic vacuoles enlarge and increase in number (Wake, 1971) (Figs. 32 and 33). In hypervitaminosis A animals, small vacuoles increase in the periphery of thin cytoplasmic processes (Wake, 1971) (Fig. 33). Both in the normal and hypervitaminosis A animals, a branching pattern and the fine structures of their dendritic processes are more clearly demonstrated with the silver impregnation method than with the Kupffer's original gold impregnation method.

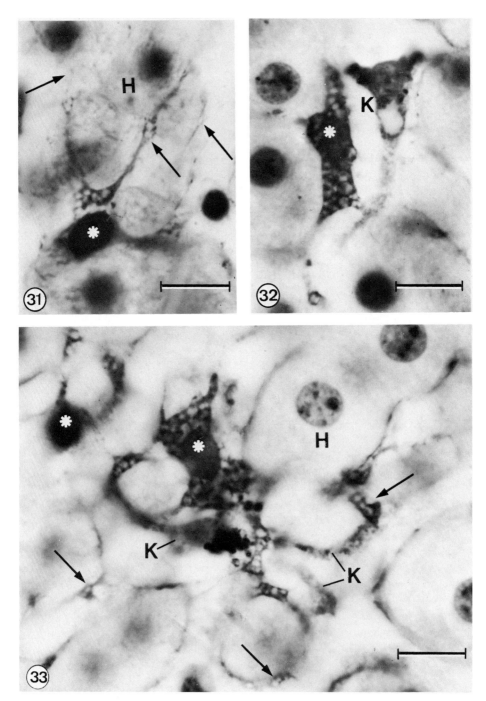

E. Distribution, Number, and Volume of the Stellate Cells

The stellate cells are distributed rather regularly in the liver lobules. Kupffer (1876) stated that the average distance between the neighboring stellate cells corresponds to 1–3 times the diameter of the parenchymal cell. Rhodin (1964a) reported that there are about 25 stellate cells per 400 μm^2 in a section of the mouse liver. Bronfenmajer *et al.* (1966) stated that there exist 3–5 such cells per 60–100 parenchymal cells. According to recent stereological examination (Blouin *et al.*, 1977; Blouin, 1977), hepatocytes constitute 78% of parenchymal volume in the rat liver, the nonhepatocytes account for 6.3% and consist of 2.8% endothelial cells, 2.1% Kupffer cells, and 1.4% stellate cells.

Stellate cells that are revealed by the gold chloride method are less dense in the central zone of the lobule than in the peripheral area and have a unit area relation of 0.4 in the central zone, 1 in the intermediate zone, and 1.1 in the peripheral zone (Wake, 1971). However, since the gold-reactive stellate cells (i.e., vitamin A-containing stellate cells) of the central zone increase in number after the administration of excess vitamin A, the stellate cells in the central zone appear to contain no vitamin A lipid droplets in the normal condition.

V. Ultrastructure of the Stellate Cells

Gold preparations studied with the help of an electron microscope reveal that metallic gold was deposited on the lipid droplets in the perisinusoidal stellate cells, i.e., "fat-storing cells" or "lipocytes" (Wake, 1971) (Fig. 25).

The nucleus in the stellate cells is oval or more or less elongated in electron micrographs (Hübner, 1968; Ito and Shibasaki, 1968; McGee and Patrick, 1972; Yamamoto and Enzan, 1975). It is frequently indented by lipid droplets in the cytoplasm (Ito and Shibasaki, 1968; McGee and Patrick, 1972) (Fig. 34). One or more nucleoli and sometimes "spheridium" can be seen (Ito and Shibasaki, 1968).

A conspicuous feature of these cells is their well-developed granular reticulum (Wood, 1963; Rhodin, 1964a; Ito and Shibasaki, 1968; Kobayashi and

---

Fig. 31.   A stellate cell from normal rabbit liver. Note fine dendritic processes of the stellate cell (asterisk) spread on the hepatic cell cord (H). Suzuki's silver impregnation method. Bar, 10 μm. ×1800.

Fig. 32.   A stellate cell (asterisk) and a Kupffer cell (K) from a liver of a rabbit 2 weeks after subcutaneous injections of excess vitamin A acetate. Lipid droplets in the stellate cells increase in number and in size. The perisinusoidal stellate cell can be clearly distinguished from the Kupffer cells by the silver impregnation method. Bar, 10 μm. ×1800.

Fig. 33.   Micrograph from the same section shown in Fig. 32. A Kupffer cell (K) is close to a stellate cell (asterisk). Note lipid droplets (arrows) within the dendritic processes of the stellate cells. Suzuki's silver impregnation method. H, Parenchymal cell. Bar. 10 μm. ×1800.

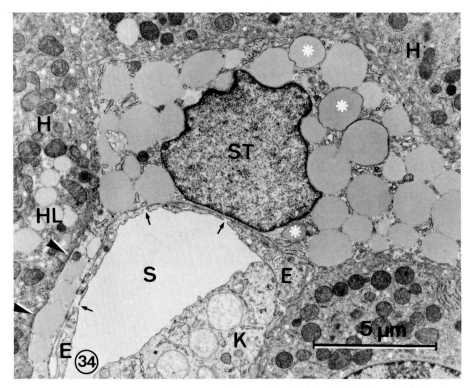

Fig. 34.   Electron micrograph of a stellate cell (ST) in the rat liver 2 weeks after subcutaneous injections of excess vitamin A, illustrating abundant lipid droplets in the cell body and its subendothelial process (arrow heads). Asterisks show membrane-bound Type I lipid droplets. E, Endothelial cell; H, parenchymal cell; K, Kupffer cell; small arrows, fenestration of the endothelial cell. ×6600.

Takahashi, 1971; Wake, 1971; Hirosawa and Yamada, 1973; Kovacs and Horvath, 1975; Yamamoto and Enzan, 1975; Yamada and Hirosawa, 1976; Tuchweber *et al.*, 1976; Blouin *et al.*, 1977; Tòth *et al.*, 1977; Wisse, 1977) (Figs. 37 and 38). The cisternae of the granular reticulum (RER) are moderately expanded and contain fine filamentous materials. The relative proportion of RER membrane of the stellate cells to that of total hepatic RER is 1.3% (Blouin *et al.*, 1977). This value is highest among the organelles of the stellate cells, suggesting active protein synthesis in these cells. Considering the findings that agranular reticulum is poorly developed (Wake, 1976a) and that these cells contain a small quantity of glycogen particles (Tanikawa *et al.*, 1965; Schnack *et al.*, 1966; Yamamoto and Enzan, 1975), these cells may not participate in active glycogen metabolism.

Several investigators have observed that the stellate cells display a conspicuous Golgi complex (Nakane, 1963; Bronfenmajer *et al.*, 1966; Hübner, 1968; Ito

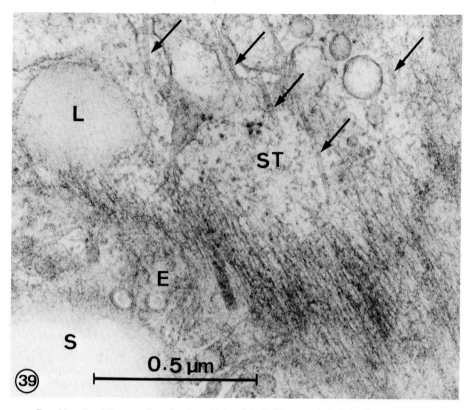

FIG. 39.   An oblique section of a sinusoidal wall including an endothelial cell (E) and a stellate
cell (ST), showing microfilaments and microtubules (arrows) in the stellate cell. L, Lipid droplet;
S, sinusoid. ×88,000.

1970; Kobayashi and Takahashi, 1971; Wake, 1971, 1974; Kobayashi *et al.*,
1973; Blouin *et al.*, 1977; Blouin, 1977; Wisse, 1977), rabbit (Matsuo, 1959;
Yamagishi, 1958; Rhodin, 1964a), and man (Ladman, 1964; Rhodin, 1964a;
Bronfenmajer *et al.*, 1966; Schnack *et al.*, 1966; Ito and Shibasaki, 1968; Muto
et al., 1977). On the other hand, no lipid droplets were reported in calf stellate
cells (Wood, 1963). In the rat, 25.3% of all the stellate cells contain lipid
droplets (Blouin, 1977). In man, the diameter of lipid droplets is about 1–2 μm
(Ladman, 1964) or 3 μm (Schnack *et al.*, 1966), and in rat they measure about 2
μm (Wake, 1964).

Bronfenmajer *et al.* (1966) stated that the lipid droplets are not surrounded by
a limiting membrane (unit membrane). However, Wake (1974) differentiated
two types of lipid droplets: membrane-bound (Type I) (Figs. 34 and 42) and
non-membrane-bound (Type II) (Fig. 34). Type I are smaller than Type II, and

the former are surrounded by a unit membrane. Type II are ca. 2 μm in diameter and are located in the cytoplasmic matrix. In the Type I lipid droplet, the unit membrane does not have any direct contact with the surface of the lipid droplet, but a light layer (intercalated layer), which is 4–8 nm thick, is intercalated between the inner leaflet of the unit membrane and the dense surface layer (6 nm in thickness) of the lipid droplet. The thick portion of the intercalated layer is characterized by the presence of a limited number of vesicles derived from that of the MVBs (see Section VI, A). Both types of lipid droplets seen by electron microscopy in stellate cells react with gold chloride (Wake, 1974).

In electron micrographs, the cytoplasmic processes that are demonstrable with light microscopy using gold and silver impregnation methods lie in the subendothelial space and in the narrow spaces between the parenchymal cells. The cytoplasmic processes are well-developed (Yamagishi, 1959; Wood, 1963; Schnack et al., 1966; Ito and Shibasaki, 1968; Blouin et al., 1977; Muto et al., 1977; Wisse, 1977) and are called "subendothelial processes" (Ito and Shibasaki, 1968). The subendothelial processes contain few to several lipid droplets (Schnack et al., 1966; Ito and Shibasaki, 1968; Muto et al., 1977), microtubules (Wisse, 1970; Yamada and Hirosawa, 1976), and 5 nm filaments (Wisse, 1970; Yamada and Hirosawa, 1976; Muto et al., 1971). Along the surface of these processes, a number of pinocytotic vesicles (Ito and Shibasaki, 1968) or caveolae (Muto et al., 1977) may be observed.

The stellate cells usually have no basal lamina (Ito and Shibasaki, 1968; Yamada and Hirosawa, 1976). However, small narrow strands of material resembling a basement membrane could be discerned underneath the endothelial lining or between the lining and the underlying subendothelial projections of the stellate cells (Ito and Shibasaki, 1968; Wisse, 1970). The stellate cells appear to adher to the wall of sinusoids by these basement membrane. This assumption is supported by the finding that, when the sinusoidal capillaries contract and become separated from the hepatic cords, the stellate cells lie on the sinusoidal wall (Kupffer, 1876, 1898; Wake, 1978).

The function of the subendothelial processes remains obscure. However, the subendothelial processes might reinforce (Ito and Shibasaki, 1968) or anchor (Wisse, 1977) the endothelial lining of the sinusoids. Moreover, increased surface of these star-shaped processes favors incorporation of vitamin A from the blood stream through the fenestrae of these endothelial-lining cells. The third

---

FIGS. 40, 41, and 42.   Developing Type I lipid droplets (L_1) in the stellate cells from a rat with hypervitaminosis A. (From Wake, 1974, reproduced by permission of the Editor of *Journal of Cell Biology*.) Figure 40: A multivesicular body (MVB) and a developing lipid droplet in the matrix of a MVB. ×65,000. Figure 41: Two small lipid droplets in the MVB. Cytoplasmic vesicles are associated with the MVB. Arrow indicates dense plaque. ×160,000. Figure 42: Section of Type I lipid droplet surrounded by the unit membrane. Note the limited number of vesicles (arrows) of MVB. L_2, Type II Lipid droplet. ×60,000.

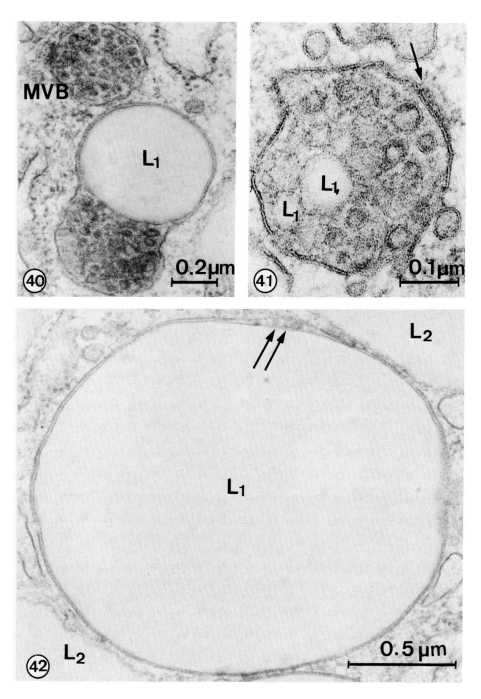

possibility that these cytoplasmic processes may enhance the efficiency of con-traction of sinusoidal capillaries to regulate the blood stream in the liver lobules cannot be ruled out, because massive 5 nm filaments (actin-like filaments) are contained in their processes.

The cytoplasmic processes or cell bodies of the stellate cells are separated from the parenchymal cells by an intercellular space of varying width in which colla-gen bundles are often seen. The narrowest cleft between cells is ca. 20 nm and is seen where the stellate cells occur apposed to the smooth surface of the paren-chymal cells (Ito and Shibasaki, 1968; Wisse, 1970; Wake, 1971). Neither desmosomes nor any other adhesive structure are observed between them.

VI. Function of the Stellate Cells

After discussing the morphological nature of the stellate cells at light and electron microscopical levels, the remaining sections will be devoted to the functions of these cells. At least two functions of the stellate cells, i.e., the storage of vitamin A and the role they play in fibrogenesis in the liver, have been definitely established by recent investigations.

A. STORAGE OF VITAMIN A

The liver is a major depot for vitamin A. Ninety-five percent of the total vitamin A in the cat, rat, and guinea pig, and 70–85% in man is stored in the liver (Wiss and Weber, 1963). In the liver, vitamin A is incorporated into the stellate cells, the Kupffer cells, and the parenchymal cells (Wake, 1975b), though the main storage site of vitamin A is the stellate cells. The author will first describe the changes in these three kinds of cells after the administration of vitamin A, and then discuss the possible mechanism of vitamin A uptake by the stellate cells.

1. Changes after the Administration of Excess Vitamin A

a. *The Stellate Cells.* Following the administration of excess vitamin A, the lipid droplets in the stellate cells increase remarkably (Figs. 32–34) and impart a strong fluorescence to them (Nakane, 1963; Wake, 1971) (Figs. 26 and 27). These lipid droplets react intensely with gold chloride and the precipitation of reduced gold occurs on their surface (Wake, 1971, 1973, 1974) (Fig. 25). They also store vitamin A analogs such as retinyl methyl ether and aromatic retinyl methyl ether (Kent *et al.,* 1977b). However, 13-*cis*-retinoic acid, aromatic retinoic acid ethylamide, aromatic retinoic acid ethyl ester, and retinyl butyl ether are not incorporated. These lipid droplets decrease or disappear during vitamin A depletion (Fig. 43).

Ultrastructural changes in the stellate cells after the administration of vitamin

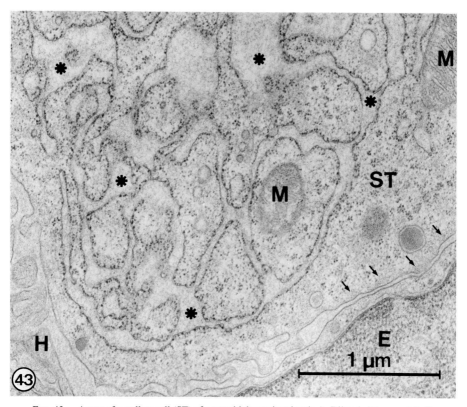

Fig. 43. A part of a stellate cell (ST) of a rat with hypovitaminosis A. Dilated cistern contain fine filamentous material. No lipid droplet is observed. Small arrow, microfilaments; E, endothelial cell; H, parenchymal cell; M, mitochondrion; asterisks, endoplasmic reticulum. ×38,000.

A have been observed in the rat (Wake, 1964, 1971, 1973, 1974, 1975b, 1976a; Kobayashi and Takahashi, 1971; Tuchweber *et al.*, 1976; Ikejiri and Tanikawa, 1977) and in the mouse (Hirosawa and Yamada, 1973; Yamamoto and Enzan, 1975; Yamamoto *et al.*, 1978). The nuclei of the stellate cells attain an indented appearance due to the increased lipid droplets, and other organelles are pushed into the narrow cytoplasm between the lipid droplets (Fig. 34). Hypertrophied stellate cells protrude into the lumen of sinusoids and are covered by a lining of thin endothelial cells (Wake, 1976a; Ikejiri and Tanikawa, 1977). This bulging of the stellate cells may disturb the microcirculation in the sinusoids (Ikejiri and Tanikawa, 1977).

In addition to the remarkable increase in size of the vitamin A-containing lipid droplets, MVBs appear to increase in number in the stellate cells after the administration of excess vitamin A (Kobayashi and Takahashi, 1971; Wake, 1974, 1975b). Small lipid droplets often appear in the MVBs (Wake, 1974,

1975b). These lipid droplets appear to fuse with one another to form larger ones (Figs. 40 and 41). They react with gold chloride and are well preserved after gold treatment (Wake, 1974). When the lipid droplets become large, the MVB vesicles in the matrix are clustered in a group (Fig. 40), and then decrease in number (Fig. 42). It has been convincingly demonstrated that Type I (membrane-bound) lipid droplets develop in the matrix of the MVBs (Wake, 1974).

Structures similar to the Type I lipid droplets (named "lypolysosomes") have been reported in the parenchymal cells of the hamster (Nehemiah and Novikoff, 1973, 1974) and of patients with Wilson's disease (Hayashi and Sternlieb, 1975). The lipid constituents of these lipid droplets are attractive problems to be investigated.

When an ordinary MVB of about 0.4 μm in diameter develops into a Type I lipid droplets measuring about 2 μm in diameter, the surface area of the surrounding membrane increases by about 25 times. This membrane appears to be formed by the fusion of cytoplasmic vesicles. Indeed, the membrane fusion of cytoplasmic vesicles with the limiting membrane of the MVBs is frequently seen (Fig. 41). These cytoplasmic vesicles might function as the vehicle not only for the formation of surface membrane but also for vitamin A or various enzymes in the matrix of the MVBs. Further investigations are, however, necessary to clarify the source of these vesicles.

Some studies on the liver during vitamin A intoxication in man have also been carried out (Lane, 1968; Rubin *et al.* 1970; Hruban *et al.*, 1974; Russel *et al.*, 1974; Farrell, 1977). These studies demonstrated many large lipid droplets in the stellate cells. For examples, Hruban *et al.* (1974) showed a large lipid droplets (ca. 18 μm in diameter) in the stellate cells of the human liver during vitamin A-intoxication. An investigation aimed at elucidating the vitamin A metabolism in Hurler's syndrome (in which case the stellate cells are filled with remarkably large quantities of lipid droplets; Callahan *et al.*, 1967) is awaited with interest.

b. *Kupffer Cells.*   Following the administration of excess vitamin A to vitamin A-deficient rats, two kinds of vitamin A fluorescent cells can be differentiated: perisinusoidal stellate cells, which emit an intense fluorescence imparted by coarse lipid droplets (Figs. 27 and 29); and the Kupffer cells, which emit a weak fluorescence from fine lipid granules present in the vacuolar system of the cytoplasm (Fig. 29). Electron microscopically, Kobayashi and Takahashi (1971) failed to demonstrate lipid droplets in the Kupffer cells after the administration of excess vitamin A. In contradiction to their observations, Wake (1971, 1975b) found numerous small lipid droplets (chylomicrons) in the phagosomes and/or phagolysosomes of the Kupffer cells (Fig. 35). These chylomicrons correspond to the source of weak vitamin A fluorescence under a fluorescence microscope.

Kupffer cells may remain in association with the perisinusoidal stellate cells (Ito and Shibasaki, 1968; Wisse and Daems, 1970; Wisse, 1974) (Fig. 36). This association becomes very intimate after the administration of vitamin A (Wake, 1976; Ikejiri and Tanikawa, 1977). The author refers to this phenomenon as a "Kupffer cell–stellate cell conjugate." The Kupffer cells migrate into the space of Disse through large pores of endothelial cells, or the processes of the Kupffer cells protrude into the endothelial pores to attain an approximation with the perisinusoidal stellate cells loaded with many vitamin A droplets. The narrowest space between them is ca. 40 nm in width. Occasionally a small strand of basement membrane-like material is intercalated between them.

The functional significance of the Kupffer cell–stellate cell conjugate is not yet understood. However, this phenomenon and the incorporation of vitamin A-chylomicrons in the Kupffer cells may suggest that vitamin A in the blood stream is incorporated into the stellate cells via the Kupffer cells (Wake, 1971). Hirosawa and Yamada (1975), however, suggest that vitamin A is directly incorporated into the lipid droplets of the stellate cells, because, after the intravenous injection of tritiated vitamin A into the mice, silver grains are found in the stellate cells as early as 10 minutes after the injection. On the other hand, the experimental results on the vitamin A storage in the liver after blockade of the reticuloendothelial system (RES) have been controversial; the RES blockade decreases the storage of retinyl ester after vitamin A injection (Lasch and Roller, 1936; Wendt and König, 1937; Thiele and Nemitz, 1939; Truscott and Sadhu, 1948; Krishnamurthy and Ganguly, 1956), but the injection of India ink *in vivo* 1 day before the perfusion experiment does not affect the uptake, esterification, or storage of [^{14}C]retinol (Zachman and Olson, 1965). Normally, the Kupffer cells contain only 4% of total vitamin A in the liver (Linder *et al.*, 1971). No esterase activity is demonstrated in the Kupffer cells (Hori and Kitamura, 1972). The discrepancies in the data reported appear to be due to the difference in dosage of vitamin A, the nature of its suspension, and the time lapse after the administration of vitamin A. Another possible function of the Kupffer cells is the destruction and removal of the hypertrophied stellate cells that are laden with many lipid droplets containing vitamin A. However, it seems obvious that the Kupffer cells play a role in the vitamin A metabolism in the liver.

c. *Parenchymal Cells.* According to Linder *et al.* (1971), more than 96% of vitamin A is stored in the parenchymal cells. However, this concentration of vitamin A estimated by them should be mostly in the third cell type, i.e., the stellate cells (Wake, 1975b).

Soon after the injection of [^{3}H]vitamin A acetate, a small amount of silver grains appears in the lipid droplets, mitochondria, nucleus, and smooth and rough endoplasmic reticulum of the parenchymal cells (Hirosawa and Yamada, 1973). Morphologically medium-sized lipid droplets appear in the peribiliary dense bodies. They emit vitamin A fluorescence and also react with gold chloride (Wake, 1975b). These lipid droplets appear to be excreted into the bile canaliculi by exocytosis, because gold-reactive lipid droplets are also present in the lumen of the bile canaliculi (Wake, 1975b). This morphological finding is consistent with the view expressed in biochemical reports (Bohdal and Hruba, 1962; Olson, 1968). When radioactive retinol, retinal, or retinoic acid in micellar solution is injected intraportally, 25, 35, or 60% respectively, of the administrated radioactivity is excreted in the bile within 24 hours (Zachman and Olson, 1965; Zachman et al., 1966). Five days after the intraportal injection of [6,7-^{14}C]retinoic acid, 95% of the radioactivity is recovered from the bile in one case (Zachman et al., 1966).

2. Mechanism of Vitamin A Uptake in the Stellate Cells

Since vitamin A-storing cells in the liver have long remained neglected, the mechanism of vitamin A uptake at the cellular level has just begun to attract attention.

In the liver, retinyl ester is hydolyzed into retinol, which passes through the plasma membrane and is reesterified (preferentially as palmitate) (Mahadevan et al., 1963, 1964; Lawrence et al., 1966). In the rat liver, 97% of stored vitamin A is retinyl ester (Futterman and Andrew, 1964). The reesterification of retinol occurs in the membrane fraction (microsome) at pH 4.5 and pH 8.2 (Futterman and Andrews, 1964). Since recent histochemical investigations using fluorescence microscopy (Nakane, 1963; Wake, 1964, 1971, 1975b), the gold chloride reaction (Wake, 1964, 1971, 1974, 1975b), and autoradiography (Hirosawa and Yamada, 1973) have revealed that the stellate cells are the main storage site of vitamin A in the liver, the above biochemical data can now be employed to explain the mechanism of vitamin A uptake and storage in the stellate cells.

As already mentioned, vitamin A lipid droplets develop in the matrix of the MVBs in the stellate cells. In general, the cytoplasmic staining reactions exhibited by the MVBs indicate the presence of various hydrolases (Friend and Farquhar, 1976; Holtzman et al., 1967; Friend, 1969) and MWBs are considered to be a kind of lysosome. All lysosomal hydrolases have an acid pH optimum (de Duve and Wattiaux, 1966). Thus, it is reasonable to assume that at pH 4.5 reesterification of retinol occurs in the MVBs in the stellate cells. On the other hand, reesterification of pH 8.2 may correspond with the development of Type II lipid droplets (non-membrane-bound). It is likely that Type II lipid droplets are derived from Type I (membrane-bound) after losing their surrounding membrane (Wake, 1974). Thus, the reesterification at pH 8.2 appears to be essential for further development of lipid droplets in the cytoplasmic matrix.

It is interesting to note that vitamin A is incorporated into the lysosomal system both in the stellate and the parenchymal cells. On the basis of comparison with the findings that the vitamin A droplets contained in the lysosomes of parenchymal cells are excreted into the lumen of the bile canaliculi, a possible mechanism of vitamin A storage in the stellate cells has been suggested (Wake, 1975b): retinol is reesterified in the MVBs of the stellate cells to form lipid droplets (Type I). These lipid droplets also might be capable of being discharged into the space of Disse by the fusion of the surrounding membrane with the cell membrane. After losing the surrounding membrane, however, the lipid droplets (Type II) cannot be excreted, and are stored in the cytoplasmic matrix.

B. Role in the Fibrogenesis in the Liver Lobules

The role of the perisinusoidal stellate cells in the fibrogenesis in the liver lobules has received much attention in recent years. There is a general agreement in the literature that the stellate cells are closely associated with the collagen bundles (Ito and Nemoto, 1952; Schmidt, 1960; Schnack et al., 1966; Stockinger, 1967; Motta et al., 1978), contain extensive granular endoplasmic reticulum (Wood, 1963; Ito and Shibasaki, 1968; Kobayashi and Takahashi, 1971; Wake, 1971; Hirosawa and Yamada, 1973; Yamamoto and Enzan, 1975; Tuchweber et al., 1976), and lack basal lamina (Ito and Shibasaki, 1968; Yamada and Hirosawa, 1976). On the basis of the presence of these fibroblast-like structures of the stellate cells, synthesis of collagen fibers in these cells was postulated (Wood, 1963; Rhodin, 1964; Bronfenmajer et al., 1966; Stockinger, 1967; Schnack et al., 1966, 1967; Ito and Shibasaki, 1968). Histochemically Tanaka et al. (1976) demonstrated glutamyl transpeptidase, which appears to take part in the synthesis of fiber protein containing hydroxyproline, glycine, proline, and glutamic acid in the stellate cells. Kawanami (1973) showed aggregation of fine granules impregnated by the periodic acid methenamine silver (PAM) stain in the cytoplasmic processes of the stellate cells, as well as in the cytoplasmic processes of the fibroblasts in the connective tissue in the liver. These data indicate that the stellate cells can assume morphological and functional characteristics of fibroblasts.

Stenger (1965) pointed out that fibroblastic proliferation and new collagen formation are significant factors in the development of fibrosis induced by CCl_4. Popper and Udenfriend (1970) also stated that new formation of fibers is of major importance in the hepatic fibrosis rather than fiber accumulation from collapse as has previously been emphasized. The findings that the fibroblastic cells are derived from the perisinusoidal stellate cells have stemmed from the development and application of techniques for the identification of stellate cells, e.g., fluorescence microscopy, the gold chloride reaction, and administration of excess vitamin A. It has been reported that an accumulation of stellate cells occurs in the necrotic region of the liver lobules following CCl_4 intoxication (Bahu et al., 1975; Yamamoto and Enzan, 1975; Kent et al., 1977a,b). The incorporation of [3H]thymidine in the stellate cells of these regions also increases remarkably (McGee and Patrick, 1972; Yamamoto and Enzan, 1975; Kent et al., 1977a,b).

Autoradiographic studies designed to evaluate the synthesis of collagen or mucopolysaccharide have been conducted. Uptake of [3H]proline (McGee and Patrick, 1972) and [35S]sodium sulfate into hepatic mucopolysaccharide (McGee and Patrick, 1969) is markedly increased by the induction of acute and chronic liver injury with CCl_4. The majority of these cells do not appear to be sinusoidal lining cells, but they are the perisinusoidal stellate cells, because they are located around sinusoids and are intercalated between the vascular lining and necrotic hepatocytes. Kent *et al.* (1976), using an immunofluorescence technique, showed that the accumulation of stellate cells in the necrotic region is associated with the appearance of Type III collagen. Shaba *et al.* (1973) demonstrated both protocollagen proline hydroxylase and the substrate for the enzyme in the crude mesenchymal cell populations containing various cell types isolated from Pronase digests of mouse liver. The substrate is increased in cells derived from animals with acute CCl_4 liver injury. Such investigations will be promoted by the development of the method to separate the perisinusoidal stellate cells purely from the liver tissue.

Now it raises a question: how the two functions (i.e., vitamin A storage and fibrogenesis) correlate with each other in the stellate cells. Although no direct evidence has so far been presented, it is reported that excess vitamin A results in the development of liver fibrosis in man (Muenter *et al.*, 1971; Hruban *et al.*, 1974; Russel *et al.*, 1974) and that vitamin A stimulates the formation of collagen in the skin (Ehrlich *et al.*, 1973; Lee *et al.*, 1973). Hassel *et al.* (1975) showed a stimulation of glycopeptide synthesis in the liver with excess vitamin A. Sudhakaran and Kurup (1974) pointed out that there is a possible correlation between the vitamin A status and the lipid and glycosaminoglycan levels in the liver and aorta in rats.

The stellate cells are thought to be precursors of the collagen-forming cells (i.e., the fibroblasts) and, upon injury, they transform into fibroblasts that are responsible for fibrogenesis (Schnack *et al.*, 1967; Popper and Udenfriend, 1970; Yamamoto and Enzan, 1975; Yamamoto, 1975; Kent *et al.*, 1976, 1977a,b).

VII. Origin and Differentiation of the Stellate Cells

The stellate cells are derived from the mesenchymal cells in the subendothelial space of the embryonic liver (Tahira, 1958; Yamagishi, 1958; Ito *et al.*, 1960b). The mesenchymal elements of the perisinusoidal space are formed from the mesenchyma in the septum transversum in the 7-week human fetus (Yamamoto and Enzan, 1975).

According to Naito (1976), in the 13-day rat embryo, the walls of sinusoids consist of two layers: endothelial cells and pericytes. The latter cells contain electron-lucid nuclei, a few glycogen granules, and lipid droplets. Their cytoplasmic processes extend between the parenchymal cells. These dendritic cells give rise to the stellate cells. Purton (1976) described another type of mesenchymal cell in the avian liver. They have the ability to take part in the formation of a bile canaliculus with the parenchymal cells, and he further suggested that they may be able to differentiate into the perisinusoidal stellate cells.

In both the fetal and neonatal periods, no transitional stages between different types of sinusoidal cells are seen, but mitoses of the stellate cells are observed, which indicate that these cells are independent and self-proliferating at early stages in the development of the liver (Naito and Wisse, 1977).

The state of development of the lipid droplets present in the stellate cells varies from species to species. In the 5-month human fetus liver, small lipid droplets occur in the cytoplasm of stellate cells (Tahira, 1958). In the rabbit liver, the number of lipid droplets increases as early as 10 days after birth (Yamagishi, 1958). On the other hand, the lipid droplets in the chick stellate cells are extremely few in number even 120 days after hatching (Ito et al., 1960b). Whether the limited number of lipid droplets in the stellate cells in the fetal and early developmental stages after birth depends upon the blood content of vitamin A, or on the functional potentialities of vitamin A uptake of these cells, remains obscure.

VIII. Vitamin A-Storing Cells in Extrahepatic Organs

Since the gold chloride reaction was found to demonstrate vitamin A histochemically, the classical investigation by Rothe (1882), who demonstrated gold-reactive cells in the cat stomach and duodenum, is noteworthy. Utilizing the fluorescence microscope, Kudo (1938) reported the existence of vitamin A-containing cells in the spleen, lung, kidney, lymph node, and thymus, as well as in the liver. He thought that vitamin A is stored mostly in the reticuloendothelial Kupffer cells of the liver, and the remaining quantity of vitamin A is incorporated into other reticuloendothelial organs of the body and excreted from the kidney. However, due to the existence of vitamin A in the testis, ovary, eye, and hypophysis (Popper and Greenberg, 1941; Popper, 1944; Tomita, 1955), it has been suggested that the vitamin A-containing organs do not always correspond to the reticuloendothelial system. As already mentioned, it has been clearly demonstrated by Nakane (1963) and by Wake (1964), independently, that vitamin A is not stored in reticuloendothelial Kupffer cells, but in the perisinusoidal stellate cells. A detailed report on the vitamin A-containing cells in the liver (Wake, 1971) has stimulated several researchers' interest in similar kinds of cells in the extrahepatic organs (Hirosawa, 1976; Yamamoto et al., 1977, 1978; Hirosawa, 1977). The same methods used for demonstration of the stellate cells in the liver have been applied to the identification of vitamin A-storing cells in various other organs. These methods include the increase of fluorescence and number of lipid droplets (Yamamoto et al., 1977; Kusumoto and Fujita, 1977), the gold chloride method (Kusumoto and Fujita, 1977), and autoradiography of [³H]vitamin A (Hirosawa and Yamada, 1974, 1975; Yamada and Hirosawa, 1976; Hirosawa, 1977).

Vitamin A-storing cells and/or cells that are capable of uptake of vitamin A following administration are distributed in the following organs.

Lung.   Vitamin A is incorporated in the lipid droplets of septal cells of the alveolar septa (Yamada, 1974; Hirosawa and Yamada, 1974; Yamada and Hirosawa, 1976; Yamamoto *et al.,* 1976, 1978; Hirosawa, 1977). Kusumoto and Fujita (1977), however, report that the vitamin A fluorescence is more intense in the subpleural and perivascular connective tissues than in the alveolar septa.

Digestive Canal.   These cells are distributed throughout the digestive canal extending right from the esophagus up to the large intestine and are seen in the lamina propria, especially in the villi, submucous layer, and adventitia (Hirosawa and Yamada, 1975, 1977; Yamada and Hirosawa, 1976; Hirosawa, 1977; Kusumoto and Fujita, 1977; Yamamoto *et al.,* 1978).

Spleen.   Vitamin A is reported to be contained in the reticular cells (Yamamoto *et al.,* 1977) or in the cells of the Billroth cords of the red pulp (Kusumoto and Fujita, 1977). A few cells occasionally occur in the white pulp also.

Adrenal.   Vitamin A-containing cells are located in the perisinusoidal spaces of the cortex, as well as in the medulla (Hirosawa and Yamada, 1975; Yamada and Hirosawa, 1976; Hirosawa, 1977; Kusumoto and Fujita, 1977; Kent *et al.,* 1977b).

Ductus Deferens and Uterus.   This kind of cells has been reported to exist in the ductus deferens and in the uterus (Hirosawa, 1977).

Skin.   After the administration of excess vitamin A, vitamin A fluorescence has been shown to appear also in the fibroblasts of the dermis (Yamamoto *et al.,* 1978).

Other Organs and Tissues.   The cells in which Sudan III-positive lipid droplets occur after the injection of vitamin A are distributed in lymph nodes, thymus (capsule, cortex, and medulla), reticular parenchyma of the bone marrow, adventitia of the aorta, lamina propria mucosae of the trachea, the oral mucosa, and tonsil (Kusumoto and Fujita, 1977.)

It is now believed that the vitamin A-storing cells in different organs have a number of features in common with those of the stellate cells in the liver. These cells are irregular in shape, have slender cell processes, and possess vitamin A-containing lipid droplets that increase with hypervitaminosis A. No basal lamina is found around the cell surface, but well-developed granular endoplasmic reticulum and Golgi complex are present. Filaments that are 5 nm thick are seen along their plasma membrane. Microtubules are frequently present. The cells are located in the connective tissue space and exhibit a close relationship with both the endothelium of the vascular vessels and various epithelial tissues. Yamada and Hirosawa (1976) proposed a common classification for these cells under the name "vitamin A-storing cell system." In contrast to this view, Kusumoto and Fujita (1977) pointed out that a close relationship between the "vitamin A-uptake cells" and the capillaries is not always observed and cited instances of these cells in the subpleural or perivascular connective tissue of the lung. On the other hand, Yamamoto *et al.* (1978) consider that both perivascular vitamin A-storing mesenchymal cells and fibroblasts may be categorized into one cell type of fibroblastic cell origin.

IX. Concluding Remarks

The present article traces the long, complicated history of the stellate cells and has emphasized that these cells were discovered by Kupffer in 1876. Their perivascular localization, stellate form, and the lipid droplets in their cytoplasm were recognized even during the first period (1876–1897). According to Kupffer's later concept in 1898, however, these cells were the integral elements of the wall of the sinusoids, and that notion has introduced deep-rooted misconceptions in liver histology. The stellate form, the Verfettung, and vitamin A-storage were thought to be features of the reticuloendothelial Kupffer cells during the second period (1898–1970). In 1971, the stellate cells were rediscovered by the present author, and the perisinusoidal cells, which were reported by others under various names, such as "granular cells," "fat-storing cells," "interstitial cells," "lipocytes," or "perisinusoidal connective tissue cells," were confirmed to be nothing but the stellate cells. Conclusively, there exist four kinds of cells in the liver lobules: the parenchymal cells, the endothelial cells, the phagocytic Kupffer cells, and the perisinusoidal stellate cells.

In the last several years, our understanding of these cells has greatly improved. In this article, the morphological characteristics of these cells, which were demonstrated by various methods and techniques, have been surveyed and summarized in order to establish the present state of our knowledge of the morphological criteria of these cells.

The stellate cells are distributed only in the lobules of the liver and constitute 1.4% of parenchymal volume in the rat liver. They are stellate-shaped, are located in the space of Disse, and adhere to the sinusoidal wall. Their dendritic processes spread over the outer surface of the sinusoids and encircle the sinusoids or run longitudinally along them. Along the surface of these processes, a number of micropinocytotic vesicles or caveolae are observed. No basal lamina is discernible. The stellate cells are separated from the parenchymal cells by intercellular spaces of varying dimensions in which collagen bundles were often seen.

The nucleus is round or oval, and "spheridium" are frequently seen. The stellate cells display conspicuous granular reticulum, Golgi complex, and lipid droplets. They contain poorly developed agranular reticulum and a small quantity of glycogen particles in their cytoplasm. Mitochondria are small. The lysosomes noted in these cells are mostly multivesicular bodies. Microfilaments are distributed in the subsurface cytoplasmic matrix which are apposed to the endothelial cells. Microtubules and 10 nm filaments are also contained. Centrioles are associated with the Golgi complex and a cilium projects into the perisinusoidal space.

The stellate cells in the liver of various mammals might be morphologically classified into two types: in the first type, branching of the cytoplasmic processes is well developed many medium-sized lipid droplets are apparent; in the second type, cytoplasmic processes are not so conspicuous and one or two large lipid droplets are located in the vicinity of the nuclei. The first type of droplets are observed in cat, rat, rabbit, and man, whereas the second type occur in pig, hamster, and bull. The above morphological criteria might contribute to the identification of similar kinds of cells widely distributed in the extrahepatic organs, as well as in cells in the developmental stages.

Application of the fluorescence microscopy, the gold chloride reaction, and autoradiography for [^{3}H]vitamin A for the histochemical demonstration of vitamin A have shown that vitamin A is stored in the lipid droplets of stellate cell cytoplasm. A remarkable increase of these lipid droplets, both in number and size, is noted in the stellate cells of the liver in experimental animals administered with excess vitamin A and in patients with acute and chronic vitamin A-intoxications. Two kinds of lipid droplets are differentiated: membrane-bound Type I and non-membrane-bound Type II. On the contrary, in hypovitaminosis A livers, these lipid droplets decrease and disappear. The mechanism of this selective uptake of vitamin A into the stellate cells gives rise to many fascinating speculations. It has been suggested that the multivesicular bodies in the cytoplasm play a role in the development of lipid droplets containing vitamin A. At least Type I lipid droplets are derived from the multivesicular bodies. Participation of reticuloendothelial Kupffer cells, as well as hepatic parenchymal cells in the metabolism of vitamin A, has also been discussed.

The fibroblastic activity of the stellate cells are not so well defined. However, proliferation of these cells and the incorporation of [^{3}H]thymidine and [^{3}H]-proline into these cells during liver injury strongly suggest the active role of the stellate cells in the hepatic fibrogenesis, and probable synthesis of Type III collagen, in the liver lobules. The role of vitamin A in fibrogenesis remains yet to be explained in the metabolic sense. The concept of the existence of a "vitamin

A-storing collagen-producing cell system'' in the liver and extrahepatic organs has been given up in the last few years.

REFERENCES

Asch, E. (1884). Inaug. Dissertation, Bonn University.
Aschoff, L. (1924). *Ergebn. Inn. Med. Kinderheilk.* **26**, 1.
Aterman, K. (1958). *Acta Anat.* **32**, 193.
Aterman, K. (1963). *In* "The Liver. Morphology, Biochemistry, Phisiology" (Ch. Rouiller, ed.). Academic Press, New York.
Aterman, K. (1977). *In* "Kupffer Cells and Other Liver Sinusoidal Cells" (E. Wisse and D. L. Knook, eds.). Elsevier, Amsterdam.
Bahu, R., Minick, O. T., Inouye, T., Kent, G., and Popper, H. (1975). *J. Cell Biol.* **67**, 15a.
Berkley, H. J. (1893). *Anat. Anz.* **8**, 787.
Blouin, A. (1977). *In* "Kupffer Cells and Other Liver Sinusoidal Cells" (E. Wisse and D. L. Knook, eds.). Elsevier, Amsterdam.
Blouin, A., Bolender, R. P., and Weibel, E. R. (1977). *J. Cell Biol.* **72**, 441.
Bohdal, M., and Hrubá, F. (1962). *Acta Biol. Med. Ger.* **8**, 60.
Browicz, T. (1900). *Arch. Mikr. Anat.* **55**, 420.
Bronfenmajer, S., Schaffner, F., and Popper H. (1966). *Arch. Pathol.* **82**, 447.
Cajal, S. R. (1909). "Histologie du système nerveux de l'homme et des vertébrés," Vol. II. Maloine, Paris.
Cohn, E. (1904). *Beitr. Pathol.* **36**, 152.
Callahan, W. P., Hackett, R. L., and Lorincz, A. E. (1967). *Arch. Pathol.* **83**, 507.
De Duve, C., and Wattiaux, R. (1966). *Annu. Rev. Physiol.* **28**, 435.
Disse, J. (1890). *Arch. Mikr. Anat.* **36**, 203.
Domagk, G., and Dobeneck, P. von. (1933). *Virchow Arch. Pathol. Anat.* **290**, 385.
Ehrlich, H. P., Tarver, H., and Hunt, T. K. (1973). *Ann. Surg.* **177**, 222.
Farrell, G. C. (1977). *Am. J. Digest. Dis.* **22**, 724.
Fischer, W. (1912). *Virchow Arch. Pathol. Anat.* **208**, 1.
Friend, D. S. (1969). *J. Cell Biol.* **41**, 269.
Friend, D. S., and Farquhar, M. G. (1967). *J. Cell Biol.* **35**, 357.
Futterman, S., and Andrew, J. S. (1964). *J. Biol. Chem.* **329**, 4077.
Hassell, J. R., Jones, C. S., and Deluca, L. M. (1975). *J. Cell Biol.* **67**, 159a.
Hayashi, H., and Sternlieb, I. (1975). *Lav. Invest.* **33**, 1.
Heidenhain, R. (1880). *In* "Handbuch der Physiologie" (L. Herrmann, ed.). Vol. 5, p. 1.
Helly, K. (1911). *Ziegler Beitr.* **51**, 462.
Higgins, G., and Murphy, G. (1928). *Anat. Rec.* **40**, 15.
Hirosawa, K. (1977). *Acta Histochem. Cytochem.* **10**, 253.
Hirosawa, K., and Yamada, E. (1973). *J. Electron Microsc.* **22**, 337.
Hirosawa, K., and Yamada, E. (1974). *Acta Anat. Nippon.* **49**, 60.
Hirosawa, K., and Yamada, E. (1975). *Proc. Int. Congr. Anat. 10th* p. 482.
Hirosawa, K., and Yamada, E. (1976). *J. Cell Biol.* **70**, 269a.
Hirosawa, K., and Yamada, E. (1977). *Cell Tissue Res.* **177**, 57.
Hirt, A. and his co-worker (1929). Cited by Hirt, A., and Wimmer, K. (1940). *Klin. Wchnschr.* **19**, 123.
Hirt, A., and Wimmer, K. (1940). *Klin. Wchnschr.* **19**, 123.
Hoffmann, F. A., and Langerhans, P. (1869). *Virchow Arch. Pathol. Anat.* **48**, 303.

Holtzman, E., Novikoff, A. B., and Villaverde, H. (1967). *J. Cell Biol.* **33,** 419.
Hopwood, D., and Nyfors, N. (1976). *J. Clin. Pathol.* **29,** 698.
Hori, S. H., and Kitamura, T. (1972). *J. Histochem. Cytochem.* **20,** 811.
Horiuchi, T. (1960). *Kobe J. Med. Sci.* **6,** 185.
Hruban, Z., Russell, R. M., Boyer, J. L., Glagov, S., and Bagheri, S. A. (1974). *Am. J. Pathol.* **76,** 451.
Hübner, G. (1968). *Anat. Anz. (Verh. d. Anat. Ges. 62. Vers.)* **121,** 495.
Ikejiri, N., and Tanikawa, K. (1977). *In* "Kupffer Cells and Other Sinusoidal Cells" (E. Wisse and D. L. Knook, eds.). Elsevier, Amsterdam.
Ito, T. (1951). *Acta Anat. Nippon.* **26,** 42.
Ito, T. (1956). *Acta Anat. Nippon.* **31,** 10.
Ito, T. (1973). *Gunma Rep. Med. Sci.* **6,** 119.
Ito, T., and Nemoto, M. (1952). *Okajima Folia Anat. Jpn.* **24,** 243.
Ito, T., and Nemoto, M. (1956). *Okajima Folia Anat. Jpn.* **28,** 521.
Ito, T., Satsuki, S., and Tsukagoshi, N. (1952). *Arch. Histol. Jpn.* **3,** 239.
Ito, T., Tahira, R., and Tsunoda, K. (1953). *Arch. Histol. Jpn.* **5,** 541.
Ito, T., and Shibasaki, S. (1968). *Arch. Histol. Jpn.* **29,** 137.
Ito, T., Shibasaki, S., and Kitamura, T. (1960a). *Arch. Hisol. Jpn.* **20,** 629.
Ito, T., Tanaka, Y., and Nemoto, M. (1960b). *Arch. Histol. Jpn.* **19,** 565.
Ito, T., Watanabe, A., and Takahashi, Y. (1962). *Arch. Histol. Jpn.* **22,** 429.
Jaffé, R. H. (1938). *In* "Handbook of Hematology" (H. Downey, ed.). Hoeber, New York.
Kano, K. (1952). *Arch. Histol. Jpn.* **4,** 13.
Karrer, P. (1960). *Vitam. Horm.* **18,** 571.
Kawanami, O. (1973). *Acta Pathol. Jpn.* **23,** 717.
Kent, G., Gay, S., Inonye, T., Bahy, R., Minick, O. T., and Popper, H. (1976). *Proc. Natl. Acad. Sci. U.S.A.* **73,** 3719.
Kent, G., Minick, T., and Inonye, T. (1977a). *J. Cell Biol.* **75,** 193a.
Kent, G., Inonye, T., Minick, O. T., and Bahu, R. M. (1977b). *In* "Kupffer Cells and Other Liver Sinusoidal Cells" (E. Wisse and D. L. Knook, eds.). Elsevier, Amsterdam.
Kitagawa, T. (1960). *Arch. Histol. Jpn.* **18,** 493.
Kitamura, T., Yamagishi, M., and Uchida, G. (1956). *Arch. Histol. Jpn.* **10,** 587.
Kiyono, K. (1914). *Folia Haemat.* **18,** 149.
Kobayashi, K., and Takahashi, Y. (1971). *Arch. Histol. Jpn.* **33,** 421.
Kobayashi, K., Takahashi, Y., and Shibasaki, S. (1973). *Nature (London) New Biol.* **243,** 186.
Kovacs, K., and Horvath, E. (1975). *Res. Exp. Med.* **165,** 245.
Krishnamurthy, S., and Ganguly, J. (1956). *Nature (London)* **177,** 575.
Kudo, T. (1938). *Nisshin Igaku* **27,** 583.
Kupffer, C., von. (1876). *Arch. Mikr. Anat.* **12,** 353.
Kupffer, C., von. (1898). *Verh. Anat. Ges.* **12,** (Vers. in Kiel), 80.
Kupffer, C., von. (1899a). *Arch. Mikr. Anat.* **54,** 254.
Kupffer, C., von. (1899b). *München Med. Wchnschr.* **32,** 1.
Kusumoto, Y., and Fujita, T. (1977). *Arch. Histol. Jpn.* **40,** 121.
Ladman, A. J. (1964). *Anat. Rec.* **148,** 304.
Lane, B. P. (1968). *Am. J. Pathol.* **53,** 591.
Lasch, F., and Roller, D. (1939). *Klin. Wchnschr.* **15,** 1636.
Lawrence, C. W., Crain, F. D., Lotspeich, F. J., and Krause, R. F. (1966). *J. Lipid Res.* **7,** 226.
Levine, V. (1932). *Arch. Pathol.* **14,** 345.
Linder, M. C., Anderson, G. H., and Ascarelli, I. (1971). *J. Biol. Chem.* **246,** 5538.
Löwit, M. (1888). *Zieglers Beitr. Pathol. Anat.* **4,** 223.
McGee, J. O'D., and Patrick, R. S. (1969). *Br. J. Exp. Pathol.* **50,** 521.

McGee, J. O'D., and Patrick, R. S. (1972). *Lab. Invest.* **26**, 429.

Mahadevan, S., Seshadri-Sastry, P., and Ganguly, J. (1963). *Biochem. J.* **88**, 531.

Mahadevan, S., Deshmukh, D. S., and Ganguly, J. (1964). *Biochem. J.* **93**, 499.

Marshall, A. H. E. (1956). "An Outline of the Cytology and Pathology of the Reticular Tissue." Thomas, Springfield, Illinois.

Matsuo, U. (1959). *Bull. Kobe Med. Coll.* **15**, 265.

Mayer, S. (1899). *Anat. Anz.* **16**, 180.

Meyer, K. A., Popper, H., and Ragins, A. B. (1941). *Arch. Surg.* **43**, 376.

Motta, P., Muto, M., and Fujita, T. (1978). "The Liver. Scanning Electron Microscopy." Igaku-Shoin, Tokyo and New York.

Muenter, M. D., Perry, H. O., and Ludwig, J. (1971). *Am. J. Med.* **50**, 129.

Muto, M. (1975). *Arch. Histol. Jpn.* **37**, 369.

Muto, M., Nishi, M., and Fujita, T. (1977). *Arch. Histol. Jpn.* **40**, 137.

Naito, M. (1976). *J. Jpn. Soc. RES.* **16**, 25.

Naito, M., and Wisse, E. (1977). *In* "Kupffer Cells and Other Liver Sinusoidal Cells" (E. Wisse and D. L. Knook, eds.). Elsevier, Amsterdam.

Nakane, P. K. (1963). *Anat. Rec.* **145**, 265.

Nehemiah, J. L., and Novikoff, A. B. (1973). *J. Cell Biol.* **59**, 246a.

Nehemiah, J. L., and Novikoff, A. B. (1974). *Exp. Mol. Pathol.* **21**, 398.

Nesterowsky, M. (1875). *Virchow Arch. Pathol. Anat.* **63**, 412.

Nicolescu, P., and Rouiller, Ch. (1967). *Z. Zellforsch.* **76**, 313.

Novikoff, A., and Essner, E. (1960). *Am. J. Med. Sci.* **29**, 102.

Nyfors, A., and Hopwood, D. (1976). *Acta Derm. Ven.* **56**, 465.

Ogawa, K., Minase, T., Yokoyama, S., and Onoe, T. (1973). *Tohoku J. Exp. Med.* **111**, 253.

Olson, L. A. (1968). *Vitam. Horm.* **26**, 1.

Oppel, A. (1900). "Lehrbuch der vergleichenden mikroskopischen Anatomie der Wirbeltiere. III Teil. Mundhöhle, Bauchspeicheldrüse und Leber." Fischer, Jena.

Patrick, R. S., and McGee, J. O'D. (1967). *J. Pathol. Bacteriol.* **93**, 309.

Pfuhl, W. (1932). *In* "Möllendorff's Handbuch der mikroskopischen Anatomie des Menschen. V/2 Bd. Verdauungsapparat. Die Leber." Springer, Berlin.

Platen, O., von (1878). *Arch. Pathol. Anat.* **74**, 268.

Ponfick, E. (1869). *Virchow Arch. Pathol. Anat.* **48**, 1.

Popper, H. (1941). *Arch. Pathol.* **31**, 766.

Popper, H. (1944). *Physiol. Rev.* **24**, 205.

Popper, H., and Greenberg, R. (1941). *Arch. Pathol.* **32**, 11.

Popper, H., and Schaffner, F. (1957). "Liver: Structure and Function." McGraw-Hill, New York.

Popper, H., and Udenfriend, S. (1970). *Am. J. Med.* **49**, 707.

Purton, M. D. (1976). *Experientia* **32**, 737.

Querner, F. (1935). *Klin. Wschr.* **14**, 1213.

Querner, F., and Sturm, K. (1934). *Anat. Anz.* **78**, 289.

Redgrave, T. G. (1976). *Aust. J. Exp. Biol. Med. Sci.* **54**, 519.

Rhodin, J. (1964a). *Anat. Rec.* **148**, 326.

Rhodin, J. (1964b). *Proc. Int. Symp. RES., 4th* p. 108.

Ribbert, H. (1904). *A. Allg. Physiol.* **4**, 201.

Riegele, L. (1928). *Z. Mikr. Anat. Forsch.* **14**, 73.

Riegele, L. (1932). *Z. Zellforsch.* **15**, 311.

Rodahl, K. (1950). *J. Nutr.* **41**, 399.

Rothe, P. (1882). "Ueber die Sternzellen der Leber." Inaug.-Dissertation Munich University.

Rubin, E., Florman, A. L., Degnan, T., and Diaz, J. (1970). *Am. J. Dis. Child.* **119**, 132.

Russell, R. M., Boyer, J. L., Bagheri, S. A., and Hruban, Z. (1974). *N. Engl. J. Med.* **291**, 435.

Rüttner, J. R., Brunner, H. E., and Vogel, A. P. (1956). *Schweiz. Z. Allgem. Pathol. Bakteriol.* **19,** 738.

Satsuki, S. (1952). *Arch. Histol. Jpn.* **3,** 449.

Satsuki, S. (1954). *Arch. Histol. Jpn.* **6,** 33.

Satsuki, S. (1955). *Arch. Histol. Jpn.* **8,** 557.

Satsuki, S., Tsunoda, K., and Shindo, K. (1956). *Arch. Histol. Jpn.* **9,** 517.

Schaffner, F., Barka, T., and Popper H. (1963). *Exp. Mol. Pathol.* **2,** 419.

Schilling, V. (1909). *Virchow Arch. Pathol. Anat.* **196,** 1.

Schmidt, F. C. (1956). *Z. Mikr. Anat. Forsch.* **62,** 487.

Schmidt, F. C. (1960). *Anat. Anz.* **108,** 376.

Schnack, H., Stockinger, L., and Wewalka, F. (1966). *Wien. klin. Wchschr.* **78,** 715.

Schnack, H., Stockinger, L., and Wewalka, F. (1967). *Rev. Int. Hepat.* **17,** 855.

Shaba, J. K., Patrick, R. S., and McGee, J. O'S. (1973). *Br. J. Exp. Pathol.* **54,** 110.

Stenger, R. J. (1965). *Exp. Mol. Pathol.* **4,** 357.

Stockinger, L. (1967). *Anat. Anz. (Verh. Anat. Ges.)* **120,** 545.

Sudhakaran, P. R., and Kurup, P. A. (1974). *J. Nutr.* **104,** 871.

Sunaga, Y. (1954a). *Arch. Histol. Jpn.* **7,** 241.

Sunaga, Y. (1954b). *Arch. Histol. Jpn.* **7,** 252.

Suzuki, K. (1958). "The Experimental Therapy," No. 310–320. Takeda Pharmaceutical Ind., Osaka.

Suzuki, K. (1963). *Proc. Gen. Ass. Jpn. Med. Congr., 16th, Osaka,* **IV** 13.

Tahira, R. (1958). *Arch. Histol. Jpn.* **14,** 495.

Takahashi, T. (1959). *Arch. Histol. Jpn.* **17,** 343,

Takeuchi, T., Sasaki, M., Shiraishi, Y., Akasaka, M., and Otsuka, Y. (1967). *J. Jpn. Soc. RES.* **7,** 127.

Tanaka, M., Kosakai, M., Inomata, I., Takaki, K., and Ishikawa, E. (1976). Acta Pathol. *Jpn.* **26,** 581.

Tanaka, Y. (1960). *Arch. Histol. Jpn.* **19,** 145.

Tanikawa, K., Yoshimura, K., and Gohara, S. (1965). *Kurume Med. J.* **12,** 139.

Thiele, W., and Nemitz. K. (1939). *Klin. Wchschr.* **18,** 95.

Tomita, Y. (1955). *J. Kyoto Pref. Med. Univ.* **57,** 489.

Tòth, J., Remenár, E., Virágh, S., and Bartók, I. (1977). *In* "Kupffer Cells and Other Liver Sinusoidal Cells" (E. Wisse and D. L. Knook, eds.). Elsevier, Amsterdam.

Truscott, B. L., and Sadhu, D. P. (1948). *Proc. Soc. Exp. Biol. Med.* **68,** 255.

Tsunoda, K. (1957). *Arch. Histol. Jpn.* **13,** 583.

Tuchweber, B., Garg, B. D., and Salas, M. (1976). *Arch. Pathol. Lab. Med.* **100,** 100.

Umahara, Y. (1963). *Arch. Histol. Jpn.* **23,** 253.

Uotila, U., and Simola P. E. (1938). *Virchow. Arch. Pathol. Anat.* **301,** 523.

von Ebner, V. (1902). "A. Koelliker's Handbuch der Gewebelehre des Menschen," 6th ed., Vol. III. Engelmann, Leipzig.

Wake, K. (1964). *Proc. Annu. Gen. Meet. Jpn. Histochem. Assoc., 5th* p. 103.

Wake, K. (1971). *Am. J. Anat.* **132,** 429.

Wake, K. (1973). *In* "Electron Microscopy and Cytochemistry" (E. Wisse, W. Th. Daems, I. Molenaar, and P. von Duijn, eds.). North-Holland, Amsterdam.

Wake, K. (1974). *J. Cell Biol.* **63,** 683.

Wake, K. (1975a). *Proc. Int. Congr. Anat., 10th* p. 422.

Wake, K. (1975b). *Rec. Adv. RES Res.* **15,** 83.

Wake, K. (1975c). *Kagaku (Tokyo)* **45,** 33.

Wake, K. (1976a). *J. Electron Microsc.* **25,** 218.

Wake, K. (1976b). *Acta Anat. Nippon.* **51,** 297.

Wake, K. (1978). *Acta Anat. Nippon.* **53**, 474.

Watanabe, S. (1960). *J. Osaka City Med. Center* **9**(Suppl. 8), 99.

Watari, N. (1959). *Arch. Hist. Jpn.* **16**, 369.

Wendt, H., and König, D. (1937). *Klin. Wchnschr.* **16**, 1253.

Wiss, O., and Weber, F. (1963). *In* "The Liver. Morphology, Biochemistry, Physiology" (Ch. Rouiller, ed.). Academic Press, New York.

Wisse, E. (1970). *J. Ultrastruct. Res.* **31**, 125.

Wisse, E. (1974). *J. Ultrastruct. Res.* **46**, 393.

Wisse, E. (1977). *In* "Kupffer Cells and Other Sinusoidal Cells" (E. Wisse and D. L. Knook, eds.). Elsevier, Amsterdam.

Wisse, E., and Daems, W. Th. (1970). *In* "Mononuclear Phagocytes" (R. Van Furth, ed.). Blackwell, Oxford and Edinburgh.

Wisse, E., van't Noordende, J. M., van der Meulen, J., and Daems, W. Th. (1976). *Cell Tissue Res.* **173**, 423.

Wolf-Heidegger, G. (1941). *Z. Mikr. Anat. Forsch.* **50**, 623.

Wood, R. L. (1963). *Z. Zellforsch.* **58**, 679.

Yamada, E. (1974). *Acta Anat. Nippon.* **49**, 60.

Yamada, E., and Hirosawa, K. (1976). *Cell Struct. Funct.* **1**, 201.

Yamagishi, M. (1958). *Arch. Histol. Jpn.* **18**, 25.

Yamagishi, M. (1959). *Arch. Histol. Jpn.* **18**, 223.

Yamamoto, M., and Enzan, H. (1975). *Rec. Adv. RES Res.* **15**, 54.

Yamamoto, M., Iwamoto, T., Itagaki, T., Yamane, T., Yamashita, K., Hara, H., Iijima, S., and Enzan, H. (1977). *J. Jpn. Soc. RES.* **17**, 123.

Yamamoto, M., Enzan, H., Hara, H., and Iijima, S. (1978). *Acta Pathol. Jpn.* **28**, 513.

Yoshida, T. (1940). *Tr. Soc. Pathol. Jpn.* **30**, 188.

Zachman, R. D., and Olson, J. A. (1965). *J. Lipid Res.* **6**, 27.

Zachman, R. D., Dunagin, P. E., and Olson, J. A. (1966). *J. Lipid Res.* **7**, 3.

Zimmermann, K. W. (1923). *Z. Anat.* **68**, 29.

Zimmermann, K. W. (1928). *Z. Mikr. Anat. Forsch.* **14**, 528.

Subject Index

Contents of Previous Volumes